A CAREER IN STATISTICS

Mr Gran

Amazon.ca

A CAREER IN STATISTICS
Beyond the Numbers

GERALD J. HAHN
Retiree, GE Company, Schenectady, New York

NECIP DOGANAKSOY
GE Company, Schenectady, New York

Contributing Authors

Carol Joyce Blumberg
Leonard M. Gaines
Lynne B. Hare
William Q. Meeker
Josef Schmee

⊛WILEY

A JOHN WILEY & SONS, INC., PUBLICATION

Published by John Wiley & Sons, Inc., Hoboken, New Jersey.
Published simultaneously in Canada

For general information on our other products and services or for technical support, please contact our Customer Care Department within the United States at (800) 762-2974, outside the United States at (317) 572-3993 or fax (317) 572-4002.

Wiley also publishes its books in a variety of electronic formats. Some content that appears in print may not be available in electronic formats. For more information about Wiley products, visit our web site at www.wiley.com.

Library of Congress Cataloging-in-Publication Data:

Hahn, Gerald J.
    A career in statistics / Gerald J. Hahn, Necip Doganaksoy.
       p. cm.
    Summary: "This book serves as an excellent companion to its predecessor, The Role of Statistics in Business and Industry. In this volume, the authors help readers decide whether a career in statistics is appropriate for them and what to expect once in it. They provide insights into the work environment and how students and entry-level statisticians can best prepare themselves to succeed, offering hints for success in training, career paths, and lifelong learning. This book is a must-have for anyone considering a career in statistics, as well as for faculty who prepare students for such a career"– Provided by publisher.
    Includes bibliographical references and index.
    ISBN 978-0-470-40441-6 (pbk.)
  1.   Mathematical statistics. 2.   Commercial statistics. 3.   Statistics–Vocational guidance.
  4.   Statisticians–Vocational guidance. I.   Doganaksoy, Necip, 1960- II.   Title.
    QA276.17.H34 2011
    519.5023–dc22

                                                                                                2011008253

Printed in the United States of America

10 9 8 7 6 5 4 3 2 1

To Bea, Adrienne and Lou, Susan and John, Judy and Ben, and Zachary, Eli, Leah, Sam, and Eliza.

—G.J.H.

To my parents Sami and Mediha Doganaksoy, my wife Reyhan, and my children Hakan and Levent.

—N.D.

CONTENTS

PREFACE   XIII

ACKNOWLEDGMENTS   XVII

CHAPTER 1   *PROLOGUE: A CAREER IN STATISTICS*   1
1.1   About This Chapter   1
1.2   What is Statistics?   1
1.3   Who is a Statistician?   2
1.4   Who Employs Statisticians?   3
1.5   The Statistical Thought Process and What Makes it Special   4
1.6   Many Skills Required   7
1.7   Beyond the Workplace   8
1.8   Some Downsides of a Career in Statistics   9
1.9   The Excitement of a Career in Statistics   10
1.10   Embarking on a Career in Statistics   11
1.11   Accreditation   15
1.12   Professional Societies   16
1.13   A Preview   16
1.14   Further Reading   17
1.15   Major Takeaways   18
         Discussion Questions   19

PART I   *THE WORK OF A STATISTICIAN*   21

CHAPTER 2   *WHAT STATISTICIANS DO IN BUSINESS AND INDUSTRY*   23
2.1   About This Chapter   23
2.2   Manufactured Product Applications   23
2.3   Service Business Applications   28
2.4   Process Improvement   30
2.5   Further Applications   31
2.6   Major Takeaways   32
         Discussion Questions   32

CHAPTER 3   *WHAT STATISTICIANS ENGAGED IN OFFICIAL GOVERNMENT
                  STATISTICS DO*   35
3.1   About This Chapter   35
3.2   The Scope of Official Statistics   36
3.3   Examples of the Work   39
3.4   Challenges of the Work   41
3.5   Research Opportunities   45

3.6 Some Major Employers in the U.S. Federal Government 46
3.7 Required Credentials 47
3.8 Integration of U.S. Government Statistical Activities 49
3.9 Local Official Government Statistics Activities 50
3.10 Government Statisticians Outside The United States 51
3.11 Compensation and Other Considerations 52
3.12 Sources of Further Information 53
3.13 Major Takeaways 53
Discussion Questions 54

CHAPTER 4 WHAT STATISTICIANS DO: SOME OTHER APPLICATION AREAS 57
4.1 About This Chapter 57
4.2 Regulatory Activities 57
4.3 Health 59
4.4 National Defense 61
4.5 Other Scientific Research 62
4.6 Social and Behavioral Sciences 65
4.7 Teaching (in Nonacademic Settings) 71
4.8 Some Further Institutes for Research in the United States 71
4.9 Major Takeaways 73
Discussion Questions 73

CHAPTER 5 THE WORK ENVIRONMENT AND ON-THE-JOB CHALLENGES 75
5.1 About This Chapter 75
5.2 Receptiveness to Statistics and Statisticians 75
5.3 Where do Statisticians Fit into the Organization? 79
5.4 Two Modes of Operation 80
5.5 Grading Systems 82
5.6 Globalization 83
5.7 Hail to Our Managers! 84
5.8 Some Challenges 86
5.9 Women in Statistics 90
5.10 Major Takeaways 93
Discussion Questions 94

PART II PREPARING FOR A SUCCESSFUL CAREER IN STATISTICS 97

CHAPTER 6 CHARACTERISTICS OF SUCCESSFUL STATISTICIANS 99
6.1 About This Chapter 99
6.2 Analytical and Technical Skills 100
6.3 Communication and Interpersonal Skills 100
6.4 Ability to Size Up Problems and See the "Big Picture" 101
6.5 Flexibility 102
6.6 A Proactive Mindset 102
6.7 Persistence 103
6.8 A Realistic Attitude 104
6.9 Enthusiasm and Appropriate Self-Confidence 105

6.10 Ability to Prioritize, Manage Time, and Cope with Stress 105
6.11 Team Skills 106
6.12 Leadership Skills 106
6.13 Ability to Properly Apply and Adapt Knowledge 107
6.14 Passion for Lifelong Learning 108
6.15 Further Reading 108
6.16 As Others See It 108
6.17 Major Takeaways 111
       Discussion Questions 111

CHAPTER 7 *EDUCATION FOR SUCCESS*   113

7.1 About This Chapter 113
7.2 The Statistics Advanced Placement Course and Other
      High School Programs 114
7.3 Degrees in Statistics: The Numbers 114
7.4 How Far To Go: Bachelor's, Master's, or Ph.D. Degree? 116
7.5 Selecting the Right School and Program 120
7.6 Statistical Education: Setting the Foundations 125
7.7 Coursework in Statistics 126
7.8 Statistical Computing and Software 131
7.9 Some Other Recommendations 134
7.10 Internships and University Consulting 138
7.11 Entering Statistics from Other Fields 141
7.12 Further Resources 142
7.13 Concluding Comment: The Limits of Formal Statistical Education 142
7.14 As Others See It 143
7.15 Major Takeaways 145
       Discussion Questions 146

CHAPTER 8 *GETTING THE RIGHT JOB*   149

8.1 About This Chapter 149
8.2 Defining Career Goals 149
8.3 Identifying Opportunities 151
8.4 Résumé Writing 154
8.5 The Job Interview 157
8.6 Follow-Up 164
8.7 Providing References 165
8.8 Some Further Hints 165
8.9 Assessing Job Offers 166
8.10 Major Takeaways 170
       Discussion Questions 171

PART III *BUILDING A SUCCESSFUL CAREER
             AS A STATISTICIAN*   173

CHAPTER 9 *ON-THE-JOB STRATEGIES: PROJECT INITIATION AND EXECUTION* 175

9.1 About This Chapter 175
9.2 Project Initiation 175

9.3   Project Execution   184
9.4   As Others See It   190
9.5   Major Takeaways   194
        Discussion Questions   194

CHAPTER 10   *ON-THE-JOB STRATEGIES: COMMUNICATION, PUBLICIZING, AND ETHICS*   197

10.1   About This Chapter   197
10.2   Communication, Communication, Communication   197
10.3   Publicizing Statistics (And Statisticians)   205
10.4   Ethical Considerations   207
10.5   Major Takeaways   216
          Discussion Questions   216

CHAPTER 11   *GETTING GOOD DATA: A KEY CHALLENGE*   219

11.1   About This Chapter   219
11.2   Designed Experiments   220
11.3   Census and Random Sampling Studies of Human and Other Populations   221
11.4   Systems Development Studies   224
11.5   Observational Studies   225
11.6   Hints for Getting the Right Data   233
11.7   A Process for Data Gathering   240
11.8   Major Takeaways   247
          Discussion Questions   248

CHAPTER 12   *CAREER PATHS*   251

12.1   About This Chapter   251
12.2   Some Roles for Applied Statisticians   251
12.3   Other Major Roles to which Applied Statisticians Might Aspire   253
12.4   Some Career Paths   256
12.5   More on Statistical Leadership   258
12.6   Contributions Beyond the Workplace   259
12.7   Some More Career Examples   261
12.8   Putting It All Together   264
12.9   Major Takeaways   265
          Discussion Questions   265

PART IV   *FURTHER PERSPECTIVES: CAREERS IN ACADEMIA AND PROFESSIONAL DEVELOPMENT*   267

CHAPTER 13   *CAREERS IN ACADEMIA*   269

13.1   About This Chapter   269
13.2   What Statisticians in Academia do: An Overview   270
13.3   Types of Positions   273
13.4   Location within the Institution   277
13.5   More on Tenure   277
13.6   Life Beyond Tenure   279
13.7   Teaching Challenges   280

13.8   Research Challenges   **286**
13.9   Consulting Challenges   **292**
13.10 Administrative Service Challenges   **294**
13.11 Professional Service Challenges   **295**
13.12 More on the Academic Environment   **296**
13.13 Training to Become an Academic Statistician   **298**
13.14 Career Paths   **300**
13.15 Downsides of a Career as an Academic Statistician   **303**
13.16 Bright Sides of a Career as an Academic Statistician   **304**
13.17 A Career as a Statistician in Academia: A Summary Comparison   **305**
13.18 Major Takeaways   **305**
       Discussion Questions   **307**

CHAPTER 14   *MAINTAINING THE MOMENTUM*   **309**

14.1   About This Chapter   **309**
14.2   Some Internet Resources   **310**
14.3   Formal Education   **311**
14.4   Technical and Professional Journals   **312**
14.5   Technical Conferences   **315**
14.6   Future Opportunity Areas   **318**
14.7   Major Takeaways   **319**
       Discussion Questions   **319**

REFERENCES   **321**

INDEX   **331**

PREFACE

Many years ago, we decided on a career in statistics. It turned out to be one of the two best decisions we ever made (second, of course, only to our choices of spouses). In this book, we try to impart our enthusiasm for statistics[1] and help readers determine whether a career in statistics is for them, tell them what to expect, and provide hints for their success. We call upon our (and others') extensive practical experience to give important career guidance that goes *beyond* the technical tools discussed in most courses and texts on statistics.

This book complements *The Role of Statistics in Business and Industry* (Wiley, 2008). Our earlier book described how statistical concepts and methods help companies to improve performance and to address problems in a wide range of operations involving products and services. In this volume, we broaden our horizon to include *all* application areas of statistics and focus on helping readers build successful careers.

DISCUSSION OUTLINE

We begin with a prologue chapter that lays the foundation for a career in statistics and provides an overview of the field. The bulk of the book is organized into four parts:

- *Part I*: The Work of a Statistician.
- *Part II*: Preparing for a Successful Career in Statistics.
- *Part III*: Building a Successful Career as a Statistician.
- *Part IV*: Further Perspectives: Careers in Academia and Professional Development.

We conclude with two more chapters (Part IV). The first of these provides an in-depth examination of careers for statisticians in academia. The last chapter describes ways statisticians can stay on top of their field throughout their careers and ensure their continued effectiveness.

Each chapter concludes with a summary of "major takeaways" and a series of discussion questions. Most of these questions do not have "right" or "wrong" answers; instead, they are aimed at provoking the reader to think through the broader implications of the subject. We indicate with an asterisk those questions that do not require *any* past statistical training. Some of those that do are designed to challenge more advanced students.

---

[1] We reserve for a future volume a discussion of our enthusiasm for our spouses.

TARGET AUDIENCE

This book should be of high interest to those around the world who

- Have decided on a career in statistics and would like to prepare themselves in the best possible manner.
- Have been intrigued by statistics, perhaps after taking an introductory course in high school or college, are considering a career in statistics, and would like to know more of what it is all about.
- Are starting on their first jobs as statisticians and want to be more effective—or have been on the job for a while and want to accelerate their paths to success.

The book lends itself well as a text or supplementary text for

- A course in statistical consulting.
- A seminar course dealing with career opportunities, especially in mathematics and/or statistics.
- An independent study course preparing students for the workplace.

It is also suited, together with *The Role of Statistics in Business and Industry* (also soft-cover), for a course on statistical practice that provides aspiring statisticians insights, typically not provided in other courses, to help them succeed in their careers.

In addition, the book can serve as a supplementary text for a variety of courses to complement more technical material, for example, in conjunction with a second course on statistical methods directed at statistics majors.

Some might acquire this book to help them decide whether or not they would like to make statistics their careers. Once so decided, our comments will help them plot a path through college and beyond. And upon completion of their formal education, it should assist them in building a successful career—with the confidence of knowing at least some of the things to expect in an uncertain world.

Finally, this book can help teachers and counselors in both high schools and colleges direct and guide students who may be suited for careers as statisticians.

We assume in some of the discussion that readers have had a one-semester introductory course in statistics, but we do not require any knowledge beyond this—even though many readers will have such added knowledge.

OUR SOURCES

This book leverages insights gained during our combined 75 years of experience as applied statisticians—including, for one of us, managing a corporate statistics group for 28 years. Much of our work has been in the United States. We have, however, recently witnessed and participated in the globalization of many of our activities. We also have some academic experience gained on sabbaticals and as adjunct professors.

To broaden the scope further, we have invited five contributing authors with wide experience in academia and/or government to write two chapters and to provide us their inputs and perspectives on the others. They are

- Carol Joyce Blumberg, Mathematical Statistician, U.S. Energy Information Administration and Professor Emerita of Mathematics and Statistics, Winona State University.
- Leonard M. Gaines, Program Research Specialist, Empire State Development, New York State (Chapter 3).
- Lynne B. Hare, Principal, Statistical Strategies, LLC; Director (retired), Applied Statistics, Kraft Foods; and former Chief, Statistical Engineering Division, National Institute of Standards and Technology.
- William Q. Meeker, Distinguished Professor, Statistics Department, Iowa State University (Chapter 13).
- Josef Schmee, K.B. Sharpe Professor of Management (emeritus), Union College (Chapter 13).

We also allow readers to avail themselves of diverse viewpoints by

- Soliciting comments from 24 eminent statisticians on the traits that they regard as most important for success and on the key advice that they would give to aspiring statisticians. Their responses, appearing at the end of three chapters, reiterate or amplify points that we make and add further insights.
- Liberally quoting others, typically from previously published or presented materials; such quotes introduce many of the book's sections.
- Inviting numerous colleagues to read and critically comment on earlier drafts (see "Acknowledgments").
- Providing extensive references and suggesting web sites[2] to which readers can turn for further elaboration and viewpoints.

OTHER FEATURES

We consider topics that are rarely discussed in other books. A few examples are the downsides (as well as the excitement) of a career in statistics; the important role that our managers play in our careers; women in statistics; what courses to take; assessing job offers; a comprehensive comparison of a career as a statistician in academia versus one in business and industry and in official government statistics; different ways statisticians become involved in projects; criteria for project selection, the "elevator speech"; and a critical assessment of the yearly Joint Statistical Meetings. We also place heavy emphasis on the statistician's role in getting the right data—a fundamental subject that is often brushed aside, under the assumption that the needed data are just there (which is infrequently the case).

We use examples based on our and the contributing authors' experience to illustrate important points, and at times use the language of the workplace. Some typical examples are the use of such terms as added value, bottom line, and payoff.

---

[2] All web sites were in working condition at the time that this manuscript was in production for publication (April 2011); some may, undoubtedly, not stand the test of time.

This should help prepare readers for the jargon that they can expect to encounter during the course of their careers.[3]

With the increasing role of globalization, more and more students throughout the world will be considering statistics as a career. We take a global perspective by, for example, including a section on government statistics outside the United States and discussing effective communication in a global environment.

Wiley has created an ftp site for our book.[4] This provides further details on selected topics, as noted in individual chapters. We will also post on this site thoughts that occur to us after this book goes into production (as well as relevant new links and publications), errors and corrections (hopefully few), and reader comments (hopefully many).

AN INVITATION

In our continuing search for improvement (in possibly future editions and, in the meantime, for posting on the book's ftp site), we invite and welcome reader comments and suggestions. We can be reached at gerryhahn@yahoo.com and necipdoganaksoy@yahoo.com.

Gerald J. Hahn
Necip Doganaksoy

Schenectady, New York

---

[3] The serious reader is referred to Dilbert™ cartoons as the ultimate guide.

[4] ftp://ftp.wiley.com/public/sci_tech_med/career_statistics.

ACKNOWLEDGMENTS

We have benefited greatly from the critical assessments and helpful inputs of numerous colleagues, friends, and (even) relatives and we thank them all for their interest and contributions.

This book would not have been possible without the work of contributing authors Carol Joyce Blumberg, Leonard Gaines, Lynne Hare, William Q. Meeker, and Josef Schmee.

We were also most fortunate to have received detailed and highly useful inputs on extensive parts of the manuscript from Rich Allen, Michael Chernick, Ron Fecso, and Gipsie Ranney.

Many others provided us valuable comments. We thank Christine Anderson Cook, Tom Boardman, Christy Chuang-Stein, Cynthia Clark, Keith Crank, Len Cook, George Domingo, John Espy, Cathy Furlong, Martha Gardner, Adrienne Hahn, Bea Hahn, Judy Hahn, Ted Hardwicke, Bill Hill, Roger Hoerl, Bob Hogg, Yili Hong, Ronald Iman, Ron Kenett, Tim Keyes, John Kovar, Diane Lambert, Jerry Lawless, Vincent Lee, Ramon Leon, Alson Look, Juanita Lott, Fang Liu, Bill Makuch, Harry Martz, Diane Michelson, David Moore, Carolyn Morgan, Sally Morton, Vijay Nair, Lloyd Nelson, Margaret Nemeth, Barry Nussbaum, William Parr, Charles Pfeiffer, Antonio Possolo, Charles Sampson, Nozer Singpurwalla, Edward Spar, Bob Starbuck, Stan Strauss, Katherine Wallman, Ron Wasserstein, and Joanne Wendelberger for their contributions.

We thank Jackie Palmieri and Steve Quigley, our editors at Wiley, and their colleagues for their encouragement and help.

We also greatly appreciate the unwavering support and encouragement by Necip's Manager at GE, Roger Hoerl.

Finally, we thank our wives, Bea Hahn and Reyhan Doganaksoy, for their continued encouragement, strong support, and remarkable perseverance over the many hours we spent in writing this book.

CHAPTER 1

PROLOGUE: A CAREER IN STATISTICS

1.1 ABOUT THIS CHAPTER

We begin this introductory chapter with a brief examination of what is statistics, who is a statistician, and who employs statisticians. We then comment on the statistical thought process and what makes it special, the many skills required to be a successful statistician, and the role of statistics beyond the workplace. We provide "equal time" to presenting some downsides of a career in statistics and counter this with a brief summary of the excitement of such a career. We then indicate some alternative paths for embarking on a career in statistics, comment on ongoing efforts for accreditation, and review professional societies for statisticians. We conclude the chapter with a preview of what is to follow.

1.2 WHAT IS STATISTICS?

> The rational basis for change is data. Data means statistical methods.
>
> —W. Hunter

Some informal definitions of statistics, provided by various well-known statisticians, are

- The science of learning from (or making sense out of) data (J. Kettenring).
- The theory and methods of extracting information from observational data for solving real-world problems (C.R. Rao).
- The science of uncertainty (D.J. Hand).
- The quintessential interdisciplinary science (S. McNulty).
- The art of telling a story with data (L. Gaines).

We prefer the preceding over the more formal definitions found online,[1] such as statistics is "the science that deals with the collection, classification, analysis, and interpretation of numerical facts or data, and that, by use of mathematical theories

---

[1] For example, http://dictionary.reference.com/.

A Career in Statistics: Beyond the Numbers, Gerald J. Hahn and Necip Doganaksoy.

of probability, imposes order and regularity on aggregates of more or less disparate elements." And we note that Brown and Kass (2009) devote a 19-page article (including discussion) to an in-depth examination of "What is Statistics?"

Statistics has applicability in almost all areas of human endeavor. To take just a few examples:

- Economists need to understand statistical concepts to make predictions.
- Psychologists need to be able to interpret empirical relationships between variables.
- Business executives need to appreciate the role of uncertainty in decision making.
- Biologists need to understand that the reactions of organisms to stimuli are not deterministic and that there will be variation among individuals.
- Design engineers need to know how to conduct statistically valid experiments to develop the best possible products.

We also note with interest that in 2010 Britain joined other countries in asking their statisticians to develop a "happiness index" to be added to their existing national household survey.

The diversity of applications is further illustrated by the 24 sections of the American Statistical Association (ASA) as of the end of 2010. These include those that deal with biopharmaceutical applications, business and economics, defense and national security, education, environmental applications, epidemiology, government, health policy, marketing, physical and engineering sciences, quality and productivity, social statistics, and sports.[2]

We will defer discussing specifics to the next three chapters, but note that Tanur et al. (1972) and Peck et al. (2006) provide 15 and 25 (different) articles, respectively, that demonstrate the use of statistics in a wide variety of application areas.

Some areas of application have taken on lives of their own. Thus, biostatistics is the application of statistics to the analysis of biological and medical data. Going even further, actuarial science—the application of mathematical and statistical methods to assess insurance risk—has become a separate profession.

Statisticians at a particular point in their careers are commonly engaged in one or a few of these application areas. It is, however, not unusual for individuals to become involved in an appreciable number of such areas during the course of their careers.

1.3 WHO IS A STATISTICIAN?

Professional statisticians are trained in statistics and actively use statistics and statistical concepts and thinking in much of their work.

The ASA now has a program of accreditation for its members (Section 1.11) and such programs exist in various other countries. The ASA accreditation program has

---

[2] Other ASA sections focus on more specialized topics such as Bayesian statistical science, nonparametric methods, statistical consulting, statistical computing, and statistical graphics.

high requirements (e.g., 5 years of documented experience in practicing statistics) and is completely voluntary. We anticipate that there will be many statisticians in the United States, especially until accreditation becomes popular, who have not been formally accredited.

Moreover, what constitutes as adequate education in statistics to qualify an individual to be a statistician is highly job dependent (and somewhat controversial), although there are some communalities (Section 7.4). Thus, to simplify matters, we will just assume statisticians to be those who regard themselves as such.

One does not have to be a statistician to use statistics. The use of statistical methods by nonstatisticians—whom we will refer to as "practitioners"—is at an all-time high and will likely continue to increase. This places additional responsibilities on statisticians. We shall return to these topics at various junctures throughout this book.

1.4 WHO EMPLOYS STATISTICIANS?

> Data are widely available; what is scarce is the ability to extract wisdom from them.
> —Hal Varian (2010)

A simple, and perhaps somewhat exaggerated, answer to the question "who employs statisticians?" is "essentially all large and some medium-sized, and even small, organizations." Somewhat arbitrarily, we categorize employers of statisticians as follows:

- Business and industry that manufacture products and/or provide services (Chapter 2).
- (Mostly government) agencies engaged in gathering, analyzing, and reporting official statistics or in related statistical activities (Chapter 3).
- Organizations involved in various other application areas, including those engaged in regulatory activities, health, national defense, other scientific research, and the social and behavioral sciences. Employers include government, research institutes, and universities (Chapter 4).
- Self-employed (typically private statistical consultants). Discussion of these is postponed to Chapter 12 since this is a role unlikely to be taken by statisticians early in their careers.
- Academia (Chapter 13).

The ASA estimates that its membership is broken down approximately as follows:

- 47%: Business, industry, nonprofit (other than government and academia), self-employed, or other.
- 42%: Academia.
- 11%: Government (national, state, provincial, or local).[3]

---

[3] In considering the preceding numbers, note that not all statisticians in the United States are members of ASA. The numbers might, for example, overrepresent the proportion of statisticians in academia since academicians may be more likely to become members of ASA.

In any case, like other professionals, statisticians work for customers who benefit from their work. These customers might be direct, such as a client who has commissioned specific work, or indirect, such as users of the Consumer Price Index or students in a professor's class.

1.5 THE STATISTICAL THOUGHT PROCESS AND WHAT MAKES IT SPECIAL[4]

1.5.1 The Scientific Method

Almost 500 years ago, during the reign of Queen Elizabeth I, Sir Francis Bacon, an English philosopher, lawyer, and statesman, addressing the Royal Society, proposed a new way to gain increased understanding of nature. Instead of drawing conclusions from their own preconceived notions, religion, or other traditional sources stemming from Aristotelian thinking, scientists should engage themselves in observation and experience. He didn't take the idea much further than that, but others built on it, and, a few centuries later, it led to the formal development of the scientific method.

Basically, the scientific method calls for starting with a conjecture about the state of nature, cause and effect relationships, or differences among phenomena (e.g., animal typing by physical characteristics, medical treatment differences, or rates of differently induced chemical reactions). We then make observations, that is, gather data, in order to confirm or deny that conjecture. This must be done in such a way that the results are reproducible by others. Thus, we build knowledge about the state of the universe by confirming or denying conjectures.

All of this may not sound like rocket science today, but at the time it was truly revolutionary, and the results have been spectacular. The scientific method has, for example, guided the increase of crop yields to help feed starving populations. It has led medical researchers to understand the causes of diseases and find cures. And it has resulted in electrical engineers learning how to produce microchips efficiently in mass quantities, allowing access to such modern technologies as laptop computers, the Internet, computer-based controls, and safety mechanisms in planes and automobiles.

The applicability of the scientific method is, moreover, not limited to the hard sciences, such as chemistry and physics. Our understanding of sociology and psychology and other social or human sciences has relied on its use as well. The list of beneficial applications is endless. It seems safe to say that the overwhelming majority of advances in human civilization have taken place as a consequence of the application of the scientific method.

1.5.2 Where Statistics Fits In

But what does this have to do with statistics and the way statisticians think?

Put simply, we assert that statisticians, in many ways, might be considered gatekeepers of the scientific method. There are good reasons for this lofty claim. After

---

[4] This section was written by contributing author Lynne Hare.

all, a key factor in any scientific endeavor is a concern for obtaining unbiased results with a wide range of applicability.

In many, or even most, situations, it is impractical or even impossible to enumerate completely an entire population. Thus, you need to develop precise and accurate estimates from a well-selected sample. Say, for example, that we want to characterize the mean weight of salmon in a lake. You can't get all the salmon in the lake and weigh them—but you can take an appropriately selected sample and use its mean weight as an estimate of the mean weight of all salmon in the lake. But to draw correct conclusions, you must concern yourself with the representativeness (as well as the size) of the sample. Does it, for instance, include only farmed or only Northern Pacific salmon? If so, the estimate is clearly biased with regard to determining the lake's entire salmon population and applies only to the limited portion of the population under examination. So, when statisticians are called upon to propose a sampling study, they ask all sorts of (both) broad and specific questions, which may perhaps initially seem impertinent, and then use the answers to help develop a plan that is maximally informative under the circumstances. When the results of the study become available, statisticians then quantify the uncertainty in the findings and provide warnings about the generality of the results. And this is where the gatekeeper status comes in.

Of course, statistics and statistical thinking are more complicated than that. Many studies are observational, and others involve designed experiments (Section 11.2). But because statistics—in light of its reliance on scientific sampling— is, in a large part, the science of making decisions under uncertainty introduced by the sampling process, and because sampling applies to almost all intellectual scientific pursuits, one could assert that statistics is relevant to just about every discipline.

1.5.3 A Peek into How Statisticians Work

It All Starts with the Theory. Statistics, just as other scientific disciplines, has theoretical components, many of which have roots in mathematics. Theoretical statisticians conduct important work in developing new methods, improving on existing ones, and coming up with novel ways to address applied problems. Expanding the theory is essential to the health and well-being of the field and requires strong mathematical skills. Most statisticians receive training in basic statistical theory and rely on such theory in applications—many of which are not straight-forward textbook situations—and in understanding the basic assumptions underlying the methods that they are using (even though they are generally not conducting research in theoretical statistics).

A Traditional View. Applied statisticians work primarily on using statistics to address issues in other disciplines, generally at the behest of what we will refer to as "problem owners." These may be administrators, economists, engineers, scientists, social scientists, or others, who are typically leaders or representatives of a larger project team.

In the past, problem owners often came to statisticians with a *fait accompli*; that is, they had already conducted a study and gathered data, and wanted to know what it all meant. In response, statisticians typically asked questions about the study objectives, the data gathering process, the measurement methods, and so on. This was often followed by the unenviable task of sorting through the data structure, and then the data themselves, attempting to extract meaningful findings. Sometimes, this was not possible, resulting in much wasted effort. This unfortunate situation was a consequence of a frequently held misunderstanding (often furthered by what was taught in school) of statisticians as merely appliers of a series of tools to evaluate already collected data—as opposed to being purveyors of the scientific method who concern themselves with the entire problem and the gathering of the appropriate information to address it.

A Collaborative Step-by-Step Approach. A much preferred alternative route for statisticians' participation—and one that, fortunately, is becoming increasingly prevalent—is that of collaboration. Effective collaboration takes the form of close involvement by the statistician with, and as part of, the project team from the beginning to the end of the project. In this context, the statistician is an advocate, working to assure unassailable results and aiding in the discovery process.

Contrary to popular opinion, discovery is not a one-time event. Instead, it is an iterative process progressing from synthesis to analysis to the next synthesis and the next analysis, and so on, all the while building knowledge and leading to scientific advances. Project objectives are reviewed and shared initially with all stakeholders, including the statistician, who then typically proposes sampling, data gathering, or experimentation strategies. These need to take into consideration such matters as sources of variation to be overcome and the precision and accuracy of the measurement process—in addition to numerous practical considerations, such as time, financial, and facility constraints.

Team members then review the statistician's proposed plan with regard to its feasibility, ability to meet objectives, expense, and various operational details—based on their own technical backgrounds and roles in the project. In this manner, the combined knowledge of the team is thrust at the problem and effectively utilized. This may lead to the statistician modifying the plan and the conduct of an initial "pilot study" to test out the recommended approach.

At the end of the study, and often at strategic times during its execution, the statistician conducts appropriate analyses to extract information from the data. And it doesn't end there; the entire team, statistician included, is typically involved in reporting the results to management in a readily understandable manner, quantifying benefits, explaining limitations, conducting follow-up investigations, and implementing the findings.

We discuss the need for statisticians to be proactive team members throughout this book, and especially in Chapters 5 and 9.

Gaining a Common Understanding. Statisticians strive to ensure that there is a common understanding by all involved—and most importantly by management—of the problem to be addressed and of objectives and strategies. If there is lack of agreement on these, then the team members are likely to be working at cross purposes, impairing the likelihood of success. So, in the initial discussions, the statistician strives to ensure that all involved are moving in the same direction. Do we agree on the project objectives and guidelines, including the time schedule? Do all the stakeholders see eye to eye? Do we have approval from oversight agencies, if any?

Recognition of Variability. Statisticians, perhaps more than most others, think about variation and its impact on the findings of the study. They strive to understand the overall process and especially the sources of variability.

The data gathering needs to be planned so as to apply as broadly as possible to the population or process of interest, as illustrated by the salmon sampling problem. Are the data truly representative of the population or process of interest and will all known relevant sources of variation be taken into account? And, consistent with the scientific method, can we expect the results to be reproducible?

In short, the statistician seeks to ensure that, despite the underlying variability and the study limitations, the right amount of the right kind of data will be acquired. We will return to this important topic in Chapter 11.

Statistical Thinking. The recognition of variability has, in turn, led to the concept of statistical thinking. In contrast to deterministic thinking, statistical thinking recognizes that

- All work occurs in a system of interconnected processes.
- Variation exists in all processes.
- Using data to understand and reduce variation is key to success.

Statistical thinking has especially broad applicability in situations in which there is an opportunity to reduce variation, rework, and waste and improve quality and productivity. And, even though our comments have been in terms of industrial settings, statistical thinking applies to all areas of application of statistics. See Britz et al. (2000) and Hoerl and Snee (2001) for further discussion.

1.6 MANY SKILLS REQUIRED

Many students are attracted to statistics because they like and are good at mathematics. Mathematical ability—supplemented by sound statistical training—though certainly a necessary requirement for a successful statistician is far from being a sufficient one.[5]

Successful statisticians need to be skilled in many areas. We summarize some of these briefly in this section and provide further elaboration throughout the book, especially in Chapters 6 and 7.

---

[5] Thus, we have some difficulty with a definition of a statistician that we found online as "a mathematician who specializes in statistics" (http://www.thefreedictionary.com/statistician).

1.6.1 Interpersonal Skills

Numerous interpersonal skills, starting with ability to communicate effectively, are essential for a successful career in statistics.

1.6.2 Knowledge in Related Technical Fields

Statisticians need to be comfortable with computers and computer software. Those working in business and industrial settings, especially, need to be trained in operations research/management science/decision theory and be knowledgeable in various statistics-related methods developed by engineers and computer scientists, such as artificial intelligence, knowledge discovery in databases and data mining, and neural networks.

1.6.3 Application Area Knowledge

Statistics does not exist in a vacuum. It is employed in many different areas. To be credible, statisticians need to be familiar with the application areas in which they are involved. Having a minor, or even an undergraduate degree, in the field in which you hope to be employed will provide you an important advantage.[6] Many statisticians learn about their chosen application area on the job, but the more prepared they are to get up to speed rapidly, the better.

1.7   BEYOND THE WORKPLACE

1.7.1 Impact on One's Thinking

Statistics is more than a career. It is a way of looking at situations that pervade our daily lives. Statisticians tend to think in terms of probabilities and variation, search for the data that might support specific contentions, and assess their validity. This is why some claim that statisticians—perhaps like scientists, engineers, and lawyers— are more rational (and possibly less emotional) in their lives—or, at least, their reasoning—than many in other fields.

In assessing the safety of different forms of travel, for example, statisticians might estimate the probabilities associated with each of the alternatives, such as flying commercially, driving, taking a bus, or taking a train, based upon available data, and then tailor the results to their specific situations (such as driving skill and road conditions).

Training in statistics also gets one to look critically at studies reported in the media. Many statisticians enjoy leisurely discussions of the potential pitfalls of such studies with colleagues, friends, neighbors, and family.

---

[6] Alson Look tells the following story, which he attributes to Lloyd Nelson, a one-time Ph.D. chemist who became a well-known industrial statistician. A junior statistician once came to Lloyd and showed him a designed experiment that he was proposing. After reviewing the plan, Lloyd urged him not to go forward with it. "Why not?" the statistician asked. "Because—based on my knowledge of chemistry—it might blow up the plant," Lloyd responded—while running in the other direction.

With an appreciation for statistics, there comes a passion for procuring as much useful data as possible *before* making a decision. In purchasing a car, we may avidly—some might say obsessively—seek data on repair frequency to guide our selection. Statisticians typically do not salt their meals before tasting them. You are unlikely to hear a statistician (or a mathematician) say something like "This object is perfectly round!" If told that a particular community has over 300 days of sunshine yearly, a statistician might assert that this statement has little meaning without a definition of exactly how a "day of sunshine" is defined and measured. And only a statistician, when told that a bird in the hand is better than two in the bush, might respond, perhaps somewhat facetiously, with "that depends upon the probability that you can capture the two birds in the bush and your risk function (what, for example, if the proverbial birds are actually your pair of shoes?)."

1.7.2 Promoting Statistical Literacy

Our society is flooded by statistics and claims based thereon, often by advocates who may be more interested in making a point than in a valid and fair assessment of the data upon which their claim is based. A key mistake, for example, is to infer causation from association (Section 11.5.1).

A responsibility of statisticians is to promote statistical literacy among the general public. This means, in general, helping others become more aware of the uses and misuses of statistics, and, in particular, pointing out the fallacies of claims or studies of questionable validity.

A number of books have pointed the way—even though sometimes in a somewhat negative manner—following in the steps of the classic *How to Lie with Statistics* (Huff, 1954), and with equally intriguing titles including Best (2001, 2004), Hooke (1983), and Levitt and Dubner (2005, 2009). We report on our experiences in teaching a short course on statistical literacy targeted at adults in Hahn et al. (2009).

1.8 SOME DOWNSIDES OF A CAREER IN STATISTICS

> If it moves, it's biology; if it stinks, it's chemistry; if it doesn't work, it's physics; if it puts you to sleep, it's statistics.
>
> —Anonymous student

There must be some downsides associated with being a professional statistician. What are they?

Start with the name and the general perception. Let's face it—the public's image of a statistician, though generally improving, is not in the same league as that of an astronaut, biotechnologist, or physicist. Statisticians are associated with "statistics"—and statistics are often regarded as boring or deceptive or both. It will not take long for you to hear the worn quote that Mark Twain attributed to Benjamin Disraeli that there are "three kinds of lies: lies, damn lies, and statistics," or the story about the chap who became a statistician because he did not have the personality to be an accountant.

Initial negative impressions are furthered by the often less than exciting introductory statistics course that many have taken in college. Such courses may have focused on the mechanics of the calculations or the mathematical theory, at the expense of demonstrating the broad applicability of statistical concepts. The ASA and many colleagues are working hard to help make introductory courses more appealing and to place greater emphasis on statistical concepts and statistical thinking. There has been important progress in recent years, including the publication of some down-to-earth introductory texts and the introduction in the United States of the Advanced Placement (AP) course in statistics for high school students (Section 7.2). As in any subject, knowledgeable, enthusiastic instructors are essential; see Section 13.7.1 for further discussion.

Another concern is that, under some organizational structures, statisticians are viewed as "outsiders." This can make them vulnerable to budget cuts and undermine their effectiveness. It is one further reason why we advocate throughout this book that statisticians strive to become proactive participants and team members in the activities in which they are involved.

Because the value of statistics is not universally recognized, many statisticians, especially early in their careers, spend much time marketing themselves and the added value they provide—often in a "soft sell" mode. This can be unappealing to those who prefer to focus on their technical work and development.

1.9   THE EXCITEMENT OF A CAREER IN STATISTICS

> A degree in statistics is one of the top degrees to have in terms of getting good and secure jobs. On top of that, it offers many opportunities for a challenging and rewarding career.
>
> —Sarah Needleman

The downsides of statistics as a profession warranted mention. But the perception may be changing. Lynne Hare reports "In years past, I used to get nerd-associations with statistics. Now I get mostly awe. Wow, you can understand that stuff? And you make money doing it? 'Yup,' I say. 'Not many people can do it, the demand is high, and I can make a bundle.'" This viewpoint (also expressed in the January 2009 *Wall Street Journal* article by Sarah Needleman cited at the beginning of this section) has been reinforced by recent articles in such publications as the *New York Times* (2009) and the *Washington Post* (2009). We have, in fact, found it highly gratifying to note the increased recognition that statistics has been accorded over the course of our careers.

We remain convinced that for many the negatives of our profession are far overshadowed by the excitement, opportunities, and challenges. We wake up, at least most days, eager to face the challenges of our jobs—a key criterion in selecting a career. Some of the reasons that statistics continues to excite us are

- The *diversity* of problems in which we become involved. As per our earlier discussion, statistics deals with just about everything. Or quoting a comment attributed to the famed statistician John Tukey, "the best thing about being a statistician is that you get to play in everyone's backyard. In an age of specialization, we might be the remaining scientific generalists." Most statisticians eventually focus on one or a few application areas, at least for a while—but even within these there is much diversity.

- The *intellectual challenge* of the work. Statisticians have been called data detectives (and, as per Section 1.5.2, gatekeepers of the scientific method). Problems are rarely clear-cut, and often part of our challenge is to define "the real problem." There is ample opportunity to be imaginative.

- The *importance* to our employer and often to society of what we do.

- The opportunity to *interact* with a wide variety of interesting people with different professional and personal backgrounds.

It should be no surprise, therefore, that CareerCast.com rated statistics as #4 in its 2011 best jobs ratings (of 200 jobs).[7]

1.10 EMBARKING ON A CAREER IN STATISTICS

1.10.1 Some Alternative Paths

Individuals with a variety of backgrounds choose careers in statistics. These range from people interested in the field from early on and others originally trained in mathematics seeking an applications oriented career to those whose work has led to an appreciation of the importance and excitement of statistics and the opportunities it provides.

Here are a few examples:

- Eliza enjoyed mathematics in high school. She decided early on to be a mathematics major in college. However, she did not want to go into teaching and was always driven by practical problems. Her high school mathematics teacher suggested she take the AP course in statistics. She became fascinated by the many areas in which statistics could be applied and decided to minor in statistics in college. Eventually, she decided that statistics provided the opportunity to combine her mathematics skills with her interest in real-world problems. So she enrolled as a Ph.D. student in the subject, specialized in biostatistics, took added coursework in biology and other sciences, and upon graduation accepted a job with a pharmaceutical company.

- Juan majored in electrical engineering as an undergraduate. He took an introductory course in statistics and was fascinated by its many applications. At the end of his junior year, he took a summer internship in the research and development division of a semiconductor company. He worked closely with the division statistician who helped him plan a test program to assess a proposed new transistor fabrication process and analyze the results. These experiences

---

[7] The top three jobs were software engineer, mathematician, and actuary. The three worst jobs were roustabout (performs routine physical labor and maintenance on oil rigs and pipelines), ironworker, and lumberjack. The #4 rating for statistician was an advance over the #8 rating in 2010; the rating in 2009 was #3.

made him decide to obtain a master's degree in statistics. Upon graduation, he returned to the semiconductor company as a statistician.

- Boris held an undergraduate degree in psychology with the career goal of going into human resources. On his first job with a large company, he became heavily involved in a study to gain a better understanding of characteristics of job applicants that could predict on-the-job success. Working with a statistician, he came to appreciate the statistical intricacies of conducting such a study and the many ways in which statistics can help organizations to be successful. He enrolled in a part-time program in statistics at a local university. This eventually led Boris to a master's degree in statistics, a career shift, and a position with a government agency responsible for conducting statistical surveys.

- Leah had a Ph.D. in chemistry; she had taken a couple of statistics courses. Upon graduation, she took a job with a large brewery. With some guidance from her former statistics professor, she became engaged in planning and analyzing a statistically based test to help ensure that a proposed new brew would consistently meet customer taste criteria. After a while, people began to come to her for help on statistical problems. Over a period of years, Leah became heavily engaged in the self-study of statistics and took a few courses at a local college. She started attending statistical conferences and taking short courses on selected topics, such as the design of experiments. Eventually, her knowledge matched that of a master's degree graduate in statistics. When the company set up a Six Sigma program, she was chosen to become a technically oriented black belt and was responsible for teaching the statistics part of the training program to her fellow employees. Eventually, she was officially designated the company statistician.[8]

1.10.2 How We Got to be Statisticians

And how did *we* become statisticians? We hasten to say that neither approach is suggested as an ideal, or even a recommended, path.

Gerry Hahn   As an undergraduate, I commuted to the School of Business at the City College of New York (now Baruch College). I was studying to be an accountant, like the majority of students there. But my (unofficial) mentor (2 years older than me) advised me otherwise. "Do something different and more imaginative," he suggested. (He became a successful stockbroker). Besides, there was then an oversupply of accountants; entry job accountants typically earned $55 per week (in 1950). Statisticians were making at least $60!

So I took an introductory course in statistics and found it most interesting. I decided to major in statistics, and after graduation enrolled in Columbia University's School of Business, where I received a master's degree in statistics.

---

[8] Leah's career resembled that of the early statistician William Sealy Gossett, who started as a scientist at the Guinness brewing company, where he derived the statistical *t*-test (and was identified as "Student" in publishing his work in a scholarly journal).

An advantage of living in New York City was that, while still going to school, I was able to get hands-on experience working in the market research department of an advertising agency (the now defunct Biow Company). One of my tasks was to help evaluate the effectiveness of different advertising methods in the then hot emerging field of television. In one study, we polled a random sample of New Yorkers to assess the relationship between people's recollection of advertising messages and their product use. The results were reported by the *New York Times* (the only time in my career that I made the *Times*).

Shortly thereafter, I was drafted into the U.S. Army (the Korean War ended while I was in basic training). I was assigned to the Chemical Corps and was stationed at Dugway Proving Ground in the Utah desert. I became involved in analyzing field test data to assess the impact of chemical weapons under different environmental conditions. This experience got me interested in more physically based applications of statistics. I was also recruited to teach an introductory statistics course on base.

Returning from duty and still wearing my army uniform (I had made it to the lofty rank of corporal), I stopped off in Schenectady, NY, to interview at GE. It was the company's patriotic duty *not* to turn me down. So began a 46-year career with the company! While working at GE, I was able to extend my technical knowledge in statistics, via a summer stint at Virginia Tech, a master's degree in mathematics at Union College, and, finally in 1971, a Ph.D. in operations research and statistics from Rensselaer Polytechnic Institute.

Looking back, my entrance into statistics was more a fortunate set of circumstances than a planned strategy. A significant deficiency, for which I have tried to compensate with subsequent courses and spare time reading, was my relatively limited training in the sciences.

P.S. One of my three daughters, despite the bad example set by her father, received her bachelor's degree in statistics at the School of Agriculture at Cornell University, followed by a master's degree from the School of Public Health at the University of California in Berkeley. She went to work at the March of Dimes and, subsequently, the Medical School of the University of California in San Francisco (UCSF). This provided an entry into the field of epidemiology, in which, after a number of years of part-time study, she received a Ph.D. degree from Berkeley. As a faculty member at UCSF, she is currently working on studies of HIV/AIDS (especially in Africa) and of the homeless.

Necip Doganaksoy   I did my undergraduate study at the Management Department of the Middle East Technical University in Ankara, Turkey, graduating in 1983. As part of a traditionally strong science and engineering school, the curriculum was heavily loaded with courses on quantitative topics. I was drawn to these since I liked applied mathematics and also wanted to learn more about computer applications, which were rapidly being integrated into my courses.

After an introductory course on probability and statistics, I went on to take elective courses on topics such as econometrics, forecasting, and operations research. During my senior year, I took a course on statistical quality control that was taught by a professor who had recently received his Ph.D. in statistics at Union College in

Schenectady. This course was my first exposure to the industrial applications of statistics and I found it fascinating.

One feature of the course was the additional reading material handed out by the instructor. These papers carried titles such as "Coefficient of Determination Exposed!" and "How Abnormal is Normality?" Compared to the orderly textbooks to which I had been accustomed, these papers raised some thought-provoking issues about the applications of statistics. I was destined to meet the author, Gerry Hahn, a few years later and to become his colleague.

I became increasingly interested in advanced study in statistics. A graduate degree in statistics would most likely lead me toward an academic career. Employment at a government agency would have been another possibility, but between the two, the academic path seemed more appealing to me at the time. In contrast, opportunities in statistics in the business and industrial sector in Turkey were limited.

I decided to follow the lead of my professor and apply to Union College for graduate study. Union College, at the time, had a graduate program in applied statistics and operations research. One of its major purposes was to serve the large GE engineering and scientific community in Schenectady. In addition to its own small full-time faculty, statisticians from (what is now) the GE Global Research Center were actively involved with the program, teaching courses and guiding graduate students. This program provided a nice balance between applications and theory, and suited me well.

My first 2 years at Union were very intense and led to a master's degree. In addition to the regular coursework, I undertook an aggressive study program to further my background in mathematics. As a consequence of my association with the adjunct faculty, I also became interested in industrial and engineering applications of statistics. While working on my dissertation, I started an internship with the GE Statistics Program. I became a full-time member of the group in 1990 after finishing my Ph.D. Given GE's diverse business interests, I wanted to have the opportunity to work on a wide array of projects and benefit from the experienced statisticians in the group. This required me to abandon, at least for the time being (20+ years and still counting), my plans to go into academia.

P.S. One is never too young to get started. My two young sons are already being presented with play situations to develop their statistical thinking skills, even though their true passions are space travel and driving a school bus. We will update their progress in future editions of this book.

1.10.3 Some Further Insights

In an article in *Amstat News*[9] (2009a), five statisticians (four of whom are in academia) respond to a variety of challenging questions starting with "what or who inspired you to be a statistician?" *Amstat News* and the ASA's Statisticians in the News web site[10] periodically feature articles on the careers of successful statisticians.

---

[9] The membership magazine of ASA (Section 14.4.1).

[10] http://www.amstat.org/about/statisticiansinthenews.cfm.

For example, see Bruce and Bose (2010). The story of Jim Goodnight—one statistician who made it big—is portrayed in Sidebar 1.1.

SIDEBAR *1.1*

A GOOD ROLE MODEL

Jim Goodnight, a Fellow of ASA, holds a Ph.D. degree in statistics from North Carolina State University and was a faculty member there from 1972 to 1976. In 1976, he cofounded the SAS Institute, with fellow statistician John Sall and others, to analyze agricultural research data. Today, SAS is a world-renowned software giant, with over 10,000 employees and $2.3 billion in revenues in 2009. The company culture has received much acclaim and made various "Best Places to Work" lists. Goodnight was #33 on the *Forbes Magazine* list of the 400 richest Americans in 2009 and has become renowned for his philanthropy.

Once in statistics, there are again different paths to success. We discuss some of these in Section 12.4.

1.11 ACCREDITATION

Unlike some other professions, including actuarial science, there was no formal program of accreditation or certification of statisticians in the United States for many years. In 2009–2010, however, the ASA Board of Directors endorsed a program of individual accreditation for its members and for testing procedures for its implementation; see *Amstat News* (2009b) and Johnston (2010).

The general criteria for accreditation[11] include

- An advanced degree in statistics or a related quantitative field with sufficient concentration in statistics.
- At least 5 years of documented experience in the practice of statistics.
- Evidence that the applicant's work as an applied statistician is of high quality,
- Effective communication skills.
- Adherence to the ASA's "Ethical Guidelines for Statistical Practice" (Section 10.4).
- An ongoing record of professional growth.
- At least two supporting letters from persons of substantial stature who have firsthand knowledge of the work and skills of the applicant.

Once granted, accreditation is for 5 years, after which it may be renewed.

In announcing this program, (then) ASA President Sally Morton emphasized that it is voluntary, and not the same as certification, which, in other professions, may be required before one can practice and which may require taking an exam.

[11] http://www.amstat.org/accreditation/index.cfm.

Other countries that offer accreditation programs through their statistical societies include Australia, Canada (Gibbs and Reid, 2009), and the United Kingdom (Lee, 2008).

1.12   PROFESSIONAL SOCIETIES

There are numerous professional societies worldwide that provide guidance to aspiring statisticians and opportunities to stay abreast, grow professionally, and network with others throughout their careers.

The mainstream organization of professional statisticians in the United States is the approximately 18,000-member American Statistical Association. There is also a much smaller (about 3000-member) Institute of Mathematical Statistics (IMS), made up principally of academicians.

Many other countries have their own statistical associations, such as the Royal Statistical Society (of the United Kingdom), the New Zealand Statistical Association, the Swedish Statistical Association, and the Statistical Society of Canada. There are also regional organizations for statisticians, such as the European Network of Business and Industrial Statisticians. The International Statistical Institute (ISI) is a global organization, founded in 1885, with 2000 elected members from more than 133 countries "who are internationally recognized as the definitive leaders in the field of statistics." Election to membership is based on an individual's professional achievements. In addition, there are many more statisticians that are involved in ISI through its seven sections. ISI activities, such as its World Statistics Congress and other meetings, are open to all.

Statisticians, especially in business and industry, might also be active in closely related organizations, such as the American Society for Quality and its Statistics Division,[12] the Institute for Operations Research and the Management Sciences, and the Decision Sciences Institute, or more specialized groups such as the International Biometrics Society.

Others might belong to the mainline organization in their fields of application, such as the American Association for Public Opinion Research, the American Chemical Society, the American Economics Association, the American Educational Research Association, the Institute of Electrical and Electronics Engineers, and the Society for Clinical Trials.

We discuss some ways in which statisticians avail themselves of professional societies and their offerings in Sections 14.4 and 14.5.

1.13   A PREVIEW

Is a career in statistics right for you? If so, what can you expect, how can you be best prepared, and what are some things you need to know to succeed? Our goal is to help you answer these questions.

---

[12] http://www.asqstatdiv.org/.

We have organized the book into three major parts, plus this introductory chapter and two supplementary chapters (Part IV). The three parts deal with

- Part I: The Work of a Statistician.
- Part II: Preparing for a Successful Career in Statistics.
- Part III: Building a Successful Career as a Statistician.

In Part I, we describe what statisticians do in business and industry (Chapter 2), in gathering, analyzing, and reporting official government statistics (Chapter 3), and in a variety of other application areas (Chapter 4). We then comment on the general environment in which statisticians work and the challenges that they face on the job (Chapter 5).

We begin Part II with a discussion of the essential personal traits of a successful statistician (Chapter 6). We then describe the technical knowledge that aspiring statisticians need to acquire in their education (Chapter 7) and suggest strategies for getting the right job (Chapter 8).

In Part III, we provide guidance on how to succeed on the job. We propose on-the-job strategies dealing with project initiation and execution (Chapter 9) and consider communication, publicizing, and ethical considerations (Chapter 10). The statistician's all-important, and frequently underappreciated, role in getting good data is then presented (Chapter 11). We conclude this part of the book by describing alternative career paths for statisticians (Chapter 12).

Our major focus in the preceding chapters is on a career in applied statistics.[13] Many of our comments also apply to statisticians in academia. Academia, however, also has some distinct characteristics of its own. In the first chapter of Part IV (Chapter 13), we describe these characteristics and extend our discussion from the earlier chapters to the academic world.

In the final chapter (Chapter 14), we stress the importance of maintaining momentum through lifelong learning and indicate some ways statisticians continue to stay abreast and move ahead.

1.14   FURTHER READING

The *Amstat News* provides a wealth of further information on a monthly basis. This includes the results of periodic salary surveys, such as those for business, industry, and government statisticians by Dias et al. (2009) and for statisticians in academia by Crank (2010a, 2011). The publication's Member Spotlight series is also of strong interest to aspiring statisticians.

Various other publications and web sites—including that of the ASA—provide further information (Section 14.2).

---

[13] In this book, we use the term "applied statistics" to refer to work conducted by statisticians outside of academia and refer to those who do the work as "applied statisticians." We recognize, however, that much work conducted in academia is applied and that many in academia consider themselves to be applied statisticians.

1.15   MAJOR TAKEAWAYS

- Statistics has been broadly defined as "the science of learning from data." It has applicability in almost all fields of human endeavor.
- Professional statisticians are trained in statistics and actively use statistics and statistical concepts and thinking in much of their work.
- The chief employers of statisticians are business and industry; government and other agencies engaged in gathering, analyzing, and reporting official statistics; and academia. Organizations involved in various other undertakings, such as social science research, also employ statisticians. Other statisticians are self-employed, typically working as private statistical consultants.
- Statisticians can, in many ways, be considered as gatekeepers of the scientific method. When feasible, they develop studies, often based on sampling or experimentation that are maximally informative. When the data become available, they analyze the results, quantify the associated uncertainty, and provide appropriate warnings about the generality of the findings.
- Applied statisticians work primarily at the behest of problem owners and are most effective when involved from the outset as team members in a collaborative step-by-step approach to addressing problems, beginning with gaining a common problem understanding.
- Statisticians focus on the impact of variability in all phases of a study. This has motivated the concepts that are part of statistical thinking.
- Successful statisticians need much more than mathematical ability and statistical training. They require important nontechnical skills, knowledge in related technical fields, and familiarity with their areas of application.
- Statistics is more than a career. It is a way of looking at situations that pervade our daily lives. Also, statisticians strive to promote statistical literacy among the general public.
- Statistics has traditionally not been as well recognized as some other professions and has been negatively impacted by less than exciting introductory courses. This might require statisticians to engage in appreciable "soft selling" to overcome.
- However, for the right person, the preceding downsides of a career in statistics are far overshadowed by the excitement, opportunities, and challenges of such a career. These include the diversity of problems, the intellectual challenges, the importance of the work, and the chance to interact closely with others.
- Many roads can lead to a career in statistics. Some initially set out to study statistics, perhaps spurred on by an interest in applying their mathematical training and skills. Others are led to statistics by witnessing its importance, excitement, and opportunities.

- The American Statistical Association has embarked on a program of individual voluntary accreditation for its members. Similar programs exist in some other countries.

- There are numerous professional societies worldwide that provide guidance to aspiring statisticians and opportunities to stay abreast, grow professionally, and network with others throughout their careers.

DISCUSSION QUESTIONS

(* indicates that question does *not* require any past statistical training)

1. Statistics has been defined as "the science of learning from data." What are some of the data with which citizens are confronted daily? How are such data typically obtained and analyzed?

2. In the example(s) that you provided in response to the preceding question, what were some deficiencies in the data gathering and analysis and how might these be overcome?

3. What are some of the issues that statisticians who are asked to develop a national happiness index need to address?

4. When did you/will you regard yourself to be a professional statistician? Explain.

5. Give examples of statistical practitioners that you have encountered and their work. Comment on how they benefited (or might have benefited) in their work from guidance by a professional statistician.

6. *Provide some examples of how statistics has been used to further the scientific method.

7. *Cite some problems in which you have been involved, or of which you are aware, that were addressed by an iterative approach, and describe that approach.

8. Statisticians focus on variability. Measurement error is an important source of variability in many applications. What is meant by measurement error? How can it be assessed and controlled in a particular application?

9. We assert that "initial negative impressions are furthered by the often less than exciting introductory statistics course that many have taken in college." What have been your and your friends' experiences in this regard? How can the introductory course be made more appealing?

10. Research and comment on the methodology used by CareerCast.com in its ratings of best (and worst) jobs. What are some issues that might be raised about this methodology? Does the methodology provide any clues of why statistics dropped from #3 in 2009 to #8 in 2010 and then back up to #4 in 2011?

11. *Consider the career paths of Eliza, Juan, Boris, Leah, Gerry, and Necip. What are the pros and cons of each of these paths? Which of the paths, or modifications thereof, do you feel most comfortable with?

12. *Check out and report on the current status of the American Statistical Association's accreditation efforts.

13. *Which professional society(ies) do you belong to or might you consider joining in the future? What specific benefits do these provide to students and, subsequently, statistics professionals?

PART 1

THE WORK OF A STATISTICIAN

THE **PURPOSE** of this part of the book is to provide a comprehensive description of what statisticians do in different application areas.

We describe what statisticians do in business and industry (Chapter 2), in official government statistics (Chapter 3), and in various other application areas (Chapter 4). We then comment on the environment in which statisticians work and the challenges they face on the job (Chapter 5). (We defer the major discussion of careers in academia to Chapter 13.)

Our breakdown of areas in which statisticians are engaged is somewhat arbitrary. There is also some overlap. For example, new product development conducted by business and industry (Chapter 2) involves scientific research (Chapter 4). And much scientific research leads to manufactured products (e.g., nanotechnology and molecular pathology) that are likely to be built by industry. Statisticians engaged in gathering, analyzing, and reporting official statistics (Chapter 3) are often doing so as a prelude to formulating regulations, providing information to improve public health, or supporting scientific or social research (Chapter 4). Also, applied statisticians working in any setting may develop new statistical techniques or extend existing one to address problems that they encounter.

WHAT STATISTICIANS DO IN BUSINESS AND INDUSTRY

2.1 ABOUT THIS CHAPTER

> Bear in mind that not only is the corporate world different from the academic one, but it itself also varies according to the type of company: from products to services, aerospace to commerce, manufacturing to marketing, research-oriented to results-oriented, and so on.
>
> —Arnold Goodman

Our earlier book (Hahn and Doganaksoy, 2008) describes the role of statistics in business and industry in detail. Also see the series of articles in Coleman et al. (2010). We, therefore, provide only a brief overview in this chapter of

- Manufactured product applications.
- Service business applications.
- Process improvement.
- Further applications.

2.2 MANUFACTURED PRODUCT APPLICATIONS

> Failure to understand variation is the top problem in U.S. industry today.
>
> —Lloyd Nelson

2.2.1 Evolution

Statistics first gained popularity in business and industry in the early twentieth century in what was then referred to as statistical quality control. This principally dealt with ensuring the stability of manufacturing processes (using control charts, building on the pioneering work of Walter Shewhart) and with sampling incoming materials and outgoing product to identify and act upon defective batches (acceptance sampling). Defense industry applications during World War II subsequently led to the formation of central statistical groups in business and industry. Much of the early work of these

A Career in Statistics: Beyond the Numbers, Gerald J. Hahn and Necip Doganaksoy.
© 2011 John Wiley & Sons, Inc. Published 2011 by John Wiley & Sons, Inc.

groups tended to be reactive or "firefighting," such as responding to a field performance or reliability problem.

The use of statistics, and a broader appreciation of what it can do, received a significant boost in the United States in the 1980s, as a result of the internationally recognized work of W. Edwards Deming[1] (see Sidebar 2.1). This built upon the success of his and others' work in helping Japan use statistical methods to build high-quality products.

SIDEBAR 2.1

DEMING'S CONTRIBUTION

A major thrust of Deming's message was the importance of understanding the sources of variability and its reduction through statistically based approaches (subsequently incorporated into the concepts of statistical thinking, see Section 1.5.3). Deming stressed the importance of enhancing one's process knowledge and the need for continuous improvement. He claimed that management's lack of understanding variability often leads to unnecessary tampering with the process. For example, Deming strongly criticized the overemphasis placed on the "most recent numbers" (e.g., last quarter sales by region) in management decision making.

Deming was a proponent of a proactive role for statisticians, saying, for example, "if I had been waiting for them to come for help, I'd still be waiting" (Deming, 1986). He also had strong ideas about statistical leadership roles (Section 12.5). Deming had much more to say to management—who flocked to his seminars—as laid out in his 14 points; see Deming (1986, 1993).

Further impetus to the use of statistics was provided in the 1990s by the popularization of Six Sigma (Section 5.2.2).

As a consequence of these developments, management came to realize that waiting to address quality in manufacturing (and reliability based upon field performance) is too late. Instead, the most cost-effective approach, and the one that maximizes customer satisfaction, is to build high quality (and reliability) into the *initial design* of products and processes. It became clear that statistical concepts, in general, and the statistical design of experiments, in particular, can play an important role in helping design high-quality (and reliable) products and in their subsequent validation. Thus, though firefighting problems still arise, a greater proportion of industrial statisticians' work today is focused on ensuring high quality and reliability in product or process introduction and in continuous product and process improvement.

2.2.2 Product Design, Reliability Assurance, and Scale-Up

Under the new philosophy, statisticians help design engineers develop the best possible product and guide its successful transition to manufacturing. This includes

---

[1] The quality improvement movement was significantly impacted by a number of "gurus." In the United States, in addition to Deming, these included Phil Crosby, Armand Feigenbaum, Bill Golomski, Joe Juran, and Dorian Shainin.

participation in setting the design goals, ensuring measurement capability, designing a product that meets or exceeds its performance goals, design validation, and transition to manufacturing.

Some typical applications of statistics are to help

- Determine customer needs and expectations through market research studies.
- Evaluate the impact of critical variables on product characteristics and assess their robustness to variability in manufacturing conditions and in product use through designed experiments.
- Plan validation and scale-up test programs and analyze the results.

An especially important element of successful product design is to ensure high product reliability. Reliability is defined formally as "the probability that a product will satisfactorily perform its intended function under field conditions for a specified period of time." It has been referred to, less formally, as "quality over time."

Ensuring high reliability typically includes planning statistically based accelerated test programs to provide an early indication of failures and their causes so that these can be addressed in a timely manner. For example, light bulb manufacturers might make design changes to improve average product life and/or reduce variability using the results of a statistically planned experiment that assesses the impact of different design variables on product life. Another illustration, dealing with washing machine reliability, is given in Section 11.7.

2.2.3 Manufacturing

Variations in manufacturing processes increase variability in product performance and costs and hurt productivity. Today, a major goal of statistics during manufacturing is to improve both quality and productivity by reducing variability.

Some typical applications of statistics are to help

- Assess manufacturing stability and capability.
- Evaluate and remove measurement bias and quantify and reduce measurement variability.
- Identify the reasons for deviations from target performance and act thereon.
- Devise ways to study the structure of variation, identify its sources, and act to reduce undesirable variability.
- Monitor the process and signal significant changes, identify and remove special causes of variation, and provide the path to permanent improvement.

2.2.4 Field Support

The major goal of statistics during field support is to help develop an optimum servicing and problem avoidance strategy and leverage the information on field performance to build improved product in the future.

Some typical applications of statistics are to help

- Establish proactive product servicing initiatives, such as preventive maintenance and parts replacement to avoid field failures, and mitigate the impact of those failures that do occur.

- Track product field reliability, provide early identification of failures and their causes, address these in a timely manner (including recall of selected product), and avoid their repetition.[2]

- Compare field performance of a product with that of competitive products and help make statistically valid advertising claims based on the results.

2.2.5 Some Specific Product Applications

Statistics has traditionally been applied in highly diverse manufacturing industries such as

- Discrete parts assembly, involving products or systems with varying complexity, ranging from light bulbs and motors to automobiles, electronic equipment, locomotives, airplanes, and turbines.

- Process industries, involving such products as chemicals, petroleum and plastics, foods and beverages, textiles, and pharmaceuticals.

- Products that require both discrete parts assembly and processing, such as semiconductor manufacturing.

Different statistical methods are typically used in discrete parts assembly and process industries. The approach employed often varies by type of business and manufacturing operation. We briefly describe some prominent product applications and suggest a few of the statistical challenges that they present.

Pharmaceutical Industry. The 1962 Kefauver–Harris Amendments to the Food, Drug, and Cosmetics Act of 1938 requires U.S. pharmaceutical products to be proven safe and effective to the Food and Drug Administration (FDA) before they can be put on the market.[3] To achieve such proof, preclinical and clinical trials need to be designed, executed, and the results reported. This requires the design and analysis of carefully controlled statistically planned investigations as part of a multiphase approval process. As a result, the pharmaceutical industry is one of the largest employers of statisticians in the United States today. Additional statisticians are employed by regulatory agencies such as the FDA (Section 4.2.1) and contract research organizations that perform work for pharmaceutical companies.

Statisticians in the pharmaceutical industry are engaged most extensively in the design and analysis of preclinical and clinical studies. These focus initially on the

---

[2] Such analysis of field data was traditionally a major task of many industrial statisticians. Though (unfortunately) still important today, the focus, as per our earlier discussion, has shifted to averting such problems through improved product design and by proactive product servicing.

[3] Similar initiatives have been implemented or are underway throughout the world; see Hamasaki et al. (2009) for developments in Japan.

impact of a new drug on animals and on healthy volunteers. If the preliminary results appear promising, they are followed by in-depth evaluations to assess the impact of the drug on humans and patients with the targeted disease. See Turner (2007) and Cleophas et al. (2006) for detailed descriptions of the design, methodology, and analysis methods in new drug development and of the use of statistics for clinical trials. We discuss ethical considerations in the design and analysis of clinical trials in Section 10.4.5.

Some other activities conducted by statisticians in the pharmaceutical industry include

- **Support of Basic Research**: Help identify potential new drugs from a vast library of compounds and perform *in vitro* (test tube) and *in vivo* (in the body— animal or human) studies on promising candidates.

- **Pharmaceutical Development**: Determine how to make a new chemical or biological compound and explore what process steps maximize product yield and minimize cost.

- **Marketing and Commercialization**: Conduct scientific studies to characterize a new product, survey physicians and patients, prepare publications, and give public presentations.

- **Manufacturing**: Guide the production of the drug in final market formulation, usually, in large quantities (e.g., assessing uniformity of drug content, stability, and dissolution).

- **Field Tracking**: The assessment of the field performance of approved drugs or other medications, for example, the evaluation of the long-term health impact of taking large doses of a vitamin. Such studies have become increasingly important in recent years.

See Peterson et al. (2009) for an in-depth discussion of the use of statistics in pharmaceutical applications beyond clinical trials and Buncher and Tsay (2005) and Millard and Krause (2001) for more general discussions of statistics in the pharmaceutical industry. Also, the American Statistical Association publishes a quarterly journal *Statistics in Biopharmaceutical Research* and has a Biopharmaceutical Section.

Semiconductor Industry. The semiconductor business is characterized by its many process steps and efforts to improve product yield and is consequently highly data-intensive. Some typical applications of statistics are

- Study of the impact that process variables at one stage of production have on the performance at subsequent stages.
- Monitoring and controlling critical process variables.
- Product burn-in to weed out likely early field failures.

Czitrom and Spagon (1997) and Czitrom (2003) provide detailed discussions of statistical approaches and methods dealing with process improvement for the semiconductor industry.

Food and Beverage Industry. The highly competitive food and beverage industry often calls for direct interactions with customers and searching for ways of providing them maximum satisfaction, while still maintaining a healthy profit. Some typical challenges are

- Minimizing the variability that customers experience in repeat purchases of the same product.
- Monitoring food safety and spoilage.
- Making customer preference assessments based on data that are typically ordinal (e.g., preference scale from 1 to 5) or categorical (e.g., fruity, acidity, or bitter taste).
- Obtaining appropriate data and using the results in a valid and defensible manner to make advertising claims about the product.

2.3 SERVICE BUSINESS APPLICATIONS

There has been growing recognition that statistical concepts can play an important role in business and industry beyond manufactured products, notably in the improvement of service businesses and functions and in the development and improvement of business processes (discussed in the next section).

2.3.1 Traditional Applications

Some service businesses have been using statistics for a long time. As indicated earlier, the success of the insurance business, and especially life insurance, is based upon statistical assessments and has led to the development of actuarial science as a separate profession; see Szabo (2004) and Lakshminarayanan (2008a). Airlines have, to a large degree, built their scheduling and dynamic ticket pricing strategies on statistical and related operations research methods. More recently, online retailers have used information on past purchases to identify potential future customers and to construct appropriate messages targeted at such customers.

2.3.2 Financial Services Applications

The financial services industry is ... one of the most information-driven industries, where analytics and statistical analysis play a key role in decision making every day.

—Sami Huovilainen (2010)

The large amount of money at stake in financial transactions has made finance an especially attractive area for statistical modeling. A key role of statistics in financial services is to help quantify risk and uncertainty. Some typical applications are the development of

- Credit scores to determine which future applicants for credit should be approved and, if so, for how large a line of credit.

- Default risk modeling for mortgages and commercial loans.
- Fraud detection schemes (e.g., of improper auto insurance claims) that maximize the identification of culprits, while minimizing the number of falsely suspected innocent individuals.
- Risk assessment as, for example, in loan portfolio acquisition.

The 2000s witnessed historic levels of consumer and commercial lending, a rapid proliferation of new financial products (e.g., derivatives and securitization of bonds), and the rise of "financial engineering" as a new field. During the global recession, starting in 2008, financial service companies received much of the blame and the need for improved statistical models for assessing and managing risks became evident; see, for example, Hutchinson (2010).

2.3.3 Communications Industry

The communications industry is another highly data-intensive business that has traditionally been a heavy user of statistics. Typical applications include

- Identification of peak usage times and evaluation of the capability to provide normal service during such times.
- Assessments of software reliability.
- Evaluation of the impact of alternative pricing strategies.
- Studies of Internet traffic patterns.

2.3.4 Statistical Software Development and Support

The revolutionary advances in computing technology have been a key driving force behind the use of statistics by statisticians and practitioners alike. We discuss and differentiate between different types of statistical software in Section 7.8.

The development of statistical software (and of general purpose software with statistical features) provides significant opportunities and challenges for statisticians. An obvious requirement for all such software is technical correctness—but much more is needed. Statisticians want their software to include the latest state-of-the-art methodologies and to provide the flexibility to allow them to build in their own features and to adapt existing features to address the idiosyncrasies of individual problems.

More recently, the so-called "democratization of statistics" (Section 5.2) has led to greater accessibility and use of statistical software by nonstatisticians, together with online information on how to use such software. Statisticians involved in the development of software to be used by practitioners need to ensure that the software is easy to use and that the information about it is not only technically correct but also easily understandable and helpful to those with limited background in statistics. Another goal is to make the software robust to potential misuse by those who do not have an intimate knowledge of the underlying methodology. This suggests emphasizing graphical methods (and supplementing statistical analyses with graphical displays) and, when possible, encouraging users to focus on methods that require minimal assumptions. In addition, it calls for internal checks and warning messages as well as clear statements of the statistical assumptions that remain.

In addition to the development of statistical software, statisticians provide support to users. This involves helping them apply the software successfully and responding to questions related to both technical features and product use, as well as teaching and marketing activities; see Kiernan (2009).

2.3.5 Internet Operations

> What makes a winning Google statistician? First and foremost, the love of data. What our data lacks in quality, we make up for in quantity.
>
> —Daryl Pregibon (2009)

The emergence of the Internet has led to a wealth of new applications for statisticians. Some of these, such as helping advertisers determine the optimum timing for the purchase and placement of ads, are similar to those encountered and addressed by statisticians in traditional media, that is, print, radio, and TV. They are characterized by vast databases.

The new technology has also led to a variety of new problems calling for the use of statistical and related methods, such as

- Directing advertisers to web pages based upon key words.
- Optimally updating the repository of Internet pages by "refreshing" current web pages (i.e., balancing the cost of having a stale web page against that of fetching a new page).
- Optimizing when to purchase online ads (in advance or on the spot) and how to price these.
- Helping a computer search engine provider assess the effectiveness of different search techniques, model user behavior, and evaluate the effectiveness of different advertising strategies.

Pregibon (2009) provides an overview of the use of novel experimental design approaches (e.g., "random cookie" experiments to assess the effectiveness of different advertising messages) and of observational studies (Section 11.5) in online applications.

2.4   PROCESS IMPROVEMENT

The success of any organization—be it product manufacturer or service operation—depends heavily on how well its processes perform. These may involve

- Taking and fulfilling customer orders for products and services.
- Employee recruitment, orientation, and retention.
- Accounting and financial reporting.
- Customer service.
- Procurement of goods and services.

The failures of a business process, unlike that of many products or services, may not always be immediately apparent, especially if these are not continuously tracked and measured. Customers, for example, experience problems in the form of delayed or incorrect shipments, multiple referrals to different departmental units, or inadequately addressed complaints (including those that the customers themselves did not vocalize—but will remember in making future purchases). Improving business processes increases customer satisfaction, reduces internal costs, and raises productivity.

Statistics plays an important role in improving existing business processes and developing new ones for both manufacturing and service operations. Some examples are helping

- Improve a billing process to eliminate errors and collect revenues faster.
- Increase a call center's accuracy and timeliness in addressing customer issues.
- Improve an order fulfillment process by reducing cycle time and errors in product shipments.
- Reduce placement errors in TV advertising (e.g., avoid back-to-back airing of commercials of competing brands).

See Sidebar 2.2 for another example.

SIDEBAR *2.2*

ARE THE CAPTIONS UNDERSTANDABLE?

Word captioning of TV broadcasts for the hearing impaired is mandatory in the United States. For live programming, this is typically done manually by using stenographers who, to keep up with the broadcast, sometimes need to paraphrase rather than provide verbatim captioning. Large fines may be imposed for what regulators deem to be inadequate captioning, especially during critical events (e.g., weather emergencies). Broadcasters need to measure, improve, and track the "quality" of their closed captions and to uncover the root causes of the word inaccuracy rate (i.e., fraction of incorrect words) in captions. To address these issues, we were asked by a broadcast network to develop a statistical study to (1) quantify the word inaccuracy rate and its variability for different broadcast offerings and to (2) break down the variability by contributing components, such as program, speaker, background noise, stenographer, and so on. The results were then used to reduce inaccuracy rates and to establish controls that would help minimize future inaccuracies.

2.5   FURTHER APPLICATIONS

Peck et al. (2006) include chapters showing how statistical methods are used in or apply to

- Detecting cell phone fraud.
- Reducing junk mail.

- Improving the accuracy of a newspaper.
- Assuring product reliability and safety.
- Randomness in the stock market.
- Advertising as an engineering science.

2.6 MAJOR TAKEAWAYS

- Applications of statistics for manufactured products start with product design, reliability assurance, and scale-up (increasingly important areas) and continue on to manufacturing and field support. The assessment of the efficacy and safety of new drugs is a key application area; two other important application areas are semiconductor products and food and beverage products.

- The use of statistics to support service businesses has expanded appreciably in recent years. Application areas include finance, communications, statistical software development and support, and Internet operations.

- The improvement of processes—including both internal operations and the servicing of customers—provides another fertile area for the application of statistics.

DISCUSSION QUESTIONS

(* indicates that question does *not* require any past statistical training)

1. *Check out Deming's teachings and report on how and the degree to which these are being used in business and industry (and in other places) today.

2. Statistical reliability analysis calls for approaches that differ, in a variety of ways, from those taught in elementary statistics courses. Describe these differences.

3. We assert in Section 2.2.5 that "different statistical methods are typically used in discrete parts assembly and process industries." Elaborate.

4. In Section 2.5.5, we state that product burn-in is used to weed out likely early failures for semiconductor products. How is statistics used in establishing appropriate burn-in procedures? What are the advantages and disadvantages of doing such burn-ins? What would a cost model for burn-in and product performance take into account?

5. *Report on some recent pharmaceutical studies in the news that required statistical planning and analysis and describe how statistics was used.

6. In Section 2.2.5, we suggest that statistics is used in the food and beverage industries to minimize the variability that customers experience in repeat purchases of the same product. Describe statistical approaches that are employed for this purpose.

7. Check out and describe how statistics is used in the development of scoring methods to determine which future applicants should be approved for credit and, if so, for how large a line of credit.

8. Select an area in which statistics is being used to improve Internet operations and describe the statistical approach used.

9. Determine an area in which statistics is used to help improve a process and describe the statistical approach used.

CHAPTER 3

WHAT STATISTICIANS ENGAGED IN OFFICIAL GOVERNMENT STATISTICS DO[1]

3.1 ABOUT THIS CHAPTER

> Official statistics provide an indispensable element in the information system of a society, serving the government, the economy and the public with data about the economic, demographic, social and environmental situation.
>
> —United Nations Statistics Division

Historically, governments have depended on statistics to understand their populations and the resources available to them. Early examples include a census in the Old Testament Book of Numbers for determining military resources; the New Testament's Roman census for taxation; and William the Conqueror's Domesday Book, a census designed to learn about the resources available in the recently conquered Anglo-Saxon lands. All of these are examples of official statistics.

Governments and their administrative arms need official statistics for policy development, implementation, and evaluation. The public at large has similar needs for information to be able to evaluate government policy, to ensure public accountability, and to be adequately informed about social and economic conditions in order to make informed decisions.[2]

Statisticians play a critical role in the collection, analysis, and reporting of official government statistics. In this chapter, we describe this role. Statisticians engaged in official statistics are typically government employees. There are, in addition, many other statisticians employed by government whose principal function involves duties *other than* official statistics, such as drug safety and environmental protection. We will discuss their work in Chapter 4.

---

[1] This chapter was written largely by contributing author Leonard M. Gaines. It also benefited greatly from inputs by Rich Allen, Cynthia Clark, Len Cook, Ron Fecso, Juanita Lott, and Katherine Wallman.

[2] This statement and part of the first statement in the next section were adapted from the United Nations Statistics Division Fundamental Principles of Official Statistics, //unstats.un.org/unsd/methods/statorg/FP-English.htm.

A Career in Statistics: Beyond the Numbers, Gerald J. Hahn and Necip Doganaksoy.

There are also many statisticians working outside of government—mostly in academic institutions, national laboratories, and research institutes (Section 4.8)— engaged in official statistics, especially in research activities. They may also be collecting official statistics under contract with a government agency.

In this chapter, we consider

- The scope of official statistics.
- Examples of the work.
- Challenges of the work.
- Research opportunities.
- Some major employers in the U.S. federal government.
- Required credentials.
- Integration of U.S. government statistical activities.
- Local official government statistics activities.
- Government statisticians outside the United States.
- Compensation and other considerations.
- Sources of further information.

We will describe characteristics of the job (such as ensuring confidentiality, privacy, and security in gathering and reporting official statistics) and various other considerations in subsequent chapters.

3.2 THE SCOPE OF OFFICIAL STATISTICS

> Statistics are integral to governance of the state. Federal statisticians heed that call every day.
>
> —Juanita Lott

3.2.1 Diversity of Needs

Government statisticians around the world develop and administer surveys to gather, analyze, and report data periodically (monthly, quarterly, yearly, etc.) on a wide variety of subjects. Some examples are population counts and characteristics, education performance results, disease occurrence, poverty levels, traffic and transportation patterns, and inflation and (un)employment numbers. The subjects surveyed might be people or organizations such as companies. The information obtained from such surveys is often referred to as "official statistics." These data are used, for example, to

- Shed light on economic and social conditions.
- Develop, implement, and monitor government policies.
- Inform decision making, debate, and discussion both within government and in the wider community.
- Allocate federal and state funds.

In addition to designing and implementing surveys and reporting their results, government statisticians work extensively with existing administrative records— information collected routinely as part of a program's operations, such as school enrollment records, and income and property tax records. Crime rates, based on police agency arrests, are another example of official statistics obtained from administrative records.

There are a number of official statistical data series that result from statistical models relying on both administrative records and sample data. The unemployment rates and related data produced by the U.S. Bureau of Labor Statistics (BLS) (and by state labor departments for smaller geographic areas) and the Census Bureau's Small Area Health Insurance Estimates are two of the many examples.

3.2.2 Diversity of Customers

There are typically multiple audiences, or customers, for official statistics. The immediate customers might be a government agency, often the one in which the participating statisticians are "housed"; another agency might be paying for the work; and a Congressional committee might use the results to frame legislation.

Businesses are also major users of official statistics. For example, data on population change can provide important inputs in making decisions about retail store location.

The general public, or a segment thereof, is frequently another customer— usually after filtering of the information through the media.

Most government agencies also provide public access to key data sets and reports through their web sites—such as the National Center for Health Statistics' data on birth and deaths or the National Center for Education Statistics' annual summary of education data. In addition, statisticians in government agencies are often asked to give further background information in response to direct inquiries.

In light of the wide audience for official statistics, the data and findings that are provided need to be reported in a comprehensive, transparent, and unbiased manner that is readily understood by each customer.

3.2.3 Different Types of Studies

The decennial population census, the monthly Current Population Survey (CPS), the monthly Consumer Price Index (CPI, see Section 3.3.1), and the quarterly Gross Domestic Product (GDP) are four well-known examples of U.S. official statistics. These differ from one another in that the decennial population census, conducted by the U.S. Census Bureau,[3] aims for a complete enumeration, or count, of every person residing in the nation. The CPS, on the other hand, is a monthly sampling study of people, also conducted by the Census Bureau. Meanwhile, the CPI is based upon a sampling of prices conducted by the BLS. The GDP is produced by the Bureau of

---

[3] Various studies cited in this chapter are, in fact, administered by the U.S. Census Bureau. Due to its expertise in and infrastructure for efficient data collection, the Census Bureau frequently conducts statistical surveys on behalf of other government agencies, including other statistical agencies. It also has limited access to selected confidential administrative records (for statistical purposes only).

Economic Analysis and relies on a number of statistical models, which make use of the results of various sample surveys, and on administrative records collected by several different agencies.

Special one-time or occasional surveys are undertaken to gather information to address key issues and to gain the needed understanding to take appropriate actions. For example, the U.S. Department of Labor conducts special surveys to explore characteristics of the labor force, such as the number of people with disabilities, union membership, and how respondents balance work and family life. Also, in addition to routine studies of diseases, government health organizations identify and track the spread of epidemics.

3.2.4 Importance of the Work

> Public policy decisions should be driven by evidence.
> —Peter Orszag

The important role that official statistics plays in society is emphasized in Sidebar 3.1 through quotes from two leaders in the field: Janet Norwood, a former Commissioner of the BLS, and Len Cook, former National Statistician of the United Kingdom and Government Statistician of New Zealand.

SIDEBAR 3.1

THE IMPORTANCE OF OFFICIAL STATISTICS

Janet Norwood expressed the need for government statistics as follows (Norwood, 2006):[4] "Those who make governmental decisions need to be informed of the facts; they require objective knowledge of what is actually happening, because only then can government act wisely. ... Statistical measurement is an important part of that knowledge. ...You cannot have a democracy without a sound statistical system that is objective and relevant. ... The statistical system in a democracy has a heavy responsibility. More and more government policy is run by the numbers. Voting districts depend on Census results. Social Security payments are increased as the CPI rises, and people's view of the success or failure of a government policy frequently depend on whether GDP or the unemployment rate goes up. Our view of what is happening in almost every aspect of everyday life is in some way affected by the numbers."

Len Cook adds (in a personal communication) that statisticians responsible for official statistics "work in a unique context, as they operate with a strong statutory authority to gather information about the lives of citizens, and of the nature of business and other organizations, in order to report on the condition of the population, and the nature of commerce. ... Their work is always visible in some way."

---

[4] It should be noted, however, that the main point of Norwood's article is to argue that the system, "while giving tremendous value, also shows signs of wear and tear" and many parts "need refurbishing and bolstering," requiring "sustained research and development to guide improvement and innovations." Reamer (2009) provides a number of specific examples of the system's alleged wear and tear.

3.3 EXAMPLES OF THE WORK

Statistics is a guide in the journey from evidence to policy.
—Sally Morton

While at first glance, the work of those engaged in official statistics might not seem very enticing, those actually engaged in it typically insist otherwise. As another leader in the field (the Chief Statistician of the United States) puts it (Wallman, 2008) "no matter how long I sit at the helm, I am always enchanted to learn of the fascinating activities that engage my fellow (government) statisticians." She provides the following (and other) examples:

- Gathering the needed data to help policy makers propose laws to reduce school class size. This might include data on the impact of class size on testing results.
- Using survey data, statistical regression models, and administrative records to estimate the number of poor, school-aged children for each of the nation's approximately 14,000 school districts.
- Conducting tutorials for reporters and administration officials about the statistical profiles of the income and tax distributions of individuals and households.

This section provides some further examples of official statistics and the work of official statisticians.

3.3.1 Measuring Inflation

The CPI, reported monthly by the BLS, is a statistical measure used to gauge inflation or deflation. It is based upon the market price of a fixed basket of goods and services purchased by consumers. The CPI is used to calculate cost of living adjustments (COLA) for government programs, such as Social Security, and it is often the basis for making cost of living adjustments in setting wages. The index is calculated as the weighted average of the prices of designated products from a random sample of businesses. Statisticians are actively involved in helping ensure that proper sampling methods are used and that the weights continue to reflect an ever-changing product purchasing environment (Section 3.4.3).

Similar indexes, though not always based upon random sampling, are used to calculate the cost of living in countries around the world.

3.3.2 Tracking Crimes[5]

The U.S. Department of Justice's Bureau of Justice Statistics (BJS) collects data on various aspects of the judicial system. For example, its National Crime Victimization Survey measures the national rate of personal and property crime victimization (both

[5] The discussion of the BJS data is based on a description by Sinclair (2009).

reported and unreported) and the characteristics and consequences.[6] The BJS also tracks various other statistics, such as the number of law enforcement officers, the rate of convictions, the flow of inmates through prisons, and the rate at which those released from jail return to crime. In addition, the Federal Bureau of Investigation tracks certain reportable crimes, based upon police department arrest records.

Such data are used, for example, to determine the need for new prisons, where to place police, and the effectiveness of different programs aimed at crime prevention or rehabilitation.

3.3.3 Improving Tax Collection[7]

The Statistics of Income (SOI) Division of the U.S. Internal Revenue Service (IRS) collects and disseminates information on the operation of U.S. tax laws. The Division also conducts specialized studies to help policy makers assess, for example, the impact of tax law changes and their effect on tax collections.

Statisticians in the Mathematical Statistics Section of SOI develop sample designs for studies dealing with tax collection and monitor their execution. This often involves the development of new methods targeted at improving the sample design and/or the subsequent estimation methods. Help is also provided, on occasion, to economists in data management and in the analysis of empirical studies.

Statisticians in SOI's Statistical Support Section provide statistical consultation to other divisions in the IRS in studies that are often of high interest to or affect the general public. Some typical projects are

- Study of alternative methods for filing federal taxes.
- Assessment of the quality of IRS responses to customers.
- Evaluation of proposed methods for reducing the time it takes the IRS to process taxpayer remittances.

3.3.4 Assessing Gender Pay Equality (and Other GAO Projects)[8]

The Government Accountability Office (GAO) is an independent, nonpartisan agency that works for the U.S. Congress and acts as a watchdog on how government money is spent. GAO's work is conducted at the request of Congressional committees or subcommittees or is mandated by law and provides helps in framing new laws.

One recent project was directed at the question "to what extent has the pay gap between men and women in the federal workforce changed over the past 20 years and what factors account for the gap?" GAO statisticians worked with the study team to analyze U.S. workforce pay data over multiple years and studied a cohort (Section 11.5.5) of those workers who entered the workforce in 1988. Using regression and

---

[6] The data collection for this study is conducted by the U.S. Census Bureau.

[7] This discussion is based on a description by Petska et al. (2001) of the work conducted by mathematical statisticians at the IRS.

[8] This discussion is based on information provided by Schwartz and Shipman (2009).

other methods, they quantified some possible explanations for gender pay gap differences. The results were summarized in a report to a Congressional committee and to subcommittee chairpersons Kennedy, Harkin, and Maloney.[9]

3.3.5 Some Other Examples

Some further examples of projects in which government statisticians collect and report official statistics are

- Evaluating the hazard associated with having infants sleep in an adult bed (by the Consumer Product Safety Commission).
- Tracking citizen complaints about police use of force (by the BJS).
- Developing models for forecasting future energy prices and allowing users to apply these models online to create their own forecasts, based upon up-to-date data provided by the U.S. Energy Information Administration.

On a lighter note, as noted in Section 1.2, statisticians in the United Kingdom have recently been commissioned to develop a population "happiness index" (and presumably asked to conduct a sampling study to arrive at a number).

In addition to activities at the federal level, statistical work in the United States also takes place at the state and local levels of government. Some examples are

- Collecting birth, marriage, divorce, and death statistics (by county and state health departments).
- Calculating the odds of lottery games sponsored by state governments.
- Monitoring and reporting on government operations dealing with, for example, the weekly number of new unemployment claims or monthly tax revenues.
- Evaluating the validity of standardized education tests and civil service examinations.

Summing up the roles of statisticians in the federal government, Ron Fecso, chief statistician at the U.S. Government Accountability Office, asserts (Fecso, 2008) that "work in the federal government provides you the opportunity to have an impact on . . . major (national) policy issues." A similar comment can be made about statisticians working for state or local governments.

3.4 CHALLENGES OF THE WORK

3.4.1 Systems Development and Updating

Statisticians are frequently engaged in the initial development and implementation of systems for gathering, storing, processing, and making accessible needed data. Developing statistically sound systems and ensuring that the work is conducted as

---

[9] The full report "Women's Pay: Gender Pay Gap in the Federal Workforce Narrows as Differences in Occupation, Education, and Experience Diminish" (GAO-09-279) is available at //www.gao.gov/new.items/d09279.pdf.

planned—while watching out for the unexpected—is an important responsibility of government statisticians. This can involve the development of new statistical techniques, evaluation studies during both the study planning and implementation phases, and quality assurance evaluations once systems for gathering official statistics are in place. A good background in computing is extremely helpful for such work.

3.4.2 Ensuring Statistical Validity

Most statistical methods rely on assumptions about statistical practices that do not always hold. It may, for example, not be possible to include some groups, such as the homeless, in the survey. Low response rates to surveys are another problem commonly faced by government statisticians. This becomes especially complicated when nonresponse tends to be concentrated in some segments of a sampled population such as young males, or in some types of businesses, but not in others.

When the statistical assumptions on which the methods are based cannot be met, government statisticians have to develop reasonable adjustments to the procedures or data to account for such concerns and to justify, implement, and document these.

We provide a more detailed discussion of getting good data in Chapter 11.

3.4.3 Responding to Changing Environments

The Challenge. A system that makes sense and provides useful results today may, in light of changing circumstances, no longer be appropriate tomorrow. Therefore, an important continuing task, beyond routine system development and implementation, is that of monitoring the continued adequacy of existing systems and modifying them, when needed. Precisely what the CPI is measuring, at any given time, needs, for example, to be reviewed periodically as social and economic conditions change; see Norwood (2006). Great care must be exercised in so doing to ensure, among other things, comparability of results across time.

Consider specifically the impact of the personal computer on the CPI. Since the 1980s, home computers have become increasingly common and sophisticated, while their prices have continuously declined. Thus, provisions need to be made in the CPI to appropriately add home computers—including changes in their performance and quality, as well as their price—to the index so that it continues to be representative of consumer purchases.[10]

Impact of Advances in Technology. Continuous improvements in technology provide further reasons for modifying a statistical system. The Internet has from its earliest days led to faster transfer of data to statistical agencies. It has also presented important new challenges and opportunities for data gathering and analysis.

---

[10] A description of how the BLS updates the CPI is included in Chapter 17 of the U.S. Bureau of Labor Statistics BLS Handbook (2007).

Government agencies, for example, work under tight laws to guarantee the confidentiality of the information they receive from survey participants (Section 10.4.4). The ability to assure respondents that their confidentiality would be fully protected in an Internet environment has been questioned. As a consequence, some government statistical agencies have been reluctant to use the Internet to collect data.

The Internet and the expansion of microcomputing have also had tremendous impact on the tabulation and dissemination of official statistics.[11] From the 1960s through the 1990s, many statistical summaries were distributed as tables on magnetic tapes for use on mainframe computers. The users of these data tended to be trained data analysts who had a good understanding of the underlying statistical issues. Today, many more people—often with little or none of the needed understanding of statistical subtleties—are able to access and manipulate the data. To address this situation, statisticians are spending increasingly more time working on ways to present data and background information to make these readily understandable to users with limited technical background.

Impact of Changes in Sociological Environment. The ever-changing environment, such as the growing influence of service businesses in the United States and the increasing role of globalization, creates the need for revising existing systems and developing new ones. In addition, since stability and comparability over time of the data being collected are of clear importance to all involved, there is a need to anticipate relevant sociological, as well as technological, changes. This includes the public's perception of the importance of government data gathering and reporting, and the ability to keep the data secure. Those who plan to work in these areas will benefit from an educational background that is strong in the social sciences, as well as statistics.

The U.S. Census Bureau's American Community Survey (ACS) is one example of official statistics responding to both changing societal needs and the long development times sometimes required to bring a new official statistics program to life. This survey was designed to address the increasingly rapid changes in socio-economic conditions by providing information on age, race, income, commute time to work, home value, and so on more frequently than afforded by the decennial census's long-form questionnaire.

The ACS replaces the once-every-10-year sampling study (about 18.3 million homes in 2000) with an ongoing monthly sample of 250,000 housing units, allowing yearly release of results, based on 12, 36, and 60 months of pooled observations. It calls for a new way of collecting and reporting data for very small geographic areas. Because of this and other reasons, it took from (the initial discussions about data collection and presentation methods in) the mid-1980s to 2006 for full implementation for data collection and 2010 for the initial release of small area data.

---

[11] The needs of official statistics and the changing environment have, in themselves, led to technological advance. Herman Hollerith, for example, developed punch card technology as a tool for processing the 1890 Census.

Another example of the need for federal statistical agencies (or programs) to respond to changing environments is provided by the North American Industry Classification System (NAICS); see Sidebar 3.2.

SIDEBAR *3.2*

THE NORTH AMERICAN INDUSTRY CLASSIFICATION SYSTEM

The North American Industry Classification System, a standard issued by the Office of Management and Budget (OMB) for classifying business establishments, was developed during the late 1980s and early 1990s. In 1997, it replaced the Standard Industrial Classification (SIC) system, which had been used since the early 1940s by U.S. federal statistical agencies to classify business establishments for the purpose of collecting, analyzing, and publishing statistical data related to the business economy. NAICS addressed the fact that the SIC system, with its very strong bias toward manufacturing reflecting the economy of the 1930s and 1940s, could no longer be modified to present meaningful information about the emerging service-based economy. Also with the increased economic relationships between the United States, Canada, and Mexico, there was a need for a common classification system that could be used by all three nations.

Statisticians from the three countries worked together to develop NAICS. One of its particularly useful aspects is the built-in recognition of the importance of the system remaining current—requiring that NAICS be reviewed and revised, as needed, every 5 years.

3.4.4   Serving as Official Data Intermediaries

Another major duty of many government statisticians, especially those in federal agencies not primarily charged with producing official statistics and those in state and local governments, is serving as data intermediaries. This function involves determining the needs of the ultimate user, finding and often reformatting the data, interpreting the data and the associated statistical limitations, and explaining to users how the data respond to their needs and what it all means.

The ultimate users, within the statistician's agency or elsewhere, may be looking for data to help design a program or develop some new legislation over the next few months. Or they may be senior officials requiring inputs for a speech to be given in an hour. They can also be members of the general public seeking official statistics for almost any reason imaginable, ranging from settling a bet to helping develop a business plan.

As data intermediaries, government statisticians must understand the data users' needs. They must also know where to find a wide variety of official statistics and how to access and use such data. On other occasions, they may have to compensate for a lack of official statistics by going to private data providers, such as in securing data on local home sale prices. It may also be necessary to make estimates based on available official statistics, for example, estimating the square footage of grocery stores in a local area from national information about average square footage per employee—subject to the stated assumption that the national figures can appropriately represent local conditions.

All of this requires an understanding of all aspects of the data being used, including why and how the data are collected and processed (and such subtleties as adjustments for missing data and methods for protecting respondent confidentiality) and delivered. This knowledge also helps the statistician understand and communicate the data's limitations, as well as usefulness for the purpose at hand.

This role, like others, while requiring good technical statistical abilities, also calls for a variety of other skills. These include understanding computer programs that can be used to present the data in various ways (such as spreadsheets for creating tables and graphs, and geographic information systems for mapping the data spatially), as well as outstanding communication abilities. We will return to these topics in later chapters.

3.5   RESEARCH OPPORTUNITIES

Many U.S. federal statistical agencies conduct ongoing research. Statisticians in such agencies work on conceiving new programs, improving existing programs and procedures, and making them more efficient. They also help develop and evaluate new technologies. As a result, government statisticians conduct research in numerous areas of statistics.

Common research topics include developing improved methods for creating small area estimates, evaluation of data suppression and publication alternatives (to protect confidentiality without sacrificing key information), and data mining of agency databases. Seasonal adjustment of time series data is an especially important topic—and one that is often not well understood by the general public. As a consequence, development of the best possible adjustments—and their defense— is a very active field for researchers.

Many official statistics are based on surveys and there is a constant need to improve these in response to social and budgetary pressures. Thus, many statistical agencies are actively involved in survey methodology research.

Research in official statistics can be truly groundbreaking. For example, the National Aeronautics and Space Agency (NASA) began exploring practical applications of aerospace remote sensing in 1968, some 4 years before the first earth resources satellite was launched. Today, remote sensing data are obtained by various agencies and contribute to our knowledge about such issues as climate change and land utilization.

A few government statisticians engaged in official statistics spend much of their time in conducting research. For many others, research studies provide an occasional change of pace and an opportunity to develop additional technical skills that are then leveraged in applied work. In conducting such research, government statisticians often collaborate with academic statisticians who may spend time with their agency through fellowships, sabbaticals, or other contractual arrangements.

3.6   SOME MAJOR EMPLOYERS IN THE U.S. FEDERAL GOVERNMENT

Statisticians in the United States are employed by various federal agencies that have responsibility for gathering and reporting data. Their specific number depends upon one's definitions of an agency and of a statistician. The BLS reports that in 2009 statisticians held about 4400 jobs in the federal government, with the highest concentration in the Departments of Commerce (especially the Census Bureau), Agriculture, and Health and Human Services.

FedStats[12] lists well over 100 U.S. federal agencies with statistics programs and gives brief descriptions of the statistics they provide (and links to their web sites), as well as contact information and key statistics.

Citro (2009) lists the 14 principal U.S. statistical agencies and their parent organization; see Table 3.1. These are members of the Interagency Council on Statistical Policy, established by the OMB to coordinate statistical policy and operations among agencies and to advise the OMB on major statistical policy issues. The budgets of these agencies are approved by seven different Congressional appropriations subcommittees, and their operations are monitored by a variety of Congressional oversight committees. In addition, numerous other agencies have budgets for statistical activities.

The 14 principal statistical agencies have various types of statistical activities among their primary duties. Nine of them have "statistics" in their titles. Together, they employ more than 10,000 permanent staff members, including approximately 2000 people who are classified as statisticians and many more with backgrounds in computer science, economics, and demography (Wallman, 2008).

TABLE 3.1   Principal U.S. Federal Statistical Agencies

Bureau of Economic Analysis (Department of Commerce)
Bureau of Justice Statistics (Department of Justice)
Bureau of Labor Statistics (Department of Labor)
Bureau of Transportation Statistics (Department of Transportation)
Census Bureau (Department of Commerce)
Department of Environmental Information (Environmental Protection Agency)
Economic Research Service (Department of Agriculture)
Energy Information Administration (Department of Energy)
National Agricultural Statistics Service (Department of Agriculture)
National Center for Education Statistics (Department of Education)
National Center for Health Statistics (Department of Health and Human Services)
Office of Research, Evaluation, and Statistics (Social Security Administration)
Science Resource Statistics Division (National Science Foundation)
Statistics of Income Division (Treasury Department)

From Citro (2009).

[12] //fedstats.gov/agencies/.

In addition to those in official statistics, there are many opportunities for statisticians in federal agencies that have other primary missions, such as administration of benefits (e.g., Medicare and Medicaid) and regulation (Section 4.2)—as well as in such agencies as the Federal Highway Administration, the Federal Reserve System, and the U.S. Geological Survey. These agencies are often heavy consumers of the data produced by the principal statistical agencies, combining such data with their own both for monitoring their own operations and for establishing policy.

Many, but not all, federal government statisticians are located in the Washington, DC, area. This high concentration of statisticians and of related professionals provides those involved the opportunity to be stimulated by, learn from, and interact with other statisticians from many different agencies, academia, and consulting firms and to actively participate in professional activities (see Fecso, 2008).[13] Several of the principal statistical agencies also employ statisticians in their regional offices, which tend to be in the nation's major cities or state capitals.

In addition, many consulting firms, such as Mathematica Policy Research, the National Opinion Research Center, the Research Triangle Institute, and Westat, have the federal government as a major client and hire statisticians to work on projects.

3.7 REQUIRED CREDENTIALS

The Office of Personnel Management states that, in general, U.S. federal professional positions require

> Successful completion of a full 4-year course of study in an accredited college or university leading to a bachelor's or higher degree.[14]

Statisticians responsible for the collection and reporting of official statistics often hold a bachelor's or master's degree. Those involved in applied research, however, often have a master's or, occasionally, Ph.D. degree. Federal government classification standards, moreover, differentiate between statisticians and mathematical statisticians.

Degree Requirements for U.S. Federal Statisticians. Government regulations require statisticians to hold a degree that

> Included 15 semester hours in statistics (or in mathematics and statistics, provided at least 6 semester hours were in statistics), and 9 additional semester hours in one or more of the following: physical or biological sciences, medicine, education, or engineering; or in the social sciences including demography, history, economics, social welfare, geography, international relations, social or cultural anthropology, health sociology,

---

[13] The Washington Statistical Society is the ASA chapter with the largest membership; it usually has several seminar programs every month.

[14] U.S. Office of Personnel Management. Group Coverage Qualification Standards for Professional and Scientific Positions, //www.opm.gov/qualifications/Standards/group-stds/gs-prof.asp.

political science, public administration, psychology, etc. Credit toward meeting statistical course requirements should be given for courses in which 50 percent of the course content appears to be statistical methods, e.g., courses that included studies in research methods in psychology or economics such as tests and measurements or business cycles, or courses in methods of processing mass statistical data such as tabulating methods or electronic data processing.[15]

The degrees possessed by those hired as statisticians are sometimes in subjects other than statistics, such as mathematics, applied mathematics, operations research, economics, business, computer science, sociology, or urban and environmental studies—although such individuals must still meet the cited basic requirements.

Degree Requirements for U.S. Federal Mathematical Statisticians. Most statisticians working in the major U.S. federal statistical agencies are hired as mathematical statisticians. They are expected to perform a much broader array of statistical work than statisticians. Their degree requirements are specified to include

24 semester hours of mathematics and statistics, of which at least 12 semester hours were in mathematics and 6 semester hours were in statistics.[16]

While the preceding are the minimum requirements, those with only a bachelor's degree are encouraged to earn a master's degree in statistics or in a closely related area, such as survey methodology. Such additional education is especially important for gaining promotions. (See Section 7.4 for further discussion on how far statisticians should take their formal education.)

Those hired as mathematical statisticians typically have statistics degrees or a degree in a highly quantitative field such as operations research or applied mathematics.

Citizenship Requirements. U.S. citizenship is a requirement for permanent U.S. government positions. There are, however, several hiring mechanisms for permanent residents and other noncitizens in some agencies (see O'Neill et al., 2008).

Degree Requirements for State and Local Statisticians. The requirements for statisticians working in state and local government in the United States vary greatly. New York State's requirements for statisticians, for example, are similar to the federal government's requirements. Thus, New York State requires a statistician to have

Either

1. A bachelor's degree or higher in statistics; or
2. A bachelor's degree or higher including or supplemented by 24 semester credit hours in statistics, biostatistics, computer science, economics/econometrics, mathematics,

---

[15] U.S. Office of Personnel Management. Statistics Series, //www.opm.gov/qualifications/standards/iors/gs1500/1530.asp.

[16] U.S. Office of Personnel Management. Mathematical Statistics Series, //www.opm.gov/qualifications/standards/iors/gs1500/1529.asp.

operations research, or research methods, 3 of which must be in statistics or biostatistics and 9 more of which must be from among: differential or integral calculus, econometrics, biostatistics, or statistics.[17]

Variations of these requirements are used for biostatisticians and for program research specialists.

In contrast, the State of Oregon uses the broader Research Analyst title to encompass statistical and other related positions. The minimum requirements for a Research Analyst are

College-level coursework that includes at least three quarter units in business or technical report writing, three quarter units in quantitative analysis or statistical principles, and six quarter units in computer concepts and microcomputer applications such as word-processing, spread sheet and database.

OR

One year of experience in work that included gathering data from various sources and using software applications to retrieve, edit or tabulate data into forms or reports.[18]

3.8 INTEGRATION OF U.S. GOVERNMENT STATISTICAL ACTIVITIES

In contrast to many other countries, federal statistical activities in the United States are basically decentralized; each agency has prime jurisdiction over its own activities. This decentralization may, in part, be a consequence of the fact that the system grew by adding separate agencies whenever congress and the executive branch felt these were needed (Norwood, 1995).

There are various other organizations dedicated to integrating and/or coordinating statistical activities involving official statistics in the United States. For example,

- The Federal Committee on Statistical Methodology, an interagency committee operating under the auspices of the OMB, brings together experienced representatives from different government agencies to identify and document best practices and to encourage improvements.
- The Council of Professional Associations on Federal Statistics disseminates information on developments in federal statistics and on educational programs about how official statistics are used; see Shipp and Cohen (2009, 2010).

---

[17] New York State Department of Civil Service. "Statistician/Statistician Trainee/Biostatistician/Biostatistician Trainee" Recruitment Announcement, //www.cs.state.ny.us/announarchive/announcements/25-280.cfm.

[18] State of Oregon. "Classification Specification Library." Research Specialist I, //agency.governmentjobs.com/oregon/default.cfm?action=viewclassspec&classSpecID=733810&agency=1807&viewOnly=yes.

- The Association of Public Data Users promotes communication between federal statistical agencies and data users.

See the book's ftp site for more details.

3.9 LOCAL OFFICIAL GOVERNMENT STATISTICS ACTIVITIES

In addition to the federal government, numerous state and local governments in the United States employ statisticians in the collection and reporting of statistics. They are engaged in addressing topics of local interest and in gathering and analyzing data needed to help develop agency policies.

However, unlike the federal government, most local governments do not have their own statistical agencies. Instead, their statisticians are likely to belong to a general policy and research unit or to an operational group. For example, this chapter's author works in the policy and research division of his state's economic development agency. In this position, he assists other analysts in finding data to help evaluate proposed programs or legislation or to promote the state's interests.

There are also many cooperative agreements between federal and state agencies aimed at meeting the statistical needs of both. These can take many forms, as suggested by the examples in Sidebar 3.3.

SIDEBAR 3.3

EXAMPLES OF COOPERATIVE AGREEMENTS BETWEEN U.S. FEDERAL AND STATE AGENICES

- Michigan's Department of Agriculture, through a cooperative agreement with the National Agricultural Statistics Service (NASS), collects and analyzes data on agricultural characteristics to help the state's farmers make informed decisions on the type and amount of crop to grow.[19] Many state agricultural statistical services operate under joint funding from NASS, share office space, and get directions from the same person.

- State employment security agencies collect, summarize, and disseminate data about employment, such as the number of jobs and the unemployment rate, under a contract with the BLS. The BLS determines the statistical methods to be employed, in consultation with the states' statisticians, who then implement the program.

- Each of the states and U.S. territories participates in the Census Bureau's State Data Center program. The Census Bureau provides the states ready data access. In return, the states disseminate the data and answer technical questions for a minimal charge (or no cost at all). They also provide the Census Bureau information about user needs to help it improve its data and products.

- The New York State Department of Labor, in partnership with the U.S. Census Bureau's Local Employment Dynamics program, develops in-depth information

---

[19] http://michigan.gov/mda/.

about local employment conditions and trends. Under a strict confidentiality agreement, the state provides the Census Bureau with information about state employment. The Census Bureau then combines such data with other information to produce and, after review by state statisticians, issue integrated synthetic data[20] on worker characteristics.[21]

In addition to cooperating with the federal government, state agencies often work with their counterparts in other states. Such cooperative efforts help statisticians who might otherwise be working in isolation to develop collegial relationships with their peers and to stay current.

The number of statisticians employed at the state and local levels varies greatly based on perceived needs. Also, some state and local government agencies that lack in-house statistical expertise seek help from other sources, such as statisticians at local universities or private consultants.

3.10 GOVERNMENT STATISTICIANS OUTSIDE THE UNITED STATES

National Agencies. Similar to the United States, countries around the world have agencies employing statisticians. The International Statistical Institute's (ISI) web site[22] provides one-stop access to the central statistical agencies of all countries.

There is much variation in different nations' statistical structures. Canada has a more centralized statistical system than the United States, with Statistics Canada as the primary statistical agency (see Sidebar 3.4). But, like the United States, the Canadian statistical system is divided between levels of government, with statisticians working for provinces (or territories), as well as for the federal government.

SIDEBAR *3.4*

CANADA'S STATISTICAL SYSTEM[23]

Canada's Statistics Act establishes a mandate for Statistics Canada, its statistical office, with virtually unlimited subject coverage. The Act also makes information requests by this agency compulsory unless specifically designated otherwise, providing Statistics Canada blanket access to many administrative records held by governments—including taxation and customs and court records. At the same time, the Act imposes an ironclad guarantee of confidentiality.

---

[20] The synthetic data are respondent-level data that have been altered slightly for public use to protect respondent confidentiality, but that retain the statistical distributional characteristics of the original data.

[21] http://lehd.did.census.gov/led/.

[22] http://isi-web.org/statsoc/directory.

[23] This sidebar is adapted from Fellegi (1996). Also see Denis et al. (2002) for a discussion on preparing statisticians for a career in Statistics Canada.

However, the Act does not establish a central statistical agency per se. Instead, it requires Statistics Canada to coordinate the national statistical system. As a result, Statistics Canada enters into joint collection and data sharing agreements with both federal government departments and the statistical agencies of individual provinces, subject to strict confidentiality protection.

In the United Kingdom, legislation in 2007 established an independent Statistics Board that reports directly to parliament. The Office for National Statistics is the nation's single largest producer of official statistics with approximately 4000 employees. The National Statistician oversees the decentralized government statistical service and the Office for National Statistics.

Even Singapore, a small country, established under a Statistics Act its own Department of Statistics to serve as the national authority and coordinator responsible for economic and population statistics.

Each country has structured its statistical system to be in line with its culture and data needs, and the major means of collecting data must do likewise. For example, some countries, such as The Netherlands and the Nordic countries, rely heavily on national register systems for their sources of data.

International Agencies. Many international (and "supranational") organizations, such as the United Nations and its component agencies and Eurostat (the European Union's statistical agency), have their own statistical systems. The ISI web site again provides access to UN agencies and other international statistical agencies (such as those of the International Monetary Fund and the World Bank).

Sometimes, such agencies meet administrative, as well as statistical, needs. The Harmonized System was developed by the World Trade Organization to provide all of the organization's member nations a common way to classify exports and imports. Eurostat also has numerous statistical requirements designed to integrate the use of statistics across the European Union.

Other international organizations, like the World Bank, employ data produced by different national systems (e.g., gross domestic product). This often requires working with data using different collection methods and trying to reconcile these to achieve a reasonable degree of comparability. Just as elsewhere (Section 11.6.7), this can be a difficult task.

3.11 COMPENSATION AND OTHER CONSIDERATIONS

The monetary compensation of government statisticians may not always match that provided by industry—especially at higher levels and in management positions. In addition, government statistical agencies often go through periods of virtually flat operating budgets and budget freezes, placing an added strain on operations (and even greater pressure for improved efficiency).

But there are other compensating benefits. Quoting Fecso (2008), "although the starting salary (in government) can be a little lower than what other employers offer, the leave and retirement program, steady employment, challenging problems and other pluses even the score. If you are motivated by service and want to like what you do and know it is important, then government may be the place for you." While Fecso was describing working for the federal government, the same comments apply, to varying degrees, for state and local governments.

3.12 SOURCES OF FURTHER INFORMATION

The articles by Norwood (2006) and the presentation by Fecso and Olson (2006) provide further details about the variety of work conducted by U.S. federal statistical agencies. Fecso (2008) and Wallman (2008) describe their careers culminating as chief statisticians of the U.S. legislative and executive branches, respectively. Also, Beale (2004), Clark (2008), Lott (2008), Schenker (2008), and Siegel (2005) provide insights from successful government statisticians and Fellegi (1996, 2004) comments on effective statistical systems and the pressures and challenges of official statistics. Periodic articles in *Amstat News* describe statistics in different government agencies; see, for example, Davis (2010). Also see Section 14.4.2 for information on current journals and other ongoing sources of information about official statistics.

3.13 MAJOR TAKEAWAYS

- The proper collection, analysis, and reporting of official statistics by government and other organizations is a vast, and critically important, challenge. Customers for the resulting information include government agencies, businesses, and the general public.
- Official statistics involves all areas of society. Some typical examples, in addition to the all-important national censuses, are measuring inflation, tracking crimes, improving tax collection, and assessing gender equality.
- In the United States, 14 major federal statistical agencies are heavily engaged in statistics; much additional work is conducted by various other federal, state, and local agencies.
- Similar work is conducted by governments worldwide, using structures and methods built around each nation's culture—as well as by international agencies, such as the United Nations.
- Statisticians are frequently engaged in the development, implementation, and updating of systems, or surveys, for gathering, storing, processing, and making accessible needed data.
- Many of the surveys government statisticians work on are ongoing; others are one-time. Systems development time can range from a few months to decades.
- Systems need to be monitored and modified over time to ensure that they remain relevant in the face of changing needs and accurate and efficient despite advances in technology and changes in society.

- Some statisticians, especially outside the 14 statistically focused agencies, work principally as data intermediaries.

- Statisticians involved in official statistics are engaged, to varying degrees, in conducting research to conceive new programs and improve existing ones and in developing and evaluating new technologies.

- Federal government job classifications differentiate between statisticians and mathematical statisticians. Most statisticians working in the major U.S. federal statistical agencies are hired as mathematical statisticians. They are expected to perform a much broader array of statistical work than statisticians.

- Although there is some variation, a bachelor's degree in statistics or a closely related field, including several statistics and advanced mathematics courses, is typically the *minimum requirement* for government statisticians engaged in official statistics. Some, especially those engaged in research, hold advanced degrees; mathematical statisticians are encouraged to get master's degrees.

- In contrast to many other countries, U.S. government statistical activities are basically decentralized. Coordination is provided by the OMB.

DISCUSSION QUESTIONS

(* indicates question does *not* require any past statistical training)

1. What are some challenges that face the planners of the next national (decennial) U.S. census in light of advances in technology and changes in society?

2. *What are some issues that statisticians at the U.S. Bureau of Labor Statistics are facing today (or will be expected to face soon) to ensure that the CPI continues to accurately reflect consumer prices?

3. *One of the "fascinating activities" in which statisticians are engaged, mentioned in Section 3.3, is "gathering the needed data to help policy makers propose laws to reduce school class size." Consider a law in this area that policy makers might be (or should be) contemplating. What information would you propose be obtained to help frame the law and how would you proceed to get the needed data?

4. How would you go about getting appropriate data to respond to the question concerning gender pay gaps in Section 3.3.4? Compare your approach to that used by the Government Accountability Office, as described in the web-accessible report GAO-09-279.[24]

5. In Section 3.5, we state that "seasonal adjustment of time series data is an especially important topic.... As a consequence, development of the best possible adjustments—and their defense—is a very active field for researchers."

---

[24] http://www.gao.gov/new.items/d09279.pdf.

Explain the need for seasonal adjustments in government time series data and provide some specific examples. How are such adjustments made?

6. Select one or more of the 14 principal U.S. federal statistical agencies enumerated in Table 3.1 (other than those discussed in this chapter) and research and report on the agency's major statistical tasks and challenges.

7. *Research and describe the statistical system for gathering and reporting official statistics used in the United Kingdom (or some other country than the United States and Canada).

8. Select a U.S. state, province, or other locality. Determine and report on the capacities in which statisticians are used and by which agencies. Where might they be used where they currently do not seem to be?

9. *On May 8, 2010, newspapers in the United States reported that the most recent U.S. government survey showed that employment was up by almost 300,000, but that the jobless rate also rose from 9.7% to 9.9% (giving ammunition to contrary political viewpoints). Explain how this discrepancy can happen and suggest some issues that need to be addressed in obtaining, analyzing, and reporting employment data.

10. How does the statistical system for measuring employment in the United States differ from that in other countries and to what degree are the resulting numbers comparable?

11. *Compare the jobs of statisticians engaged in official statistics with those in business and industry. What are some of the similarities and differences? Which area seems more appealing to you and why?

CHAPTER 4

WHAT STATISTICIANS DO: SOME OTHER APPLICATION AREAS

4.1 ABOUT THIS CHAPTER

In this chapter, we consider some added application areas in which statisticians are engaged: regulatory activities, health, national defense, other scientific research, social and behavioral sciences, and teaching (in nonacademic settings). We conclude with a brief description of some further institutes that conduct various types of research in the United States.

Some of the statisticians involved in these areas are government employees. But their work is often quite different from that of government statisticians who are principally involved in gathering, analyzing, and reporting official statistics, as described in Chapter 3. They are also more likely to hold Ph.D. degrees than their counterparts involved in official statistics.[1]

The application of statistics, together with other quantitative approaches in some of these and other areas, has given rise to a new nomenclature of professional fields, such as biostatistics, chemometrics, econometrics, and psychometrics.[2]

4.2 REGULATORY ACTIVITIES

Numerous government agencies are engaged in a variety of regulatory activities. In the United States, these include the Food and Drug Administration, the Environmental Protection Agency, the Federal Communications Commission, and the Consumer Product Safety Commission. Their work includes helping

- Establish laws and regulations to protect, for example, the general public (or particular segments) or the environment.
- Develop processes and procedures to assess that existing laws and regulations are being followed.
- Monitor to rapidly identify, and lead to correction of, specific violations.

---

[1] Government statisticians engaged in official statistics and those engaged in other activities also have some communalities, including the job grading system (Section 5.5) and restrictions on employment of noncitizens.

[2] Some, together with *Technometrics*, have also become the names of professional journals.

A Career in Statistics: Beyond the Numbers, Gerald J. Hahn and Necip Doganaksoy.
© 2011 John Wiley & Sons, Inc. Published 2011 by John Wiley & Sons, Inc.

Such activities are of strong interest to those directly impacted by them, especially business and industry, as well as the general public. We provide some further details about two such agencies.

4.2.1 Food and Drug Administration (FDA)

The U.S. FDA's mission, according to its web site, is to

- Promote and protect the public health by helping safe and effective products reach the market in a timely way.
- Monitor products for continued safety after they are in use.
- Help the public get the accurate, science-based information needed to improve health.

The FDA is the oversight agency to which pharmaceutical companies in the United States submit their plans for the testing of a proposed new drug, medical device, and vaccine, among others, for approval. Then, after they obtain and analyze the results, the companies submit their findings to the FDA requesting approval for market release (or further study), when this is felt to be warranted.

According to O'Neill et al. (2008), the FDA employs more than 200 Ph.D. or master's level statisticians, most of whose work is concentrated in the evaluation of applications for approval of new products. Their task is to ensure the statistical validity of the planned studies of a product's safety and effectiveness and to assess the company's analyses of the results of such studies. In this manner, statisticians play an important role in framing recommendations concerning the approval for market release of a new drug or medical product.

Statisticians' activities at the FDA are currently focused in three groups, whose titles summarize their basic missions:

- The Center for Drug Evaluation and Research.
- The Center for Devices and Radiological Health.
- The Center for Biologics Evaluation and Research.

4.2.2 Environmental Protection Agency (EPA)

The U.S. EPA's mission is to protect human health and the environment. Its tasks include developing and enforcing environmental regulations and in general studying environmental issues.

Much of the work of EPA's statisticians involves helping obtain valid measurements of hazardous exposure in water, in air, and in the ground and then analyze the resulting data to quantify the associated risks to humans, animals, vegetation, and so on. Like statisticians engaged in official statistics, EPA's statisticians are involved in the design and analysis of sampling studies. However, the sampling is typically of "things" rather than of people (or organizations). Thus, major challenges tend to deal with such factors as measurement capability, rather than questionnaire construction or the impact of nonresponse.

Specific applications, cited by Nussbaum (2008), are to

- Derive the relationship between childhood leukemia and environmental pollutants.
- Trace the movement of chemicals through the environment.
- Quantify fluid runoff from airplane deicing.
- Assess compliance with auto emission standards.
- Estimate automobile fuel economy.
- Evaluate drinking water safety.
- Measure and assess the hazards from toxic chemicals in soil.

The use of statistics to address environmental issues is not limited to the EPA. Industry employs statisticians to monitor and address potential environmental hazards; see Section 13.6.6 of Hahn and Doganaksoy (2008) and the comments on climate science later in this chapter.

4.3 HEALTH

Statisticians are heavily engaged in a wide variety of health issues. We have already suggested some of these in our earlier discussions of pharmaceutical products, official government statistics, and the work conducted by the FDA and EPA. The National Institutes of Health of the U.S. Department of Health and Human Services is a major employer of statisticians; see Ellenberg et al. (1997).

As we indicated in Section 1.2, the application of statistics to the analysis of biological and medical data is often referred to as biostatistics, and statisticians who work in this area are often referred to as biostatisticians; see Piccolo (2010).

Some further applications are suggested in this section. See Section 13.6.1 of Hahn and Doganaksoy (2008) for added examples.

4.3.1 Disease Prediction, Control, and Prevention

The prediction, control, and prevention of diseases, and especially the spread of infectious diseases, has given rise to the field of epidemiology, see Aschengrau and Seage III (2008), and presents numerous statistical challenges.

The Centers for Disease Control and Prevention (CDC), an agency of the U.S. Department of Health and Human Services, strive to protect public health and safety by appropriate and timely studies. Their concerns include infectious diseases, environmental health, and occupational safety.

According to the CDC web site, statisticians at CDC have been engaged in such projects as

- Estimating the probability of West Nile virus infection from blood transfusion.
- Monitoring the health and health care of people with HIV.
- Tracing the source of anthrax in mailings.

- Studying the effectiveness of school violence prevention programs.
- Evaluating causes of infant mortalities.
- Evaluating laboratories that test human serum and urine specimens for illegal drugs.
- Identifying genetic markers in parents of children with Down's syndrome.

4.3.2 Health Research

Important health research is conducted at major hospitals and clinics in the United States, such as the Cleveland Clinic Foundation and the Mayo Clinic, as well as various smaller institutions. O'Brien (2002) describes such nonprofit health science centers as "academic institutions, hybrids of industry and academia." Health research is also conducted at medical schools throughout the world.

Keller and Sargent (2010) state that the Mayo Clinic's Division of Biomedical Statistics and Informatics employs more than 175 statisticians—approximately 30 with a Ph.D. degree, 55 with a master's degree, and the rest with a bachelor's degree.[3] Bot (2009) indicates that problems addressed include responding to questions such as

- What is the risk of having a heart attack for a particular population?
- Can genetic information be used to give patients prognostic information and help determine which drug(s) to use?
- What is the effect of metastatic colorectal cancer on patients' well-being and ability to carry out daily activities?

Statistical modeling and analysis of DNA data is a rapidly advancing field. Genetic research experiments now routinely collect over 300,000 measurements per subject, leading to significant computational and statistical challenges; see Siegmund and Yakir (2007) and the journal *Statistical Applications in Genetics and Molecular Biology* for recent developments.

4.3.3 Health Care Improvement

Health care costs place a heavy burden on citizens of all countries. It is estimated that such expenditures in the United States account for 16% of the gross domestic product.[4] Administrators are, therefore, constantly seeking ways to improve processes and reduce costs through the use of disciplined, data-based approaches. This often provides huge opportunities for statisticians; see Valentine et al. (2010).

Some typical applications are

- Scheduling of physicians in a hospital emergency room so as to provide optimum care, while avoiding doctor fatigue.
- Comparison of the effectiveness of alternative treatments of a disease.

[3] Some of the 90 bachelor's degree level statisticians might have degrees in mathematics or computer science, rather than statistics.
[4] OECD Health Data 2010; key indicators at http://www.oecd.org/.

- Assessment of different approaches to patient care, for example, studying alternatives to C-sections and episiotomies and determining the optimum length of hospital stays at childbirth.
- Measurement of waiting times in physicians' offices and hospitals and of potential contributing factors to gain understanding of how waiting times can be reduced.
- Evaluation and reduction of errors in filling prescriptions, in administering medications, in performing surgeries, and in diagnoses.

Comparative effectiveness research is receiving particular attention (and funding) in the United States; see Morton (2009a) and Slutsky and Clancy (2009). This deals with making evidence-based determinations of which (medical) treatments work best, for whom, and under what circumstances. It might, for example, call for comparing two treatments of a disease, each of which had been compared in *separate* clinical trials against placebos.

Some health care improvement applications involve conducting a statistically designed experiment. Often, however, experimentation is inappropriate and we must instead rely on data from observational studies (Section 11.5). As always, an essential element is careful planning of the study to get good data.

Health care practitioners have come to use quality improvement methodologies, such as Six Sigma (Section 5.2.2), as well as statistical and operations research methods. Articles dealing with applications, such as those by Van Den Heuvel et al. (2005) and Bisgaard and Does (2009), appear in *Quality Engineering*, *Quality Progress*, and *Six Sigma Forum Magazine*.

4.3.4 Added Examples

Peck et al. (2006) provide examples of the use of statistics in biology and medicine in chapters that deal with

- Modeling an outbreak of anthrax.
- Understanding the human mind.
- Leveraging change in HIV research.
- Statistical genetics.
- DNA fingerprinting.
- How many genes?
- Mapping mouse traits.

4.4 NATIONAL DEFENSE

Statisticians around the world are engaged in national defense and security applications. In the United States, their employers are often one of the branches of the armed forces, a specific government agency—such as the Federal Bureau of Investigation

(FBI)—or one of the national laboratories (Section 4.8.1). Alternatively, such work might be conducted in industry or academia under contract with the government.

The work performed might range from support in the development of defense strategies to reliability assurance in the production of munitions. Some typical applications are

- Helping predict (and avert) terrorist attacks.
- Evaluating the effectiveness of a military defense system.
- Monitoring munitions stockpiles.
- Tracking global epidemics.
- Searching for potential computer network security threats.

Spruill and Wilson (2009) provide some discussion. However, the specifics of such activities are frequently government classified information.

The U.S. Army Conference on Applied Statistics (ACAS) has since 1995 served as a forum for the presentation and discussion of theoretical and applied papers related to the use of probability and statistics in solving defense problems. The American Statistical Association (ASA)'s Section on Defense and National Security sponsors various activities that spotlight the role of statistics and its applications in furthering national defense.

4.5 OTHER SCIENTIFIC RESEARCH

Statisticians contribute extensively to scientific research in numerous application areas. Early applications were in a variety of fields, but the most prominent effort was in agriculture; see Sidebar 4.1.

SIDEBAR 4.1

STATISTICS IN AGRICULTURAL RESEARCH

In agriculture, the extensive use of statistics in scientific research dates back to the first part of the twentieth century when it was introduced by such luminaries as the famed statistician/geneticist Sir Ronald A. Fisher at the Rothamsted Agricultural Experimental Station in Hertfordshire, United Kingdom.

Forward-minded farmers and researchers used statistically designed experiments to compare alternative crop growing methods, such as evaluating the impact of different fertilizers on crop yields. Various popular statistical designs—such as incomplete block, Latin square, and split plot designs—had their origins in agriculture. These designs were based upon the realization that crop yield and product quality differences were due not only to the growing methods or the fertilizers being compared, but also to natural variation in soil (even over a small area of land), weather conditions, and other extraneous factors. Statistical methods were similarly used to compare animal breeding methods.

This led to the formation of agricultural extension services in countries throughout the world and in the establishments of university departments of statistics, especially in land-grant universities in the United States. It has been credited as a significant factor in the rapid advance of agriculture in the United States and elsewhere.

The use of statistics in agricultural research continues today, but it is only one of many statistical application areas. Scientific research is conducted in academia or industry, sometimes with government support, or directly by government agencies. In addition, numerous national laboratories and research institutes conduct scientific (as well as other) research in the United States. Their activities are described in Section 4.8. Also, the ASA conducts an active program that concerns itself with national science efforts; see Pierson (2010).

We briefly describe a few current areas of application in this section.

4.5.1 Nanotechnology

Research in nanotechnology—an emerging field dealing with the study of matter at the molecular level—is directed mainly at developing new products and improving existing ones.

Statistics can contribute to such research in many ways. Lu et al. (2009) cite specific opportunities involving

- Specially tailored designed experiments to help learn more about the formulation process of nanocomposites (i.e., multiphase solid materials where one of the phases has at least one dimension of less than 100 nm).
- Modeling and analysis methods for working with high frequency and/or spatial data.
- Statistical and automatic process control methods for addressing low-quality and high-defect processes.
- Reliability analysis methods for low, and sometimes unpredictable, product reliability.

Some of this research resembles that undertaken in the past in other areas—such as the early work on semiconductors that also resulted initially in low yield (i.e., high end-of-line reject rate) processes.

4.5.2 Space Exploration

Space exploration presents special challenges because human life is frequently at stake. Ensuring high reliability for "manned" space vehicles is critical. Statisticians at the U.S. National Aeronautics and Space Administration address such problems as

- Evaluating the risk caused by external factors, such as ice and insulation foam, striking a shuttle during launch.
- Quantifying the uncertainty in the physical models that predict where a launched space vehicle will be at a particular point in time.

- Designing experiments to identify and rank the primary contributors to integrated drag during a space vehicle's ascent trajectory (Rhew and Parker, 2007).

One notorious example in which statistics was improperly applied in space exploration was in an analysis preceding the 1986 Challenger space shuttle disaster (Dalal et al., 1989). The low temperature predicted for the launch date was believed by some to be detrimental to the O-rings that were used to hold critical joints in place during launch. An analysis based on data from previous flights with O-ring failures led to the conclusion that low temperatures did not present a hazard to the O-rings. As a result, NASA proceeded with the launch the next day at an unusually low temperature of 31°F. Regrettably, the shuttle blew up and the entire crew died. The subsequent review determined that the analysis was flawed. This was because 17 additional past flights that had no O-ring incidents were incorrectly deemed to be irrelevant and were excluded from the analysis. A proper analysis of data from all previous flights showed a strong association between temperature and number of O-ring incidents, with low temperatures being particularly risky.

4.5.3 The Environment and Climate Science

The environment in general and climate science in particular are rapidly emerging areas of interest for applied statisticians. The Board of Directors of the ASA (endorsing the Fourth Assessment Report of the United Nations' Intergovernmental Panel on Climate Change) in November 2007 urged that statisticians become more involved in the assessment of climate change. They cited various technical challenges, including

- How best to combine climate data from different sources.
- How to identify biases (and adjust for them) in measurement systems.
- Estimating regional and local effects of climate change from global models.
- Evaluating the human health effects of climate change.

In November 2009, ASA President Sally Morton—based on a recommendation by ASA's Climate Change Policy Advisory Committee—joined leaders of 17 other science organizations in signing a letter to all U.S. senators summarizing their consensus views on climate change. The committee's justification for their recommendation is reviewed by Smith et al. (2010) (and commented on in various letters to the editor in the June 2010 *Amstat News*).[5]

The 2010 Joint Statistical Meetings featured invited presentations on such topics as climate change-caused shifts in forest fire ignitions, modeling of environmental extremes, and analysis of global climate change data. Also, Crawford (2008) describes her career in environmental statistics. Finally, we note

---

[5] The committee concluded, based upon their statistical evaluations, that there was "strong scientific evidence that climate change is happening and that human activities are the primary driver." Also see *Amstat News* (2010a) for a statistician's briefing to the U.S. Congress on climate change.

that the ASA has a section on Statistics and the Environment that issues news-letters periodically.

4.5.4 Added Examples

Peck et al. (2006) provide further examples of the use of statistics in science and technology in chapters that deal with

- Monitoring tiger prey abundance in the Russian Far East.
- Predicting the Africanized bee invasion.
- Combating electronically transmitted spam.
- Should you measure the radon concentration in your home?
- Statistical weather forecasting.
- Space debris assessment.

4.6 SOCIAL AND BEHAVIORAL SCIENCES

The number of statisticians currently engaged in work in the social and behavioral sciences is relatively small compared to some other areas. Their work, however, encompasses a wide variety of applications, a few of which we describe in this section. Also, many professionals in demography, economics, psychology, sociology, and other social sciences are highly trained in statistical methods and concepts, even though they may not regard themselves principally as statisticians.

4.6.1 Public Opinion Polls and Other People Sampling Studies

> Knowing statistics helps one to sort out the reliable from the unreliable information that comes at us.
>
> —Gipsie Ranney

Measuring public opinion by a scientifically (i.e., randomly) selected sample across a country, or over a particular region or for a particular demographic group, has become a big business. Studies undertaken prior to elections that try to predict the winner (see Ratledge, 2006) receive much media attention as do studies to assess the public's opinion on specific issues, such as the state of the economy. Various companies conduct such studies and employ statisticians and/or statistical consultants to help ensure their validity.

Drawing correct conclusions from sampling studies is critically dependent upon exactly how the study is conducted. A key issue is that of selecting a random sample from the available "sampling frame" (i.e., listing of the units from which the sample can be drawn) that ideally is identical to the population of interest but in practice only coincides as closely as is practically feasible, at best. Such issues as how to pose questions in an unbiased manner and how to handle nonrespondents also need to be addressed. Failure to do so can introduce bias in the results. Such bias, moreover, is *not* reflected in the calculated statistical "margin of error."[6]

When the polls make incorrect predictions, statistics is viewed suspiciously by the public; see Sidebar 4.2.

SIDEBAR 4.2

WHEN THE POLLS WERE WRONG

How far would Moses have gone, if he had taken a poll in Egypt?

—Harry S. Truman

National polls use statistical sampling to draw inferences. As a result, the findings are typically reported, subject to a stated "statistical margin of (sampling) error." When the predictions are found to be seriously in error—as the election results become known—statistical sampling, in particular, and statistics, in general, are sometimes blamed, even though generally it is not statistics per se that is at fault.

The polls have over time become increasingly accurate. This in part has undoubtedly been a consequence of learning from past mistakes such as

- In 1936, the *Literary Digest* magazine predicted that Alf Landon would be elected President of the United States. Two major problems with this poll were that it relied principally on a telephone sample (in 1936 only more affluent people had telephones) and the low response rate (together with failure to properly account for nonrespondents); see Squire (1988).

- In 1948, the polls confidently predicted that Dewey would beat Truman for the U.S. Presidency. A major contributor to this error was the fact that most of the polls were taken insufficiently close to the election and did not detect the swing in voter sentiment that occurred just prior to the election.

- In 2004, exit polls (taken as voters left the polls after voting) tended to predict a Kerry victory over Bush for the U.S. Presidency. The error was subsequently attributed, by some, to faulty interviewing techniques (lengthy questionnaires, inexperienced interviewers, etc.).

In each of these cases, it was the specifics of the survey, and not the concept of statistical sampling, that seemed to have led to the errors. This is, however, a detail of which most people are unaware.

As indicated in Section 2.2.2, sample studies of consumers are used by manufacturers of a product or providers of a service in their market research. These might have the goal of assessing potential customer interest in the product or service,

---

[6] Not all sample studies portend to be random. A notable example is polls conducted by radio or TV stations urging listeners to call in with their opinions on a stated question. The opinions of such self-selected (and typically highly motivated) individuals can differ appreciably from those of the population at large, and especially so, when a viewpoint on the issue has just been suggested by the radio or TV host conducting the survey. Likewise, respondents to Internet surveys are typically self-selected.

evaluating the attractiveness of specific offerings or features, or measuring the effectiveness of an advertising message or slogan.

In addition, television and other rating studies involve the sampling of a specified population. The results of these studies are followed intensely by TV networks and advertisers, as well as the general public.

We comment further on sampling studies in Section 11.3.

4.6.2 Education and Educational Testing

K to 12 Education. Hill (2006) describes statistical issues in studies to evaluate school choice programs, contrasting the use of observational studies with the use of randomized experiments (see Chapter 11), and describes challenges in designing and analyzing social experiments, in general.

Testing of children from kindergarten to 12th grade has become a frequently used (and highly controversial) way for evaluating, among other things, how different schools (and sometimes teachers) are faring in meeting mandated goals. Children, however, are not randomly assigned to schools; they enter different schools with varying starting abilities and are exposed to differing home environments. Making valid comparisons between schools (as well as teachers), therefore, becomes a complex statistical problem that calls for sophisticated statistical evaluations. The results of such assessments need, nevertheless, to be made transparent and justifiable to school administrators, politicians, and the public at large.

Higher Education. Educational testing has traditionally involved the development of tests used for college and graduate school admissions assessment. The Educational Testing Service (ETS) is the best known organization in this field. It is a private, nonprofit institution with primary focus on educational measurement and research through testing. In the United States, it produces and administers the SAT, GRE, and TOEFL examinations. ETS also designs and implements achievement and admissions tests in over 180 countries.

A major concern for statisticians working for ETS and similar organizations is to design standardized tests in a manner that ensures their validity, reliability, consistency, and fairness. In addition, statisticians get involved in research studies dealing with such topics as comparing performance differences between genders on standardized tests, assessing the association between college admission test scores and subsequent performance, and evaluating the effect of classroom teaching methods on student performance.

4.6.3 Sports Strategy and Assessment

Outcomes and performance in sports are expressed by numbers. It follows almost invariably that statistics plays an important role. This is heightened by the extraordinary interest in sports by much of the general public worldwide. Many think of sports statisticians in rather mundane terms as the keepers of the numbers who might be called upon during a rain break or a highly one-sided game to provide the announcers some interesting trivia. There is much more that statisticians can and do contribute, such as the development of optimum strategies in playing the game and of improved ways of measuring players' performance—whatever the game may be. Some sports teams have, in fact, hired professional statisticians for this purpose.

Baseball and Statistics. Baseball, a sport that seems to especially relish numbers, has availed itself of statistical concepts through the work of Bill James[7] and the development of what has been called sabermetrics—the analysis of baseball through objective evidence. Sabermetrics is concerned with the development of realistic measurements of the contribution of individual players. In assessing the number of games won by a pitcher or the number of runs batted in or scored by a batter, one would take into consideration the support the player received from other members of the team (such as the batting performance of other players) and the dimensional differences between ball parks. All of this has led to various books on the subject, such as Schell (2005) and Thorn and Palmer (1985).

Besides providing more incisive information to fans, the use of statistics in baseball has also impacted both the front office and the on-field playing strategy of forward-looking teams. The book *Moneyball* (Lewis, 2004) generated high public interest in describing how the Oakland Athletics built a highly competitive team on a relatively low budget by the strategic acquisition of players, based upon an analytic assessment of their past performance.

Further Examples. Albert et al. (2005) have compiled mainly previously published articles in a total of 44 chapters devoted to football, baseball, basketball, ice hockey, multiple sports, and miscellaneous sports. The titles (some erudite sounding; others down-to-earth) of some of the chapters suggest the diversity of applications:

- Did Shoeless Joe Jackson throw the 1919 World Series?
- Improved NCAA basketball tournament modeling via point spread and team strength information.
- Simpson's[8] Paradox and the hot hand in basketball.
- Adjusting golf handicaps for the difficulty of the course.
- Modeling scores in the Premier League: is Manchester United *really* the best?
- Heavy defeats in tennis: psychological momentum or random effect?
- Who is the fastest man in the world?

A subsequent more advanced volume (Albert and Koning, 2008) aims "to provide an accessible survey of current research in statistics and sports ... and to explain how statistical thinking can be used to answer interesting questions about a particular sport." Its chapters deal with such subjects as

- Modeling the development of world records in running.
- Odds betting market in Scottish league football.

---

[7] A well-known U.S. baseball writer, historian, and statistician.

[8] Not OJ's.

- Measurement and interpretation of home advantage.
- Myths in tennis.
- Optimal drafting in hockey pools.

Some further interesting applications of statistics in sports deal with

- Assessment of nationalistic favoritism in judging the 2000 Olympic games (Emerson et al. 2009).
- Assessment of the impact of racial bias on basketball betting (Larsen et al. 2008).
- Monitoring batting performance in cricket using control chart methods (Bracewell and Ruggiero, 2009).
- Applying Markov Chain models in predicting tennis game outcomes (Newton and Aslam, 2009).

Other Sources of Information. The ASA Section on Statistics in Sports organizes sessions at the Joint Statistical Meetings and publishes a newsletter. The International Statistics Institute also has a Sports Statistics Committee. The journal *Quantitative Analysis in Sports* has been published since 2004.

4.6.4 Legal Applications

Statisticians are frequently called upon to provide legal support, ranging from commenting to "their side's" legal staff on the efficacy of the available data to giving courtroom evidence, possibly as expert witnesses subject to cross-examination. We will discuss the role of expert witnesses (and the ethical issues presented by them) in Section 10.4.6. Some typical topics in which statisticians might become involved are the evaluation of statistical evidence concerning the commitment of a crime (see Sidebar 4.3), assessment of alleged workplace discrimination, and claimed negligence in ensuring consumer safety for products ranging from toasters to automobiles.

SIDEBAR 4.3

GUILTY OR INNOCENT?

Lucia de Berk, a Dutch nurse, was found guilty of murder and attempted murder and received a life sentence in prison. The evidence against her was principally statistical: her physical presence at a series of suspicious deaths and near deaths in hospital wards in which she had worked. The prosecution successfully claimed that the odds of this occurring by coincidence were extraordinarily low (one in 342 million) and de Berk was sentenced to life imprisonment. Subsequent examination questioned the validity and relevance of the statistical evidence against de Berk and the case was reopened in 2008 by the Dutch Supreme Court, receiving much public attention. De Berk was set free and retried, and in April 2010 was exonerated. See Buchanan (2009) and relevant web sites[9] for further discussion and updates. Also see Cobb and Gehlbach (2006) for a somewhat similar example.

---

[9] For example, Wikipedia.

Books about statistics and the law include DeGroot et al. (1994), Finkelstein (2009), Gastwirth (2000), Good (2001), and Kadane (2008). Also see Gastwirth and Pen (2010) for a discussion of statistical reasoning presented to and used by the U.S. Supreme Court in cases dealing with equitable jury representation.

4.6.5 Some Further Applications

There are numerous other applications of statistics in social science and behavioral research; see Sidebar 4.4 for an example of a political analysis.

SIDEBAR 4.4

U.S. HEALTH CARE, POLITICS, AND STATISTICS

An article entitled "The Senate's Health Care Calculations" by Gelman et al. (2009) appeared prominently on a *New York Times* Op-Ed page. (The lead author is a Professor of Statistics and Political Science at Columbia University.) The article dealt mainly with Congressional support for and opposition to the then pending health care bill in the U.S. Congress. It provided statistical evidence to support the claim that "in general, senators seem to be less interested in what their constituents, old and young, rich and poor, might think about health care, and more interested in how they feel about President Obama." The analysis rated individual senators on their level of support for the pending proposed health care initiative on a scale from 1 to 5 (based on their public statements and their committee votes). It then related these evaluations to their constituents' polled views on health care reform and to President Obama's margin of victory (or defeat) in their home states during the 2008 election. The authors' evaluation was based on what was described in the article as "a statistical method called multilevel regression and post-stratification" and included incisive graphical displays of the results.

Some other examples that illustrate the wide applicability of statistical methods in the social and behavioral sciences are

- Dating medieval manuscripts (Feuerverger et al. 2008).
- Studying the impact of religion in everyday life (Emerson and Sikkink, 2006).
- Understanding gender birth disparity in China (Hesketh, 2009).
- Estimating the damage to homes and displacement of people resulting from the 2010 Haitian earthquake (Cochran, 2010).
- Identifying changes in the incidence of crime in London (Spiegelhalter and Barnett, 2009).
- Reforming U.S. foreign assistance policy (Asher, 2010a).

Gelman and Cortina (2009) describe the use of quantitative methods in history, economics, sociology, political science, and psychology, considering such topics as

- Historical data and demography in the United States and the Americas (e.g., quantifying the slave trade).

- Econometric forecasting and the flow of information.
- Explanation of the racial disturbances of the 1960s.
- Modeling strategy in Congressional hearings.
- Formulating and testing theories in psychology.

4.7 TEACHING (IN NONACADEMIC SETTINGS)

Statisticians are almost invariably involved in teaching. We discuss the teaching of statistics in academia in Chapter 13.

Teaching is, however, not limited to academic settings. Applied statisticians are frequently involved in various teaching activities. Some of these are highly informal and part of the normal communication process—typically requiring technical concepts to be made understandable to those with limited training and/or interest in the subject. In addition, statisticians outside of academia are also frequently called upon to give talks or teach short courses on topics of interest, such as sampling studies, the design of experiments, categorical data analysis, and the use of a statistical software package. These activities may be closely integrated with specific on-the-job issues. For example, design engineers concerned with product reliability might be exposed to a short course on life data analysis.

Unlike academia, teaching statistics in business and industry and in government generally requires only a master's degree in statistics.

4.8 SOME FURTHER INSTITUTES FOR RESEARCH IN THE UNITED STATES

In this chapter, we have described a wide variety of application areas in which statisticians are engaged. Much of this work is conducted in academia, by business and industry or by government agencies. Some of it is undertaken by statisticians in national laboratories and research institutions. These tend to conduct a mix of scientific and social research and provide an environment that combines features of academia, business and industry, and government. We describe statistical activities in national laboratories and research institutions in the United States briefly in this section. We also comment on the National Institute of Standards and Technology (NIST), which, although a government agency, functions in many ways in a manner similar to national labs and research institutes.

4.8.1 National Laboratories

The U.S. Department of Energy funds and oversees various national laboratories, such as Brookhaven, Lawrence Livermore, Los Alamos, Oak Ridge, Pacific Northwest, and Sandia, for the purpose of advancing science and helping promote economic and national defense interests. Most of these laboratories are administered and operated in conjunction with universities or corporations, in which case those doing the work are typically employees of these institutions rather than the government.

Statisticians support such research in partnership with scientists and engineers. Examples of the work conducted at one of these laboratories (Los Alamos) include helping in the design of surveillance systems, biological risk assessment, development of energy data collection systems, and various national defense projects.

4.8.2 Research Institutes

The online encyclopedia Wikipedia lists a multitude of nonprofit research institutes in the United States, some of which are affiliated with universities. Some of these organizations, such as the Electric Power Research Institute, principally support specific business segments. Others are engaged in a wide variety of scientific and social research, usually sponsored by industry or government. Many of these institutions have statistical groups or employ statisticians. Some examples are the Battelle Memorial Institute, the Research Triangle Institute (RTI),[10] the RAND Corporation, and the Southwest Research Institute.

The National Institute of Statistical Sciences (NISS) was established in 1990 as an independent research institute by various U.S. statistical societies and the (North Carolina) Research Triangle universities to identify, catalyze, and foster high-impact, cross-disciplinary research involving the statistical sciences. Its research encompasses such areas as bioinformatics (the application of statistics and computer science to molecular biology), data confidentiality, statistics integration, data quality, information technology, the environment, educational statistics, and large and complex databases; see Pantula (2010a).

A few examples of research studies in which statisticians at such institutions are involved are

- Coordination of a large multiyear observational patient registry study of a new cancer treatment (RTI).
- Measuring the incidence and effects of sexual assaults in correctional facilities (RTI).
- Providing comprehensive and reliable data on early child development (RTI).
- Quantifying normal variation in EKGs (NISS).
- Assessing online reading comprehension (NISS).
- Developing reliable ways of measuring highway travel time (NISS).

4.8.3 National Institute of Standards and Technology (NIST)

NIST is the United States' national metrology institute. Statisticians typically support the development of measurement standards, calibration services, and studies that substantiate claims published in calibration and measurement certificates. Extensive work goes into quantifying the uncertainty in standards.

---

[10] RTI lists Statistics Research and Survey Research & Services as two of its 11 key activity areas.

Other statistical activities involve the support of a wide variety of NIST activities in physics, chemistry, information technology, and other scientific endeavors.

4.9 MAJOR TAKEAWAYS

- Statistics is playing an ever-increasing role in regulatory activities, such as in the work conducted at the U.S. Food and Drug Administration and the Environmental Protection Agency.

- Statisticians are heavily engaged in a wide variety of health issues, in addition to the assessment of proposed pharmaceutical products. These include disease prediction, control, and prevention; health research; and health care improvement.

- Statisticians are involved worldwide in national defense and security applications in work ranging from the development of defense strategies to the production of reliable munitions.

- Scientific applications of statistics started with principally agricultural applications. Today, statistics contributes to scientific research in many areas including nanotechnology, space exploration, and environmental and climate science studies.

- The social and behavioral sciences provide challenging opportunities in public opinion surveys, education and educational testing, sports strategy and assessment, legal applications, and many other areas.

- Some applied statisticians are heavily involved in teaching, both informally and through short courses and similar undertakings.

DISCUSSION QUESTIONS

(* indicates that question does *not* require any past statistical training)

1. What would you expect to be some major technical and nontechnical challenges faced by statisticians working for the FDA?

2. Select one or more of the applications of statistics at the EPA cited by Nussbaum (2008) and suggest a statistical approach for addressing the problem.

3. Select one or more of the examples provided in Section 4.3 and suggest how statistics might be used in addressing the question posed.

4. Significant effort is currently underway to have health care providers (i.e., physicians, hospitals, etc.) "digitize" their records dealing with patient care in a manner that permits ready and useful retrieval of information, while protecting confidentiality. What are some of the statistical evaluations to help improve patient care that such digitization suggests? Elaborate on how you

might go about such assessments. Also, indicate some likely obstacles and how you would try to address these.

5. Research and report on recent statistical evaluations that have been conducted on climate change. What were the methods used and what did the studies conclude?

6. A major concern in traditional landline telephone sampling for public opinion polls is the increasing number of homes that rely on cell phones exclusively. Why is this a concern? How is this concern being (or might be) addressed? What are some of the challenges?

7. *How have public opinion polls forecasting election outcomes fared recently? Have there been any seriously incorrect predictions? If so, to what might these be ascribed?

8. A sophisticated statistical approach for making comparisons between schools (and between teachers) based on student test results, developed originally for the State of Tennessee school system and gaining national attention, is known as value-added modeling. Research and report on this approach and comment on how it tries to address the controversial issues raised by K to 12 educational testing.

9. Look into one or more of the topics dealing with sports applications of statistics suggested in Section 4.6.3. Report on how the problem was addressed statistically.

10. Assume that you have been asked to develop a short introductory course for practitioners in one of the areas mentioned in Section 4.7 or some other area that might be of interest to a (potential) employer. Prepare a course outline.

11. Some of the examples given in this chapter involve the use of operations research and other analytical methods, as well as statistics. Give some examples of uses of operations research, indicating the specific approach and how it might be employed.

CHAPTER **5**

THE WORK ENVIRONMENT
AND ON-THE-JOB CHALLENGES

5.1 ABOUT THIS CHAPTER

In this chapter, we describe the environment in which applied statisticians work. We begin with a brief discussion of the varying degree of receptiveness to statistics and statisticians in different segments of the workplace. We then describe where statisticians fit into organizations, contrast the consulting and team member modes of operation, and overview job grading systems. We then briefly discuss the impact of globalization on our work and the important role that managers play in our lives. We consider next a variety of challenges that contribute to making the work of statisticians both challenging and interesting. We conclude the chapter by commenting on the role of women in statistics.

5.2 RECEPTIVENESS TO STATISTICS
AND STATISTICIANS

5.2.1 A Mixed Bag

The value of statistics, unlike that of many other fields from accounting to physics, is not always evident to management and to our nonstatistical colleagues and varies widely.

A few arenas, such as gambling and insurance (fields that have some commonalities), have their fundamental roots in probability and statistics. To remain competitive and profitable, both require an understanding of the probabilities associated with events of interest—such as the rate of occurrence and severity of accidents and natural disasters in the case of insurance and the odds associated with games of chance in gambling. Also, for some government agencies, such as the U.S. Bureau of Labor Statistics and the Census Bureau, the gathering, interpreting, and reporting of data is a prime mission and, therefore, the need for statistics is unquestioned in such agencies.

In some other fields, the use of statistics is, to a large degree, the consequence of mandates by government or other regulatory agencies. In the United States, for example, as a result of the 1962 Kefauver–Harris Amendments (Section 2.2.5), pharmaceutical products must undergo rigorous statistically based testing before they can receive approval by the Food and Drug Administration (FDA). Statistics, thus, provides the

A Career in Statistics: Beyond the Numbers, Gerald J. Hahn and Necip Doganaksoy.
© 2011 John Wiley & Sons, Inc. Published 2011 by John Wiley & Sons, Inc.

fundamental underpinning for most pharmaceutical testing. The pharmaceutical industry is, in fact, one of the largest employers of statisticians. Some other data-intensive businesses, such as semiconductors (Section 2.2.5) and communications (Section 2.3.3), also use statistics extensively. Banking and finance is another important area driving the demand for statisticians over the past 15 years.

In many other fields, the acceptance of statistics is based upon the perceived added value that it provides and the hiring and retention of statisticians needs to be justified (and rejustified) by tangible payoffs. In these areas, especially, statisticians need to sell themselves and their contributions.

5.2.2 Trends

The use of statistics and the recognition accorded to statisticians have sharply increased over the course of our careers.

Government has been a user of statistics since at least the early 1800s. Manufacturing industries have made some use of statistics since the early 1900s, based, to a large measure, on the work on control charts by Walter Shewhart and the development of acceptance sampling concepts by Dodge and Romig.

Interest across the board, however, has soared since World War II and the emergence of high-speed computing capabilities. As a consequence, today

- Many popular statistical methods are computer-intensive. Simulation, bootstrapping, and modern Bayesian analysis are some typical examples. Increasingly faster computers allow the use of powerful new statistical methods.

- Massive data sets are becoming increasingly commonplace, especially in such areas as financial services, genomics, and measurement-intensive manufacturing environments, such as in the process industries.

- Data exploration, especially through the use of graphical displays, is becoming more and more commonplace.

In addition, the emergence of new technologies, ranging from modern genetics to nanotechnology, has provided important new application areas of statistics.

The attention accorded to improving the quality of products and services in the 1980s, with statistician W. Edwards Deming as one of its leading and most articulate proponents, provided additional impetus to the use of statistics, especially in business and industry. Subsequently, the Six Sigma initiative, mounted by various large companies, with its (adopted) slogan "In God we trust—all others bring data," gave statistics a further boost (see Sidebar 5.1).

SIDEBAR 5.1

SIX SIGMA

Over the years, various strategies for deploying quality improvement—with varying degrees of emphasis on statistics—have commanded attention. The Six Sigma initiative is one of the most far-reaching and influential of such approaches. Introduced at Motorola in the mid-1980s, Six Sigma achieved national prominence about a decade later when Honeywell, GE, and various other large companies publicly embraced it. The Six Sigma effort is perhaps best known for its laudable—but not always attainable or even practical—goal to reduce the level of defective products and services to "3.4 defects per million opportunities."

More importantly, Six Sigma provides a highly disciplined approach involving a series of steps (see Hahn and Doganaksoy, 2008, Sections 3.2.2 and 6.5) for achieving quality improvement. It also places heavy emphasis on basing decisions on data and the use of statistical techniques. Employees are trained in Six Sigma methodology and upon completion of training designated as Six Sigma green belts, black belts, master black belts, or champions.

Six Sigma gained further momentum through its integration with lean manufacturing (sometimes referred to simply as "lean"). Lean manufacturing aims to identify and minimize process activities that do not directly add value to the final product (e.g., storage, transportation, rework, duplicate data entry, and testing). Its integration with Six Sigma calls for using Six Sigma concepts and tools, including statistical methods, to implement lean manufacturing; see Snee and Hoerl (2007).

In many companies, the initial focus is on applying Six Sigma in manufacturing (often because of the frequently rapid and quantifiable payoffs). Based on successful results, its use is then extended to other operations, such as product design, information systems, business processes, and customer service.

The use of Six Sigma has spread beyond business and industry into numerous other areas, such as health care and education, and even into some government organizations. For example, the city of Fort Wayne, IN, claims that its use of Six Sigma helped reduce lot trash pickups by 50%, avoided $1.7 million wastewater plant expenditures, and reduced the time to issue building permits from 47 days to 12 days.

The *Six Sigma Forum Magazine* (SSFM) regularly publishes articles that describe new developments in Six Sigma, as well as successful applications and lessons learned.

Over time, the work conducted by statisticians in many organizations has become more problem-oriented with major research efforts being left principally to academia. In this vein, Hoerl and Snee (2010a, 2010b) call for greater emphasis on what they term statistical engineering over statistical science. Statistical science focuses on advancing the fundamental laws and knowledge of statistics. Statistical engineering, in contrast, concentrates on how to best utilize the principles and techniques of statistical science for the benefit of humankind.

5.2.3 Statistics is for Everybody

> Some of the most innovative and important new techniques have come from researchers who would not identify themselves as statisticians.
> —Brown and Kass (2009)

Today, easy-to-use statistical software and online tutorials are readily accessible to practitioners and are often used in introductory courses in statistics. Web-based guidance can respond to technical questions about statistics. On-the-job training programs, such as Six Sigma, provide further statistical instruction to many. As a result, nonstatisticians now have the basic tools, motivation, and self-confidence to conduct their own statistical analyses[1] and often do so. This has been referred to as "the democratization of statistics" and has led to what some refer to as parastatisticians—individuals who without formal academic degrees in statistics frequently employ statistical methodology and software in the workplace.

In addition, many prominent tools currently used in data analysis, such as neural nets, data mining, Taguchi methods, Kalman filtering, and fuzzy logic, have been developed, principally by engineers and computer scientists (as suggested by the quote at the beginning of this subsection).

5.2.4 Impact of Democratization of Statistics

The ever-increasing use of statistical tools by nonstatisticians has, indeed, made this a golden age for statistics. But it is not necessarily a golden age for statisticians (as my manager reminded one of us when I asked for a salary increase). The employment of professional statisticians in business and industry, for example, does not seem to have increased in recent years—with some exceptions, such as pharmaceuticals and finance. The democratization of statistics has, nevertheless, created exciting and important new opportunities for our profession.

Freedom from the nitty-gritties of conducting routine statistical analyses allows us to work on more complex problems, and especially on those projects that are most important to our organizations. It provides us the opportunity to look at problems more holistically (see Snee, 2008) and to focus on projects, at their most important stages—such as in the planning of the data acquisition and in reviewing the conclusions. The new environment also provides statisticians opportunities for greater involvement in the development of organization-wide solutions as opposed to responding to narrow technical questions (see Sidebar 5.2).

SIDEBAR 5.2

AN ANALOGY BETWEEN COMPUTER SCIENTISTS AND STATISTICIANS

The field of computer science provides an analogy to the statistics profession. At one time, most software development was limited to specialists in computer programming. With the development of more user-friendly programming tools, this has changed. Today, much software is developed by practitioners, rather than professional computer programmers. This has allowed computer programmers to become computer scientists, providing guidance to users on higher level technical issues (such as system architecture, network reliability, and enterprise-level system security) and staking out greater leadership roles for themselves. Such individuals remain in high demand.

In a similar manner, statisticians today can leave many of the more routine tasks, especially of data analysis, to nonstatisticians and, instead, focus on higher level and more challenging, more proactive, and more visible work than before. We elaborate on this broader perspective for statisticians throughout this book.

---

[1] A possible exception is the drug approval process, an area in which most statistical analyses are still conducted by statisticians.

The democratization of statistics carries significant risks of misuse if the tools are applied uncritically. Thus, as statisticians, we need to focus on providing statistical methods, often in graphical form, that are robust to possible misuse by those with less technical knowledge than ourselves. We need to help ensure that practitioners have an appreciation of the basic underlying concepts—and especially the inherent assumptions and pitfalls—of the tools they may be using. At the same time, we should strive to collaborate with engineers and scientists involved in the development of data analysis methods to help make such methods better.

5.2.5 Public Exposure

The public visibility accorded to statisticians varies among application areas. Statisticians in business and industry may have appreciable exposure within their own companies, but typically not externally. Exceptions include testifying in a public lawsuit, such as an alleged discrimination case, or responding to a Congressional committee that is developing legislation to change the regulations for drug approval.

Government statisticians work for the community at large. Many of their reports are of strong public interest and are heavily scrutinized by the media. The release of periodic economic indexes is awaited with much anticipation, not only by economists and financial analysts but also by the general public, and can have an important impact on the stock market and the economy. It is not unusual for senior statisticians to be interviewed and questioned by the media about the results and implications of recently released official statistics.

5.3 WHERE DO STATISTICIANS FIT INTO THE ORGANIZATION?

Some statisticians are "one of a kind" in their workplace. This is especially the case in small organizations. We discuss "isolated statisticians" (mostly from the perspective of academia) in Section 13.12.3. Others may be one of a (possibly large) number of statisticians in an organization. They may be bunched together organizationally in a statistics or general support group. Or they may be dispersed throughout the organization.

5.3.1 Statistics (and General Analytical Support) Groups

Sometimes, all or most of an organization's statisticians are members of a statistics group. Thus, pharmaceutical companies, large government agencies, and national laboratories frequently have statisticians working together in the same group. Such groups often service the entire organization, even though individual departments within the organization may themselves have one or a small number of statisticians. The authors' company (GE) currently has in its Global Research Center a 14-person centralized statistics laboratory in the United States and a similar size sister organization located in India. These two groups, in addition to participating in projects within their own organizations, support the entire company and especially those departments that do not have their own professional statistician(s).

Other organizations may have many or all of their statisticians "housed" together in a general analytical support group or information technology section. In this case, the group may include other professionals with backgrounds in, for example, mathematics, operations research, or possibly such areas as artificial intelligence, computer science, or machine learning.

5.3.2 Statisticians in Functional Groups

In some organizations, statisticians are members of the group to which they are most likely to contribute, such as a quality control group in industry or in that part of a government agency that is directly engaged in survey studies. At Google, for example, there is no statistics department per se; instead, statisticians work on project teams (Pregibon, 2009).

In some of these cases, statisticians work mainly on tasks within the organization in which they are housed. For example, statisticians located in engineering, manufacturing, and marketing units in a company may work principally on projects involving product development, quality control, or consumer survey projects, respectively. In many other cases, such statisticians support other organizations within or even outside their own departments. A statistician in a research and development (R&D) group may, for example, be asked to work on a manufacturing problem or even on an employee opinion survey.

5.3.3 Local Statistical Experts

There are in addition individuals (previously referred to as parastatisticians) with a strong statistical orientation, such as Six Sigma black belts or master black belts, typically spread throughout the organization in business, industry, and government. In some cases, these are professionals who have acquired an interest in statistics and serve as local statistical experts, even though this may not, at least initially, be part of their formal job descriptions. They are an important resource for furthering statistical thinking, as well as providing statistical support, especially for organizations that do not have a professional statistician. Some of these individuals may, over time, become professional statisticians; Leah's career path summarized in Section 1.10.1 provides an example.

5.4 TWO MODES OF OPERATION

> A simple test (if your name is Joe): There is a very tough investigation coming up, and engineers and scientists say, "We've got to have Joe on the team," then you have arrived.
>
> —George Box

5.4.1 The Consulting Mode

Traditionally, statisticians have served as "consultants" on projects. In this mode, they are asked to provide expertise on specific, often technical, questions. These questions might be relatively minor or they may have important implications for the project. Occasionally, consultants are also asked to conduct a broad project overview and make open-ended recommendations. The responsibility for deciding whether or not a consultant's recommendations are followed and their actual implementation resides with the project owners. Consultants are often minimally involved in the day-to-day conduct of the project.

The consulting mode is a common model for statisticians who are in organizations different from the one from which the project originates. College professors and retirees often serve as external consultants.

There is an aura of prestige associated with the title "consultant." This tends to increase the further removed the consultant's home base happens to be from the project and the further the person is removed organizationally. This is somewhat ironic, since geographical proximity (and maybe even "rubbing elbows") can be a distinct advantage in successful project work.

5.4.2 The Team Member Mode

As their roles and contributions become more evident to the project team, and depending upon organizational and geographical considerations, statisticians may be invited to become members of the team.

Hoerl (2008a) has suggested the following characteristics of team members that differentiate them from consultants:

- Attend team meetings.
- Are on the "To" list, rather than the "cc" list for project mailings.
- Are coauthors of project reports, rather than acknowledged contributors.
- Contribute in multiple areas rather than only statistics.
- Share in the rewards for success and the consequences of failure.

Hoerl summarized the difference by saying that under consulting one thinks of a project as belonging to somebody else, as opposed to being an equal partner in its implementation.

The team member mode tends to be especially prominent in the pharmaceutical industry. As part of the team, statisticians typically participate in specific deliverables, such as the planning, analysis, and reporting of the results of a pharmaceutical trial.

The increased involvement that comes with being a team member can lead to better understanding, compared with the consulting mode, and, thus, enhances statisticians' ability to contribute. It also makes it easier to be proactive and sometimes even to assume a leadership role. Thus, you can frequently be more effective as a team member than as a consultant.

5.4.3 Fusion of Roles

There is not always a clear distinction between the consulting and team member modes. An individual may be considered to be part of a team, and provide technical guidance, but, unlike some other team members, may not be held responsible for the project outcome. Also, a particular individual may take on different roles on different projects.

5.5 GRADING SYSTEMS

Most jobs, at least in large organizations, come with official job descriptions. Many also have an associated job grade that reflects the responsibility, and often the salary range and privileges, associated with the position. The importance of these varies from one organization to the other.

Governments typically place heavy emphasis on grading systems. The U.S. government assigns grade levels to its professional employees as follows:

- Grade levels 5–7: Apprentice.[2]
- Grade levels 9–12: Journeyman.
- Grade level 13: Team leader.
- Grade levels 14 and 15: Manager.
- SES: Senior Executive Service.
- S/L: Senior Level Technologists.

The description of the tasks associated with each of these grade levels is somewhat subjective, but responsibilities and required substantive knowledge increase with the grade level. Typical entry positions for those with bachelor's degrees are at levels 5–7 and for those with master's degrees are at levels 7–11. At least at one government agency (National Institutes of Health), some new Ph.D.'s start at a grade level 13. Current salaries[3] and descriptions of the responsibilities[4] that are associated with different grade levels can be found at the U.S. Office of Personnel Management web site.

Government statisticians' positions and the associated responsibilities are defined within this system for each position. Statisticians and mathematical statisticians (Section 3.7) tend to be at grade levels 5–7 and grade level 11 or above, respectively. As new jobs arise, these are assigned grade levels and their responsibilities stated.

State governments in the United States use similar grading systems, but with different numbering systems to identify the grades. For example, the New York State structure for statisticians is

- Grade 14: Entry-level statistician (apprentice).
- Grade 18: Senior statistician (journeyman).
- Grade 23: Associate statistician (team leader).
- Grade 27: Principal statistician (manager).
- Grade 31: Chief statistician.

[2] Professional grade levels start with level 5.

[3] http://opm.gov/oca/10tables/index.asp.

[4] http://www.opm.gov/fedclass/.

Pay scale ranges, as well as various privileges, are typically associated with each grade level.

Large businesses have established somewhat similar, but often less formal and less emphasized systems. A typical career progression in an R&D organization, for example, involves the categories scientist, lead scientist, senior scientist, principal scientist, and chief scientist.

5.6 GLOBALIZATION

Companies in the United States and in other industrialized countries came to realize some time ago the cost advantages of having repetitive work performed in less developed countries. The resulting globalization has changed the working environment of many professionals, including statisticians, especially in business and industry.

5.6.1 Evolution

The offshore outsourcing of statistical work initially focused on routine analyses. Financial services companies pioneered moving simple data analysis projects, typically involving large data sets, to low labor cost countries. The globalization of many manufacturing activities, and subsequently some design and technology, also led to the associated statistical support becoming outsourced. The tasks that have been sent offshore have expanded in scope over time to include more challenging activities.

5.6.2 Some Consequences and Impact

Globalization provides important new opportunities for statisticians in developing countries around the world. It also presents some challenges and opportunities for statisticians in developed countries. Some of these are similar to the impact of democratization in that the relief from more routine tasks that globalization provides allows statisticians to work more broadly and on more important issues.

Another result has been that statisticians are frequently involved in projects that require integration over continents and working closely with their offshore counterparts. This often requires the bridging of language, culture, and time differences. See Sidebar 5.3 for a recent example that illustrates the scope and cooperative nature of a globalized project and Lakshminarayanan (2008b) for further discussion.

SIDEBAR 5.3

STATISTICS 24 × 7

This project (described in greater detail in Sidebar 2.5 of Hahn and Doganaksoy, 2008) involved the development of a web-based system to enable a (job shop) plastic pellet manufacturer's design teams to develop product manufacturing specifications and fill requests for parts with specified characteristics faster and more efficiently. This was to be accomplished by the sharing of past product data, gathered globally through statistically designed experimentation, and then estimating relationships between properties from the resulting data.

The business had design sites and manufacturing plants in the United States, Germany, Japan, and China. The core project team consisted of

- Two statisticians—one in the United States (who was also the team leader) and another in Bangalore, India.
- A product design expert in the United States.
- A system architect and two software engineers in Bangalore.

In addition, product experts from various global business sites were consulted and participated, as needed.

The core team and key stakeholders from the business met in Germany for the project kickoff. Agreement was reached on project goals, assignment of responsibilities, timing of deliverables, and a plan for communicating among team members and with stakeholders.

During the course of the project, the team kept in close contact through frequent interactions via teleconferencing and electronic file exchanges. At each critical project milestone, the project status and next steps were reviewed with stakeholders at the various business sites by videoconferencing. The project continued over a period of 2 years and led to a successful implementation of the web-based tool.

Time differences (Bangalore runs 10.5 hours ahead of the Eastern United States) resulted in teleconferences taking place at the beginning of the workday in one country and at the end of the day in the other and required some adjustments of work schedules.

One issue raised by globalization that sometimes needs to be addressed is the difference in regulations in different countries. Export control laws might prohibit the sharing of some data and software that are felt to be sensitive. The project leader needs to clear up such issues early on and ensure adherence throughout the project.

A frequent benefit of global interactions is the diversity of ideas that it generates—often as a result of the camaraderie established among participants; this expedites the successful implementation of such efforts.

5.7 HAIL TO OUR MANAGERS!

Companies are not democracies.

—Popular saying

If you plan to be a self-employed consultant, you may want to skip this section.[5] It also has less applicability to those in academia. But all others need to recognize the high impact managers have on their professional lives. Managers determine, or have an important say in, what you work on (especially early in your career), what you get paid, how rapidly you advance, and your overall job security. They are generally

[5] Keep in mind, however, as discussed in Section 5.4.1, that self-employed consultants have managers also, namely their clients.

second only to the work itself in determining job satisfaction (perhaps, closely followed by compatibility with colleagues). Good managers challenge you to do better, help you in doing so, and can play a role in advancing your career.

5.7.1 The Performance Evaluation

In most organizations, managers meet periodically with those who work for them to discuss their performance—when this is not the case, it could be a reason for concern. The evaluation usually involves a verbal discussion, often accompanied by a written statement that becomes part of the employee's record. The specifics vary from one organization to the other but typically include

- Various aspects of performance (e.g., successful projects, teamwork, bringing in contracts) and success in meeting previously stated objectives (often based upon feedback from customers). In some organizations, these are summarized in a report card format on, say, a scale from 1 to 5.
- Future objectives.
- Personal strengths and weaknesses.
- Long-term goals.
- Follow-up actions.

Employees are often asked to prepare a similar statement with their self-evaluation.

Formal evaluations are usually conducted yearly. Some managers follow these up with less formal discussions during the course of the year, especially with new employees. The performance evaluation typically forms the basis of subsequent training, project assignment, and salary action.

Some statisticians, such as Deming, dislike the subjective nature of performance appraisals. Still, they are part of the work culture in most large organizations, provide you a realistic assessment of where you stand, a path to self-improvement and advancement of your career goals, and an opportunity to raise questions and issues.

5.7.2 Many Managers

In large organizations, your manager is, in turn, dependent on his or her manager, and so on. Often the written evaluation of your performance is reviewed and approved by your manager's manager. As a consequence, you are impacted not only by your immediate manager but by all above you in the management chain. This requires you to know and be attentive to the goals and expectations of management up the line.

Especially as you take on increasing responsibility, your day-to-day work might be led by somebody other than your direct manager. A "matrixed" structure is commonplace in many organizations. Those working on multiple projects might well have multiple project managers. Others have direct contact with the ultimate customers of the product or service being provided. Thus, to be successful, and build a high professional reputation, you need to keep many masters happy. Change in management along the line can, moreover, have an important impact on you (Section 5.8.2).

5.8 SOME CHALLENGES

In this section, we briefly describe some of the things that make a career in statistics especially challenging and interesting.

5.8.1 Time Accounting

In some organizations, statisticians are paid fully out of an organization's overhead funds. In many others, especially in business and industry, statisticians charge most of their time against the specific projects on which they are working; see Sidebar 5.4 for a comparison of these two modes of operation.

SIDEBAR 5.4

WORKING ON OVERHEAD VERSUS CHARGING TIME TO PROJECTS

Working on overhead relieves you from the responsibility of having to devote time to develop formal work proposals. It usually provides you (and your management) a greater say in the specific work in which you are engaged. This allows you to focus on those activities that are in the organization's best long-term interest. It avoids the cost and bureaucracy involved in moving money from one pocket to another within the same organization.

Requiring you to charge your time against specific projects focuses your attention on the areas for which your guidance, at least in the short run, is felt to be most needed (and often most appreciated). You will usually have clearly defined customers and they are likely to give greater attention to your results than they would if they did not have to pay for them. Also, since your customers have a greater stake in your being successful, they are more willing to gather and provide you the data and other relevant information that you need.

Having you charge time against projects forces you to focus on, and stay abreast of, customers' key objectives. It also makes you more conscious of the need to quantify the specific benefits resulting from your participation. Finally, this mode of operation is likely to provide you greater job security since overhead activities are especially vulnerable to cutbacks in times of economic uncertainty.

The process of obtaining authorization for the time (and other costs) to be charged against a particular project might be quite informal, such as the person requesting your support simply providing you a charge number. At many other times, it requires your developing a formal work proposal and securing formal approval before your work can begin (see Section 9.2.5).

5.8.2 A Dynamic Work Environment

> Change is the name of the game.
> —Popular saying

Statisticians in academia can, by and large, expect a reasonably stable work environment (see Chapter 13). This is hardly ever the case in most other organizations.

A day (and, even more so, a month or year) in the life of a statistician in business and industry is often highly unpredictable, due to both changes of people and changes in the business or political environment. The dynamics of the work environment in government are typically somewhere between that encountered in business and industry and that in academia, involving a fair amount of stability and some significant unpredictability.

People Changes. Your manager may change overnight. One of us (GH) had 17 immediate managers during his 46-year career—without himself changing jobs. Similarly, project leaders or key working partners, or even entire organizations (both ours and our customer's) may change suddenly. Such changes are frequently accompanied by a major recalibration of focus and objectives, reflecting the views of the new leaders.

Statisticians working in government witness major changes in direction with the appointment of a new agency head; see Sidebar 5.5.

SIDEBAR 5.5

PEOPLE CHANGES IMPACTING FEDERAL STATISTICIANS

Most U.S. federal government statisticians are career civil servants who work for the agency (or department) that employs them (and, ultimately, "the people" whose taxes support the agency). Various procedures are used for selecting the heads of federal statistical agencies. The Director of the U.S. Census Bureau, for example, serves at the pleasure of the President, subject to Senate confirmation, and is likely to change with change in administration.[6] The heads of various other U.S. agencies are appointed by the President, again subject to Senate confirmation, for specified terms (typically 5 years).

The U.S. Chief Statistician (Section 3.8) and the heads of some other agencies are also appointed by the President and again require confirmation by the U.S. Senate—but then serve permanently without the need for reappointment. Selection of yet other agency heads follows the selection processes of their parent organizations; such agency heads are often civil servants, rather than political appointments.

Statisticians working in government often witness major changes in direction, and sometimes funding, with the appointment of a new agency head. See Wallman (2008) for a description of the impact of reorganizations on the career path of one eminent government statistician. This situation is not much different from the impact on company statisticians of a change in, say, top management or division leadership.[7]

More locally, our colleagues and working partners, as well as our customers, also invariably change over time, due to promotions, resignations, retirements, and so on, requiring us to adapt to new personalities and work patterns.

---

[6] This led to the nomination, and subsequent confirmation of a new director of the Census Bureau in July 2009—only months prior to the start of field activity for the 2010 census.

[7] The unpredictability in government may, however, be somewhat more "predictable" than that in business and industry because the timing of a key underlying factor (at least in the United States)—the date of the next election—is known.

Changes in Business and Political Environment. Goals and priorities also change as a result of various other factors beyond our control, such as economic conditions, competitive pressures, new organizational initiatives, reorganizations (again both at our end and by the customer), or some external happening. The work of statisticians engaged in air traffic control problems in the United States was appreciably altered overnight by the decrease in air traffic resulting from the events of September 11, 2001. Statisticians working in the investment community are especially impacted by the rise and fall of the stock market. Businesses are often reorganized, merged, or bought out, and as a result, their mission is altered. Changes in the economic climate and budget cuts may have a profound impact on your work. And even without such evident changes, priorities may be altered simply by the emergence of the latest "hot problem" or directive from top management, or a change in direction by a key customer.

One of us (ND) notes that, even while writing this book, his project portfolio has changed drastically largely as a direct consequence of external events. My company has significantly extended its presence in the health and biomedical sciences and in clean energy through both growth and acquisitions. These changes were accompanied by expansion of the company's global reach through increased worldwide manufacturing operations and broader international markets. At the same time, the company significantly reduced its involvement in some other areas, including those that provided sustained support of my activities for many years. These changes required me to become knowledgeable in new application areas and technologies— and to understand how a statistical approach can benefit them.

Finally, changes in budget can be a constant problem for government statisticians, especially when budgets are cut for statistical activities to release funds for politically more popular, and perhaps more visible and more immediate, programs.

5.8.3 Deadlines

There is generally much pressure on statisticians to get answers in a timely manner. As a result, you will often be subject to significant time pressures. U.S. federal agencies that have data series designated as "Principal Federal Economic Indicators" are required (by the Office of Management and Budget) to have a calendar of release dates and to carry out a number of best practices for informing the public. The Bureau of Economic Analysis provides a release schedule in which it posts dates and times at which it will release specified reports, such as quarterly estimates of gross national product. These releases are watched very closely by the Congress, media, and the business world in general.

Businesses frequently strive to shorten lead times in new product introduction so as to beat the competition. Thus, top management might set new product release dates based heavily upon business as well as product considerations. Since these dates are usually known in advance, it is possible to plan for them. Meeting required, and not always reasonable, deadlines can nevertheless get hectic. In dealing with manufactured products, time-urgent "crises"—such as premature product failures—arise unexpectedly and need to be addressed urgently. Time pressures may also be created by sudden requests to provide inputs for a forthcoming presentation to a company

CEO or testimony to a Congressional or other governmental committee. As a consequence, statisticians' findings are often needed by a date set principally by circumstances; this timing infrequently coincides with the one that is most conducive for an optimum analysis.

In addition, government statisticians are often required to produce data for policy analysis on short notice, possibly only hours. Also, extraneous events or emergency situations sometimes lead to the need to obtain and report data rapidly. A typical example is the assessment of housing needs as a consequence of a hurricane or other natural disaster.

All of this requires statisticians to work, at least on occasion, long hours to meet, especially unanticipated, deadlines.

5.8.4 Multiple Projects

Statisticians frequently work on multiple projects at the same time. This is one consequence of an iterative approach to problem solving. There may be relatively quiet times on a particular project while the experiment or sampling study that you helped to plan is being conducted or the findings that you provided are being digested. During such times you may work on other projects. But you might be interrupted unpredictably by a new issue arising in the conduct of the experiment or sampling study or by a request from management for a project update. It is not unusual for the urgent needs of multiple projects to coincide in time, requiring you to "juggle many balls in the air" simultaneously.

Working on a single major project allows you to focus your efforts on that project. Multiple projects, however, often can add interest to your work and might provide greater job security in times of change. Also, working on multiple projects sometimes allows you to see and act upon a synergy between projects to the benefit of all concerned.

5.8.5 Statisticians' Roles as Technical Arbitrators

Statisticians are often asked to be technical arbitrators. When a controversy arises that involves a probabilistic or data-driven issue, the statistician may be asked to help decide who is "right." Thus, statisticians may find themselves in the center of disputes. Sometimes, these are within their own organizations; at other times, an external organization, such as a supplier, customer, or government organization, is involved.

Here are two examples:

- A thermoplastic resin manufacturer made regular product shipments to one of its major customers, an automotive parts producer. Both the manufacturer and the customer conducted their own testing on the same product properties. In general, the two sets of test results were in close agreement. Over a short time period, however, the customer returned three shipments that, though passing the manufacturer's test, were found to be unacceptable on the customer's tests. A statistician was asked to participate in a team effort to uncover the source of the discrepancy.

- A financial lending business was made up of eight different operations (e.g., small business lending, consumer lending, etc.). Individual business units had adopted different methodologies to score the risk levels of their customers. Management wanted to unify the scoring system across the businesses. Each business provided a credible reason why their method was best suited. It was not clear if one of the existing methods could be used for this purpose. A statistician was asked to actively participate in a study to evaluate the relative merits of the methodologies and make a recommendation toward unifying these.

Often, rather than being the arbitrator, the statistician is in the employ of one of the parties to a dispute. We consider in Section 10.4 such situations and the ethical challenges that they present.

5.9 WOMEN IN STATISTICS

5.9.1 Looking at the Numbers

In the Old Days. In the past, women statisticians were vastly outnumbered by their male counterparts. Starbuck (2010) reports, for example, that of current American Statistical Association (ASA) members who joined the ASA prior to 1984 and prior to 1999, only 17% and 23%, respectively, are women. As a consequence (and in light of the criterion for election of sustained contribution to the profession), only 20% of ASA Fellow nominations and 22% of the actual awards in the years 2004–2010 went to women.

In addition, Crank (2010b) reports that in a salary survey of biostatistics faculty, broken down by professor, associate professor, and assistant professor and years of rank, full professors with 7 or more years in rank (who presumably joined the workforce some years ago) were the least gender-equal group; only 17 of the 90 professors were female. The median salary for these 17 women, moreover, was 13% below that for the 73 men.

Some women statisticians still managed to become distinguished—even in the early days. An ASA flyer on women in statistics points to Florence Nightingale (the first female member of the Royal Statistical Society), Gertrude Cox (an eminent statistician in academia, elected to fellowship in 1944, and the 1956 ASA President), and Janet Norwood (the first woman commissioner of the Bureau of Labor Statistics, serving from 1979 to 1991, and President of ASA in 1989).

The Situation Today. Figure 5.1 displays the percentage of degrees in statistics (and biostatistics) conferred to women from 2000 to 2009. This plot is based on summaries by the ASA from data obtained—sorry for the mouthful—by the National Center for Education Statistics, via their Integrated Postsecondary Education Data System.[8]

---

[8] http://nces.ed.gov/ipeds/.

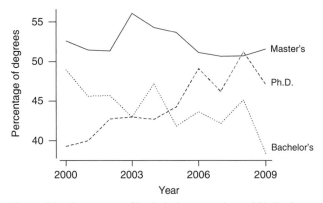

Figure 5.1   Percentage of bachelor's, master's, and Ph.D. degrees in statistics conferred to women in the United States: 2000–2009.

This plot shows that, of the degrees in statistics awarded in 2009, women received

- 38% of the bachelor's degrees (dropping from near 50% in 2000).
- 52% of the master's degrees (dropping slightly over the years).
- 47% of the Ph.D. degrees (up from below 40% in 2000).

We also note that an approximately equal number of males and females currently participate in the high school advanced placement course in statistics.

The ASA estimates that in 2010 approximately one-third of its membership, and about 40% of the new members over the past 5 years, are women.

The previously cited salary survey (Crank, 2010b) also shows assistant professors with 1–3 years in rank (presumably the most recent to join the workforce) to be the most gender-equal group with 36 women and 41 men. The salary statistics for these 36 women, moreover, were essentially identical to those of their male counterparts.

A woman president of ASA is no longer an unusual happening; three of the four presidents from 2006 to 2009 were women. We also note—although we have no numbers to substantiate it—that women statisticians have played especially important leadership roles in U.S. government official statistics. This includes Katherine Wallman, the current Chief Statistician of the United States, who has held this position since 1992.

5.9.2 Some Personal Assessments

The topic "Being a Female Industrial Statistician" was discussed at a roundtable luncheon at the 2008 Joint Statistical Meetings (JSM). We interviewed the discussion leader (Diane K. Michelson) immediately after that gathering. From her comments, we prepared the summary, shown as Sidebar 5.6, of views expressed by the nine women (from industry and students) who participated in this informal meeting.

SIDEBAR 5.6

VIEWS EXPRESSED AT 2008 JSM ROUND TABLE ON "BEING A FEMALE INDUSTRIAL STATISTICIAN"

- Women have a greater need than men to prove themselves and to gain the respect of colleagues, customers, and management.
- It is more difficult for women than for men to get merited promotions.
- Women have an advantage over men in that they tend to be more talented in communication, organization, multitasking, problem-solving, and people skills—all highly important to success as a statistician.
- A particular challenge is to dispel the assertion that women are not good at mathematics, a notion that was found to be incorrect in a recent study.

Most of these observations apply to women in various professions and are not unique to statisticians. However, they may be especially relevant for women statisticians, since statistics is not as well recognized a profession as some others (making it more difficult to break the barriers).

5.9.3 Some Comments and Tips from Women to (Mostly) Women

We assemble a variety of quotes from accomplished women statisticians aimed at helping other women. Many of these were in response to our request for comments on women in statistics. Some, like those in Sidebar 5.6, are possibly controversial:

- For every bad experience I've had with a male colleague, there have been 90 good ones. . . . I have run into situations . . . where being an assertive female has caused male colleagues to be uncomfortable or to turn away or to be, frankly, rather rude and almost abusive. You have to let these go and keep doing the work you think is important because there are plenty of colleagues that won't treat you that way. (Jana Asher)
- Be ready to be a little aggressive and make your opinions known. (Martha Gardner)
- Learn to keep your emotions in check. (Leslie Fowler)
- Dress appropriately. (Martha Gardner)
- Seek out good mentors. (Joanne Wendelberger)
- Have confidence and don't be critical of yourself. Don't constantly seek external approval or affirmation. We give others the benefit of the doubt; why not do so for ourselves? (Christy Chuang-Stein)
- Don't be offended if you are mistaken as a secretary—but don't be afraid to push back if inappropriately asked to do secretarial tasks. (Christine Anderson Cook and others)
- Men often talk easily to other men as colleagues, but talk to women about family or other nonwork topics. Learn how to graciously and smoothly

convert a nonwork conversation into a work-related one. (Christine Anderson Cook)

- Female statisticians are often challenged to demonstrate that they are "statistically significant." To successfully address the challenge even the most confident and talented person must find a supportive and mentoring environment at work and home. (Carolyn Morgan)
- Don't compromise your statistical integrity to be agreeable. (Christine Anderson Cook)
- Always behave as a professional person, which is not the same as a professional woman. (Gibbons, 2009)
- Help promote other women statisticians. (Christy Chuang-Stein)

Despite these challenges, most commentators agreed that statistics is a profession that offers women many and increasingly attractive opportunities.

5.9.4 Further Reading

In a general vein, Valian (1999) talks about the many small areas in which women were felt to be at a disadvantage in the workplace and cites scientific studies to show that these are real.

Jean Dickinson Gibbons reflects on her experiences as a female statistician (Gibbons, 2009). She also authored a chapter on statistics for a 1973 book that dealt with what were then regarded as nontraditional careers for women (Splaver, 1973).

Asher (2008) offers some hints for "all those statistician mom and dads ... contemplating a short-term or several-year 'vacation' from gainful employment." Her message is that "life at home with the kids does not mean that life as a statistician ends."

The Caucus for Women in Statistics is a source for current information. According to its web site,[9] this group was formed in 1970 to focus on specific problems associated with the participation of women in statistically oriented professions. Its activities include the issue of a quarterly newsletter (including job wanted advertisements) and the organization of technical sessions and social gatherings at statistical meetings.

5.10 MAJOR TAKEAWAYS

- The value of statistics is well recognized in some fields, such as pharmaceutical studies and official statistics, but needs to be sold to nonstatistical colleagues and management in many other areas.
- The use of statistics has sharply increased over the years with the rise of high-speed computing capabilities. The work of Deming and the emergence of Six Sigma have also been important factors contributing to the increased use, especially in business and industry.

---

[9] http://caucusforwomeninstatistics.com/.

- Vastly improved accessibility to statistics and statistical software has resulted in the "democratization of statistics." The use, and even development of, statistical methods is no longer limited to mainly statisticians. This has created new challenges and opportunities for statisticians to work on more complex and more important problems and to look at situations more holistically—as well as to guide practitioners and help them avoid pitfalls.

- Statisticians fit into organizations in a variety of ways. Some serve mainly one or only a few functions and are often organizationally members of one such function. Others may be part of a central statistics or analytical group serving multiple functions in the organization.

- Traditionally, statisticians have worked as consultants providing technical expertise. Increasingly, they are striving for and becoming project team members with broader responsibilities.

- Jobs in many large organizations are categorized by formal grades with official job descriptions. Grading systems are especially important in government positions.

- Globalization has impacted statisticians, especially in industry, and has provided new opportunities around the world. It has also changed the way many statisticians in outsourcing countries work.

- Managers play an important role in your on-the-job well-being, especially early in your career, and in your success and are responsible for the periodic evaluation of your work performance.

- Some challenges faced by statisticians in the workplace are
  - The need for charging their time against projects (and securing the authorization to do so).
  - A dynamic work environment.
  - The need to meet sometimes tight and unanticipated deadlines.
  - The need to juggle multiple projects.

- Statisticians sometimes serve as arbitrators on contentious issues.

- Women statisticians encounter some special challenges. The statistics profession, however, offers many attractive opportunities for women.

DISCUSSION QUESTIONS

(* indicates that question does *not* require any past statistical training)

1. Describe your experience to date of the receptiveness to statistics and statisticians.

2. The emergence of Six Sigma was suggested as providing impetus to the use of statistics in the workplace. Research and report on how Six Sigma is being used directly and indirectly today.

3. Today statistics is used in business and industry more extensively than ever before, but often without the involvement of statisticians. This has elevated the role of statisticians and allows them to engage in higher level efforts. Discuss and provide some examples in actual or hypothetical situations.

4. We indicated in Section 5.2.4 that "the democratization of statistics carries significant risks of misuse if the tools are applied uncritically." Describe one or more examples from your or others' experience or from the media in which this has been the case.

5. Explore further the analogy between computer scientists and statisticians in broadening their horizons as a consequence of the popularization of their fields. What are some key similarities and differences and how might statisticians address and leverage these?

6. What are the pros and cons of having statisticians organized in a centralized statistics group versus being embedded in individual operations of an organization? If possible, provide some examples.

7. Discuss the pros and cons of involvement in a project as a statistical consultant versus a team member, using a specific application in which you may have been involved or of which you are aware. How could the statistician's involvement have been improved?

8. *Consider managers that you have had on jobs to date. How have they impacted your accomplishments, your enjoyment of the job, and the degree to which the job was (or is) a learning experience? What did you learn from this experience?

9. *Give one or more examples of how things changed unexpectedly during the (perhaps short) time you were on a job. How did it impact you and how did you react?

10. *Assess your ability to handle unexpected changes, meet deadlines, and juggle multiple tasks. What can you do to improve on these?

11. Consider the example in Section 5.8.5 dealing with the situation in which a supplier and customer disagree on whether or not a product is meeting specifications, based upon their respective data analyses. What might be the reason(s) for these two analyses disagreeing? How would you find out? Suggest a specific action plan to help resolve the disagreement.

12. To what degree do you (as a woman or a man) feel the personal assessments and tips for success for (mostly) women, in Section 5.9, still apply today? Add to our discussion based upon your own experiences.

PART II

PREPARING FOR A SUCCESSFUL CAREER IN STATISTICS

THIS **PART** of the book provides guidance on how to best prepare for a successful career in statistics.

We describe the essential personal traits required of a successful statistician (Chapter 6) and the technical knowledge that aspiring statisticians need to acquire through their education (Chapter 7). We then suggest various strategies to employ in getting the right job (Chapter 8).

CHAPTER 6

CHARACTERISTICS OF SUCCESSFUL STATISTICIANS

6.1 ABOUT THIS CHAPTER

> Your hard skills will get you an interview, but it is your soft skills that will land you a job.
>
> —Sastry Pantula (2010b)

> Statistical leaders possess the so-called soft skills in abundance.
>
> —Ron Snee (1998)

Do you have the makings of a successful statistician? This chapter should help you decide.

We begin by amplifying on the obvious. You must have strong analytical and technical skills (Chapter 7). This is, however, only a necessary, and not a sufficient, condition. To succeed (and be happy) on the job, you also need to possess a variety of the so-called "soft skills." We will discuss the importance of communication and interpersonal skills; the ability to size up problems; flexibility; a proactive mindset; persistence; a realistic attitude; enthusiasm and appropriate self-confidence; the ability to prioritize tasks, manage time, and cope with stress; team and leadership skills; the ability to properly adapt and apply knowledge; and passion for lifelong learning. We supplement our comments with the observations of some other statisticians.

Most of these traits do not apply to statisticians alone; they are required for success in many professions. But they are especially important for statisticians.

Our comments are made principally with applied statisticians in business, industry, and government in mind. However, they also have much applicability to academia, as further discussed in Chapter 13.

Expecting any single individual to possess *all* of these traits is unrealistic. However, the more of them that you have, the better will be your chances of success.

A Career in Statistics: Beyond the Numbers, Gerald J. Hahn and Necip Doganaksoy.
© 2011 John Wiley & Sons, Inc. Published 2011 by John Wiley & Sons, Inc.

6.2   ANALYTICAL AND TECHNICAL SKILLS

A successful statistician needs to have an analytical mind. You require the ability to quickly grasp and properly apply highly technical concepts and be able to carefully examine, absorb, and sometimes question what is presented to you.

Mathematics provides the foundation of the theory of statistics. You need to like mathematics and be good at it. In your training you will be exposed to much mathematics while learning the fundamentals and the nitty-gritty of statistical methodology. On the job, you will often need to bend or extend a particular method to fit the problem at hand. This requires mathematical skills.

Agility on the computer is a further technical skill important for successful statisticians. This requires, for example, familiarity with common database formats and an ability to write custom codes for data preprocessing and nonstandard analyses.

In addition, you will need to rapidly gain an understanding of the application area in which you are involved. Thus, the ability to quickly learn, at least, the fundamentals of a new field and becoming conversant in it is another important requirement for success.

We will elaborate on these comments in Chapter 7.

6.3   COMMUNICATION AND INTERPERSONAL SKILLS

> Good presentations are a key component of almost any success story.
> —David Banks (2008)

Your customers, and sometimes even your management, often start with little understanding and a limited view of the potential contributions of statistics and statisticians. This requires you to teach and "sell" the value of statistics—as well as of yourself. Throughout your career, there will be a need to explain concepts and methods in laymen's terms. To do so effectively, you must have excellent oral and written communication and interpersonal skills.

You must speak in the language of your customers—and not expect them to be proficient in yours—statistical jargon must be subverted in favor of your clients. You need to initially assess their level of sophistication and then calibrate what you say accordingly. You will have to be able to get across key ideas, conclusions, and recommendations succinctly and effectively in both informal one-on-one or small group settings, in more formal presentations, and in written communication.

A genuine interest in others, an outgoing personality, and diplomatic skills are also highly important. So is the ability to network with others with backgrounds and training that may be quite different from your own.

The ability to be "quick on your feet" is another important element of communicating effectively. This is especially important in such settings as fielding questions from your CEO, agency head, or department head; when, as an expert witness, you are under cross-examination; or when, as a government statistician, you need to respond to questions from the media or the public.

We elaborate on the preceding points and provide hints for success in Section 10.2.

6.4 ABILITY TO SIZE UP PROBLEMS AND SEE THE "BIG PICTURE"

> Statisticians must grit their teeth and also become practitioners. Only then will they discover where the truly novel problems are.
>
> —George Box

You need to be good at sizing up and diagnosing problems, appreciating their context and broader implications and assessing their importance.

Most problems that will come your way will *not* be well defined or articulated. Often, we encounter narrowly phrased technical questions, such as

- "How large a sample do I need to ..."
- "What is good software to ..."
- "How come my regression analysis yields an R^2 of 0.9, but I still get a very wide prediction interval?"

As another example, we were often asked early in our careers to "design a Latin square experiment." Later requests were for "Taguchi" experiments. Most problems have much broader implications than suggested by their initial statements; see Section 9.2.1 for an example. It is your job to ferret out these implications and to understand the real problem and its significance to the organization.

Occasionally, you may be asked questions that are of little importance to anybody other than the person posing the question. (Of course, if this person happens to be your CEO or agency head that likely makes it an important question per se). Or you may be called upon to give "statistical blessing" to a fait accompli, and an objective evaluation is less than welcome. It is important for you to appreciate such situations and act accordingly.

The ability to size up a situation astutely requires you to be able to rapidly gain an understanding of the underlying politics and history and to have a "good nose" for sensing management interest and support. It calls for an inquisitive mind and the ability to frame—and the confidence to (politely) ask—fundamental questions that might challenge underlying, and often not stated, assumptions, and listening closely to the answers. It may also require some independent digging and keen evaluation, as illustrated by the example in Sidebar 6.1.

SIDEBAR *6.1*

DEVELOPMENT OF A SALES FORECASTING SYSTEM

A service business relied on a sales forecasting system to make inventory and pricing decisions. The existing forecasting tool was cumbersome to use, poorly documented, and prone to providing inaccurate forecasts. Our initial study indicated that markedly better results could be obtained by an alternative approach based on a combination of statistical techniques. The need to replace the current method seemed evident to us. There were some technical challenges to overcome, but we were confident of our ability, with some effort, of addressing these satisfactorily.

Further discussion revealed that a key and highly placed manager who had championed the development of the existing method thought otherwise and felt that all that was needed were a few "fixes." Selling an improved new approach called for more than demonstrating its technical superiority. It required us to establish a close relationship with key direct reports of the manager and to gain their help in securing eventual project approval.

A further important element of sizing up a problem successfully is your assessment of the chances that you will be able to make a significant contribution and that the project itself will succeed (Section 9.2.4).

6.5 FLEXIBILITY

Don't put all your eggs in one basket.

—Popular saying

Applied statisticians work in a dynamic environment, as discussed in Section 5.8.2. The strategic importance of the work we are doing may be downgraded (or upgraded) at any time due to, say, a change in management or in the business climate. You need to be prepared for things to change abruptly, to anticipate and recognize change, and to have the vigor to roll with the punches.

Therefore, while weighting heavily the demands of your current customers, managers, and projects, you need to frame your work to make it as robust to change as possible. This requires a good understanding of the business environment and the information (and imagination) to recognize how this environment might change. It makes it important for you to have some knowledge of, and possibly involvement in, a variety of projects and opportunities beyond those in which you are currently engaged—so as to be able to switch over to, if need be. In other words, always have a Plan B in mind.

Some of us enjoy change and thrive on it. But it is not everybody's cup of tea. Those who would like their work to be fully predictable need to think twice before embarking on a career as an applied statistician.

6.6 A PROACTIVE MINDSET

Merriam-Webster defines proactive as acting in anticipation of future problems, needs, or changes. The democratization of statistics—resulting in today's statisticians being relieved of much routine number-crunching activities—and the dynamic environment make it essential for statisticians to be proactive. As a statistician, you need to actively seek and leverage opportunities for improvement and to engage in out-of-the-box thinking. This requires you to search for the important opportunities that are lurking around the proverbial corner and to identify, assess, and communicate your potential role and contributions.

Once on a project, a proactive mindset will push you to look at things holistically—and to seek out important aspects of the problem and novel ways of addressing it to attain the best possible results. On a technical level, this may call for you to creatively apply existing methodology in a nontraditional area or manner or even to develop a new approach to fit the situation, as in the example in Sidebar 6.2.

SIDEBAR 6.2

EARLY IDENTIFICATION OF POPULAR VIDEO CLIPS

Businesses are becoming increasingly interested in the number of visitors to their web sites and the number of clicks that a particular link (e.g., for an item on sale or a request for further information) will attract. A media company had accumulated data on the number of viewers of various video clips at its web site. There were, at any time, hundreds of such clips available for viewing. While a handful of these attracted many viewers, the large majority laid essentially dormant.

The business sought our help in using the data to identify the factors that accounted for the large variability in video view counts. The findings were to be used in site redesign and management decision making, with the ultimate goal of identifying and increasing the number of "hit" videos.

A surprising discovery was made in the early stages of our analysis: the eventual success of a video clip could be determined with high accuracy within a few days, if not hours, after posting. This potential capability was highly appealing. We, therefore, proposed a new project to develop, validate, and implement methodology for early prediction of video clip popularity—in place of the originally commissioned work. The new project was approved and successfully conducted. The resulting advantage to the business far exceeded that achievable by the original approach.

6.7 PERSISTENCE

> Statisticians should be adept at removing barriers that impede success.
>
> —Peterson et al. (2009)

Statistical concepts—because they tend to be "different" from the norm of deterministic thinking—often require reinforcement at strategically selected times before they take hold. Once on a project, action by others is often required for you to be able to make meaningful contributions. You typically need to rely on working partners to provide existing data, to collect new information, or, perhaps even, to execute designed experiments.

You need to persist in driving toward what you believe to be in the best interest of the project and the organization, and not give up easily when you are convinced you are on the right path. At the same time, you need to appreciate the fine line between persistence (or tenaciousness) and obstinacy. You need to listen carefully why others might think that what you are advocating might not work or be practical, and consider modifying your ideas accordingly so as to still achieve your major goals.

6.8 A REALISTIC ATTITUDE

> The best business solution is more important than the best statistical solution, and you need to know the difference.

—Roger Hoerl

You need to focus on both the immediate requirements of the project and the long-term goals of the organization, and not let marginal issues divert you.

It might seem cool to try out a new method that you learned in school, heard about recently, or even developed yourself—but you should do so only to the degree that this is relevant and useful for the problem at hand. You may determine that the available data are inadequate to serve the immediate needs of the project, but are tempted to apply advanced modeling with the thought that this might, just possibly, provide a rescue. But your time and efforts might be better spent in working toward procuring improved data.

We all derive satisfaction from a job well done. In a result-oriented environment, however, you can be a perfectionist only up to a point. Cost and practical considerations dictate how far to take a project. In the example presented in Sidebar 6.2, the methodology proposed to predict the ultimate success of a video clip was based solely on using data on view counts over time. More accurate and timely predictions could be obtained by utilizing further data from other sources (e.g., the genre of the clip, video quality, and language). However, this would require substantial methodological development and be more difficult and expensive to implement. It would also be harder to explain to management. The potential benefits of a more advanced approach did not seem to justify the time required and cost that would be incurred for development and implementation. Consequently, it was decided against moving beyond the initially proposed methodology.

You need to have the ability to carry your work forward only as far as it is of practical usefulness and to be able to recognize and to let go when you have reached a point of diminishing returns. Those who tend not to be satisfied until they have driven a problem to its ultimate optimal solution need to learn how to adjust their thinking so as to accommodate the practical needs of the problems that they encounter; see Sidebar 6.3.

SIDEBAR 6.3

SIMPLE SOLUTIONS VERSUS HIGH-POWERED STATISTICS

High payoff applications are not necessarily the ones that involve the most advanced or complex statistical techniques. There may be little, if any, correlation between the sophistication of the tools used for analysis and the payoff from their use. There are many situations for which simple and easy-to-understand methods—and often elementary statistical thinking—do the job.

At the same time, there are other cases for which simple methods alone are insufficient. A plot of the residuals from a fitted model may, for example, be critical to gaining understanding. This, however, may have to be preceded by fitting a sophisticated statistical model from which these residuals are obtained.

You need to understand the arsenal of tools available to you so that you can properly leverage these when appropriate—often adapting them to the problem at hand—and dismiss them when they are not. However, even when advanced tools are being used, the results need to be presented in a maximally understandable manner. This frequently calls for incisive graphical displays. In addition, although the technical details of the underlying method generally do not need to be explained to your customers (except, perhaps, in a technical appendix of the project report), the key underlying assumptions and their consequences need to be clearly conveyed.

We continue this discussion in Section 9.3.

6.9   ENTHUSIASM AND APPROPRIATE SELF-CONFIDENCE

> A certain amount of self-promotion is an important part of building a career.
>
> —Joanne Wendelberger

A key challenge facing many statisticians is to sell themselves and statistics. A prerequisite to making others enthusiastic about what you do is for you to be enthusiastic and have a positive "can do" attitude, conveying passion for your work. This calls for a high level of self-confidence.

At the same time, you need to be able to differentiate self-assuredness from arrogance. Not taking yourself too seriously, and especially maintaining a sense of humor, is helpful. So is an appreciation of the ego of others.

Even the most successful statisticians—and especially the most resourceful ones—encounter occasional setbacks. You may find yourself proposing (what you regard to be) a great idea but meet pushback from colleagues or management. The resistance may have technical merit or there may be other things going on behind the scenes, as in the example in Sidebar 6.1.

At other times, you may be doing outstanding work, but get little recognition for it—perhaps, because the overall project did not succeed. Or a promising project might be cut off in midstream due to extraneous factors. You need to be able to cope with such occasional setbacks, learn from them, and move forward. On the other hand, if you find yourself consistently not succeeding, you need to take a close and candid look, perhaps with a trusted friend or mentor, at how you go about doing things and carefully consider what changes you need to make.

6.10   ABILITY TO PRIORITIZE, MANAGE TIME, AND COPE WITH STRESS

The fact that you are frequently working on multiple projects, and the need to respond to unanticipated crises and requests, can result in an overload work situation. This makes it especially important for you to be able to manage and allocate time effectively. Unanticipated demands can be better met, and the resulting stress reduced, by scheduling yourself to meet with time to spare those demands that are known or can be readily anticipated. You need to be able to skillfully prioritize tasks—based upon their importance, their deadlines, and the time required to do the work—and be ready to reprioritize as situations change.

If you find that your proverbial plate seems to be flowing over all the time, we propose that you take a close look at your work patterns. You may be overcommitting yourself or being too generous in taking on nonessential tasks or not sufficiently delegating work that can be delegated. Or perhaps you are not working efficiently or allowing yourself to be diverted from the main goals of a project by tangential issues. A look at whether your colleagues have similar issues, and how they cope with the demands on them, might be instructive.

6.11 TEAM SKILLS

> In the long history of humankind (and animal kind, too) those who learned to collaborate and improvise most effectively have prevailed.
>
> —Charles Darwin

Statisticians frequently work as members of a project team or collaborators—a mode that we strongly advocate (Section 5.4). In this capacity, you need to provide important added value to the team and at the same time be easy to work with. It might require you to suppress some of your personal aspirations for the sake of team harmony and success.

Team efforts can present thorny issues that you need to address. For example, how do you handle a potential problem that you feel merits being brought to the attention of management, but the team feels it can fix on its own? Or what should you do when a team member, whose inputs are critical to the work, is consistently missing deadlines (and make sure that you are not that person)? There are no set answers to such questions. Each needs to be addressed on its own merits and good judgment applied.

6.12 LEADERSHIP SKILLS

> Don't allow the label of "statistician" limit your thinking about how you can contribute and add value to the organization.
>
> —Roger Hoerl

The previously discussed evolution of statistics, and the resulting more proactive roles proposed for statisticians, calls for them to exert greater leadership than in the past. This may be informally or through a specific role, such as project team leader or manager of an organization. This requires you to have the strong personal and organizational skills that characterize a successful leader.

We discuss the concept of the statistical leader in more detail in Chapter 12.

6.13   ABILITY TO PROPERLY APPLY AND ADAPT KNOWLEDGE

> You don't need to be an expert in everybody else's field, but some level of detailed understanding is critical.
>
> —Martha Gardner

The unique contribution that statisticians make is, of course, their knowledge of statistical concepts and tools and their ability to apply these effectively.

Raw knowledge in itself, however, is not enough. You need to integrate and synthesize what you know and apply it appropriately to the problem at hand. In a classroom setting, you can expect the problems at the end of a chapter (or on the final exam) to deal with topics that are discussed in the chapter. But on the job, you do not know to which chapters of which book—or to which course, if any—the problem that you are facing pertains. It is your job to select—from your reservoir of knowledge— the appropriate technical approach to use in a given application. Frequently, a "canned" method fails to address the specific needs of a given application. It is your job to tailor existing methods—or to develop an appropriate new method and, perhaps, the associated software—to fit the problem.

Consider the following example. The managers of a web site desired to estimate the total number of users who downloaded a particular file over a 1-month period. Unfortunately, the available data on file download activity were sparse and do not readily yield the desired information. Instead, new data about user characteristics were collected and recorded on a random sample of 50 users over 1 hour time periods. In this manner, the same user could, however, be sampled more than once and, thus, multiply counted—unless appropriate adjustments are made. It was noted that this sampling scenario closely resembles the capture–recapture approach commonly used to estimate the number of wild animals in a forest (or the number of fish in a lake); see Sidebar 6.4. The methodology used for estimating animal populations was, therefore, adapted to provide an estimate of the number of users who downloaded the specified file during the course of a month.

SIDEBAR *6.4*

CAPTURE–RECAPTURE METHODOLOGY TO ESTIMATE THE NUMBER OF WILD ANIMALS IN A FOREST

In its basic form (see Lohr, 2010), the goal of the capture–recapture methodology is to estimate the total number (N) of animals in a defined area. A random sample of n_1 animals is captured, tagged, and released back into the forest. In a later time period, n_2 animals are captured. Suppose that m out of these n_2 animals are found to be carrying the tag from the previous sample. N is then estimated by equating the (unknown) proportion of marked animals in the population to those in the second sample and solving for N, that is, setting $(n_1/N) = (m/n_2)$ and then estimating N as $\hat{N} = n_1 n_2/m$. This method assumes a well-mixed static population during the sampling period. The application of the methodology to changing populations (i.e., due to death, birth, and migration) often involves several cycles of capture and recapture.

6.14   PASSION FOR LIFELONG LEARNING

The statistics profession is constantly changing, as are the application areas in which many of you will be involved. You need to keep abreast of the latest developments in both. We suggest some ways for doing this in Chapter 14.

6.15   FURTHER READING

For further discussion on what makes a successful statistician, see Banks (2008), Hoerl et al. (1993), Kenett and Thyregod (2006), Kotz et al. (2005), McDonald (1999), O'Brien (2002), Snee (1998), and Chapters 3 and 4 in Coleman et al. (2010). Also, the *Amstat News* frequently carries articles on this subject.

In addition, various books consider some of the topics discussed in this chapter. For example,

- Boen and Zahn (1982), Cabrera and McDougall (2002), and Derr (2000) focus on statistical consulting.
- Spurrier (2000) considers the practice of statistics.
- Szabo (2004), though directed at actuaries, provides comments with general applicability.

6.16   AS OTHERS SEE IT

To supplement our comments, and to get more inputs, we asked some eminent applied statisticians:

- What, in your opinion, are the most important traits of a successful applied statistician?
- What key advice would you give to an aspiring statistician planning or embarking on their career?

We present below responses (some abridged or edited) to these questions. More responses are shown at the end of Chapters 7 and 9 and in quotes in various other chapters. Some of these responses bring out new points (or subtleties) that add to our comments. Other responses reiterate points that we make and may be somewhat repetitive. But, as one of our respondents suggests, if it's that important, it might warrant repetition.[1]

---

[1] Many of the respondents have held multiple positions during their careers. We generally show their present (or last prior to retirement) positions.

Successful statisticians need good communications skills in both oral and written form and a keen interest in helping others understand how statistical thinking applies to all aspects of the enterprise.

—Thomas J. Boardman, Professor Emeritus, Colorado State University; Consultant

To be successful, an applied statistician has to have strong negotiation and communication skills, a firm grasp of the scientific method, a shameless curiosity and a willingness to walk miles in the shoes of clients.

—Lynne Hare, Director (retired), Applied Statistics, Kraft Foods

Bottom line to being successful is a keen sense of customer focus and returning value for their investment.

—William Hill, Six Sigma Leader (retired), Center for Applied Mathematics, Honeywell Company

Ability to work as an equal partner on a team with non-statisticians; desire to become a subject matter expert in the domains where statistics is being applied.

—Bruce Hoadley, Research Statistician (retired), Fair, Isaac and Co. Inc. (FICO)

Successful statisticians require solid technical skills in broad areas of applied and theoretical statistics, and the inclination to learn new areas. Necessary soft skills include the ability to team and communicate with non-statisticians; leadership ability to determine what needs to be done; and the ability to energize yourself and others to get it done.

—Roger Hoerl, Manager, Applied Statistics Laboratory, GE Global Research Center

The most important trait is the ability to communicate clearly the spoken and written word to your customer.

—Ronald Iman, President, Southwest Technology Consultants; past President of the American Statistical Association

I have never felt limited by a lack of understanding of statistics. The larger challenge was translating statistical findings into concise information that is easily and accurately understood.

—William Makuch, Managing Director (retired), Structured Transactions and Analytical Research, Wachovia Corporation

An applied statistician must be able to see and appreciate a client's problem from the client's viewpoint. The less you appear to be and act like a statistician, and the more you appear to be intimately familiar and interested in the client's problem area, the more successful the interaction will probably be. Three things are extremely important: The ability to listen and to ask the right questions, so that the client's statistical needs become crystal clear; the ability to choose and use the simplest, most appropriate statistical tools for addressing the problem; and the ability to translate the statistical results into the practical, action-oriented (non-statistical) implicative language of the client. Finally, when interacting with clients don't forget to be friendly, pleasant and to smile—you'd be surprised how much this ensures success.

—Harry Martz, Laboratory Associate in Statistics (retired), Los Alamos National Laboratory

Everyone says that communicating (listening, cooperating, asking and hearing, speaking and writing) are important. But it's true even though everyone says it, so I will say it again.

—David S. Moore, Shanti S. Gupta Distinguished Professor (retired), Purdue University; past President of the American Statistical Association

Successful statisticians need strong knowledge and understanding of statistics; the ability to apply their knowledge and training to a diverse array of problems and applications; the ability to interact well with others; good computer and mathematics skills; good oral and written communication skills; and the ability to work well with others. They should be prepared to work in teams consisting of other scientists, engineers, managers, etc.

—Carolyn B. Morgan, Professor, Hampton University; formerly, Statistician, GE Corporate Research and Development

Communication, communication, communication! This includes being a good listener; the ability to ask the right questions; and the ability to communicate statistical concepts at the client's level. Ability to adapt to changing situations. Willingness to learn and apply new methods, and to look at problems outside the realm of their current knowledge.

—Margaret A. Nemeth, Statistics Center Lead, Regulatory, Monsanto Company

I want the kind of applied statistician who realizes that nature abhors a vacuum. He/she is part of a team trying to attain an objective in the subject matter, not sitting in a lonely corner analyzing data. The good applied statistician puts the emphasis on "applied" and not on the "statistics."

—Barry D. Nussbaum, Chief Statistician, Environmental Protection Agency

A successful applied statistician is a good communicator. This includes being a good listener, as well as being able to communicate ideas effectively to non-statisticians.

—William C. Parr, Professor, China Europe International Business School, Consultant

While there are many necessary traits for success, there are only two that really discriminate good and excellent applied statisticians: proactive leadership and strong interpersonal skills. I view ultimate success as a demonstration of the footprint that an applied statistician has left behind in integrating statistics into the culture, strategies and decision-making of an organization. Large footprints invariably require initiative and influence.

—Charles G. Pfeifer, Principal Consultant, DuPont Quality Management and
Technology

Have broad interests. Maintain high self esteem in difficult situations. Be an
excellent communicator and thought of as one more good mind on a project team
(rather than some sort of mathematical geek kept to run the formulas).

—Charles B. Sampson, Director of Statistics (retired), Eli Lilly and Company

A successful statistician needs a sharp mind, an absence of rigidity, the ability to
wink at p-values, and an appreciation of the foundations of statistics.

—Nozer D. Singpurwalla, Professor, The George Washington University;
Consultant

6.17   MAJOR TAKEAWAYS

Technical knowledge is a necessary but not sufficient condition for the success of an
applied statistician on the job. Successful statisticians also need to possess important
soft skills:

- Strong communication and interpersonal skills.
- The ability to size up problems and see the big picture.
- Flexibility: the ability to anticipate and adapt to change.
- A proactive mindset.
- Persistence in pursuing what is right for the organization.
- A realistic attitude and readiness to quit at the point of diminishing returns.
- Enthusiasm and appropriate self-confidence.
- The ability to prioritize, manage time, and cope with stress.
- Team skills.
- Leadership skills.
- The ability to properly apply and adapt knowledge.
- A passion for lifelong learning.

DISCUSSION QUESTIONS

(* indicates question does *not* require any past statistical training)

1. *Based on the discussion in this chapter, what would you say are the three most
   important traits of a successful applied statistician? Are there any traits—or
   degrees of emphasis—which the respondents to our questions in Section 6.16
   appear to disagree on, either among themselves or with the authors?

2. *Rate yourself (e.g., on a score from 1 to 5) on each of the characteristics of a successful statistician described in this chapter. What can you do to improve yourself on characteristics on which you do not score well?

3. *Describe situations in which you have been involved, either in a personal or in a professional context, in which you needed to do one or more of the following: be fast on your feet, size up the situation differently from that described to you, be flexible, be proactive, and be persistent.

4. *Consider a work or school (e.g., your thesis) situation in which you are currently or have recently been involved. What changes, beyond your control, happened (or might happen) to impact your work and its success? How can you make sure that what you are doing is as robust as possible to such changes?

5. *Consider a specific project in which you have been involved. How can you determine when your work has reached a point of diminishing returns?

6. *How good a job do you think you are doing in prioritizing your work, managing your time, and coping with stress? How can you do better?

7. *Do you enjoy and thrive in working on team projects or do you feel more comfortable doing things on your own? What might you do to become a better team player?

8. *You are collaborating with a human resources specialist to design and conduct a survey to assess employee job satisfaction and will need to jointly present your findings to management by a specified date. Your working partner seems unable to find the time needed to devote to this project. He consistently apologizes and promises to get his part done right away, but does not come through. What approaches might you use to foster progress, without unduly creating ill feelings?

9. Elaborate on the application of the capture–recapture methodology described in Sidebar 6.4 and the underlying assumptions.

CHAPTER 7

EDUCATION FOR SUCCESS

7.1 ABOUT THIS CHAPTER

> There is no substitute for knowledge.
>
> —W. Edwards Deming

We stress throughout this book that successful statisticians require competencies that stretch far beyond technical knowledge of statistics. That, however, in no way detracts from the need for them to have sound in-depth understanding of their subject. Competence in statistics is, after all, the key added value that statisticians provide.

Our nonstatistical colleagues, as a consequence of the democratization of statistics, have become increasingly familiar with statistical concepts and methods, and are building self-confidence in performing their own statistical analyses. When they turn to professional statisticians, they expect them to be more knowledgeable than ever.

You need to plan in advance to ensure that you acquire the right education so that—when the time comes—you will be in a position to put together an attractive résumé and embark on a rewarding career. Needless to say, this calls for such basics as working hard and striving to get top grades. But it also requires you to make good decisions on a variety of matters—such as selecting the courses that are best for you, building evidence that you possess broad personal and leadership skills, and, if possible, gaining relevant job experience.

In this chapter, we provide guidance to help you acquire the educational foundation that best meets your needs for a successful career in statistics. We consider

- The Statistics Advanced Placement course and other high school programs.
- Degrees in statistics: the numbers.
- How far to go: bachelor's, master's, or Ph.D. degree?
- Selecting the right school and program.
- Statistical education: setting the foundations.
- Coursework in statistics.
- Statistical computing and software.
- Some other recommendations.
- Internships and university consulting.

A Career in Statistics: Beyond the Numbers, Gerald J. Hahn and Necip Doganaksoy.
© 2011 John Wiley & Sons, Inc. Published 2011 by John Wiley & Sons, Inc.

- Entering statistics from other fields.
- Further resources.
- The limits of formal statistical education.
- As others see it.

7.2 THE STATISTICS ADVANCED PLACEMENT COURSE AND OTHER HIGH SCHOOL PROGRAMS

We urge students to avail themselves of opportunities to learn about statistics even before entering college.

In the United States, many are introduced to statistics in high school through the Advanced Placement (AP) course on the subject, and take the nationally offered AP Statistics exam at the end of the course. Typically taught as a 1-year offering by the Mathematics Department, students who complete the course and pass the exam may receive college credit for a one-semester introductory statistics course or advanced placement (i.e., waiving of requirement to take the introductory statistics course, possibly as a prerequisite for more advanced courses).

Started in 1997, in 2009 almost 100,000 students took the AP Statistics exam and it was the eighth most frequently taken of the 37 AP courses.[1] Not all students take the exam, making the total number taking the course even higher. About three-quarters of those who take the course do so in their senior year in high school. See The College Board (2010) for a more detailed description.

Some high schools offer a "university in the high school" program as an alternative to the AP program. Students pay tuition to a local college or university that works with the high school to provide courses that meet the institution's academic requirements. Upon successful course completion, students receive credit from the institution, which may be transferable to whatever college or university they choose to attend.

Many who study statistics in high school do so because of their curiosity about the subject and/or their interest in mathematics and do not, at least initially, think of it as a career opportunity. However, we hope and expect that at least some will be turned on sufficiently to want to consider statistics as their potential profession. See Patterson (2009) for further discussion.

7.3 DEGREES IN STATISTICS: THE NUMBERS

7.3.1 The Big Picture

Figure 7.1 shows the number of degrees in statistics conferred at the bachelor's, master's, and Ph.D. degree levels from 2000 to 2009. These graphs were prepared

---

[1] U.S. history, English literature and composition, English language and composition, calculus AB, U.S. government and politics, biology, and psychology were the seven most popular courses.

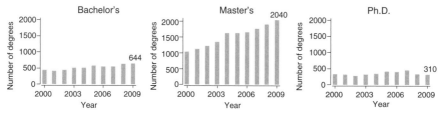

Figure 7.1   Bachelor's, master's, and Ph.D. degrees in statistics in the United States, 2000–2009.

from data compiled by the American Statistical Association, based on numbers obtained from the National Center for Education Statistics (NCES).[2]

The same person can appear multiple times in these plots—as a recipient of a bachelor's degree goes on to get a higher degree or a Ph.D. earns a master's degree along the way.

We note from Figure 7.1 that the preponderance of degrees in statistics is at the master's level. Also noteworthy is the approximate doubling in master's degrees from 2000 to 2009. During the same period, the number of bachelor's degrees in statistics rose only modestly, and the number of Ph.D. degrees remained about the same, though fluctuating over the period. Landes (2009), moreover, shows that an increasing proportion of, at least, Ph.D. degrees awarded in the United States are conferred to non-U.S. residents.[3]

It should be noted that just as it is difficult to measure precisely how many practicing statisticians there are at any given time in the United States, so it is also hard to estimate how many degrees are conferred yearly in the subject. This is due, in part, to differences in opinion as to exactly what constitutes a degree in statistics.[4]

7.3.2 A Further Breakdown

A further study, the National Science Foundation's (NSF) Survey of Earned Doctorates (using somewhat different program inclusion criteria and data gathering methods[5]), arrived at a total of 579 Ph.D. degrees in statistics in 2007 and 510 in 2008.[6]

[2] http://nces.ed.gov/ipeds/.

[3] The proportion of Ph.D. degrees awarded to non-U.S. residents increased from under 40% to about 70% from 1997 to 2006, whereas the actual number of such degrees granted to U.S. residents over the same years decreased.

[4] The ASA compilation used general and other statistics, biometry/t ' statistics, educational statistics and research methods, health and me statistics, and probability. It did not include econometrics, manageme statistics, and psychometrics.

[5] For NCES, the institution granting the degree provides the informatior forms that Ph.D. students are required to submit to their institution befor how different data gathering methods can result in different findings.)

[6] There were an additional 142 and 156 Ph.D. degrees in managemen statistics and 30 and 35 Ph.D. degrees in econometrics in 2007 and 2008

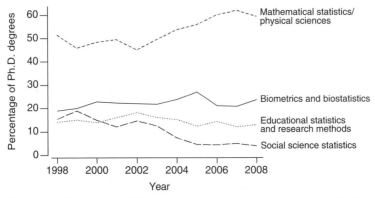

Figure 7.2   Percentage of Ph.D. degrees in statistics by area of specialization.

This survey, moreover, broke down the 510 Ph.D. degrees awarded in 2008 by area of specialization (or school granting degree); see Figure 7.2.

We note from Figure 7.2 that

- The percentage of degrees in mathematical statistics and the physical sciences has constantly been around 50% or above, and is on the rise (59% in 2008).

- The percentage of degrees in biometrics and biostatistics has been about 20% and has been rising slowly, but not consistently (24% in 2008).

- The percentage of degrees in educational statistics and research methods has been around 15% and is dropping slowly (13% in 2008).

- The percentage of degrees in social science statistics has dropped sharply to below 5% (4% in 2008).

7.4   HOW FAR TO GO: BACHELOR'S, MASTER'S, OR PH.D. DEGREE?

How far should you, as a potentially aspiring statistician, take your formal education? The answer depends heavily on your specific career aspirations, and especially whether your long-term goal is pursuing an increasingly more responsible technical career as a statistician in an application area or academia (for both of which in-depth technical training will be especially important) or a broader career leading to a position in management or in an application area (see Chapter 12).

7.4.1   Bachelor's Degrees in Statistics

Students that take a bachelor's degree in statistics fall roughly into three categories:

- Those who consider it as their terminal degree and seek employment in the field, or related fields, after graduation.

- Those who seek employment upon graduation, but are planning to continue their education part-time, possibly with financial (and other) support from their employers (and, in some cases, eventually switch to full-time study to complete their degrees).

- Those who are planning to go on to graduate school upon getting their bachelor's degree and want to build an early solid technical foundation in statistics.

Undergraduate programs in statistics are generally highly applied, but can cover the key methods and concepts. In addition, some 4-year colleges have programs to provide students with research experience. Such programs are especially attractive to those who aspire to continue their education beyond a bachelor's degree; see Bryce et al. (2001) and Tarpey et al. (2002) for curriculum guidelines.

Those who do seek a job upon getting an undergraduate degree in statistics (the first two categories) are typically asked by their employers to conduct statistical analyses using statistical software. A good understanding of such software and competence with computers, in general, is therefore especially important (Section 7.8).

The ASA conducted a symposium on "Improving the Work Force of the Future: Opportunities in Undergraduate Statistics Education" in 2000; see Bryce (2002).

7.4.2 How Far Will a Bachelor's Degree in Statistics Get You?

Many students go to work after completing their bachelor's degrees for financial or personal reasons—or because they are tired of school and want to explore the world beyond. In some cases, an undergraduate degree in statistics may be a stepping-stone for a career in an application area.

But is an undergraduate degree, with a major in statistics, sufficient for a career in applied statistics? This is a somewhat controversial subject.

In Favor of a Bachelor's Degree. Bryce et al. (2001) maintain that "Many employers have discovered that graduates of strong BA/BS programs (in statistics) possess skills that make them valuable entry-level employees." As we noted in Section 4.3.2, approximately 90 of the more than 175 statisticians employed by the Mayo Clinic hold only bachelor's degrees (although not all of those degrees were in statistics). Also, many of the statisticians engaged in official statistics have only bachelor's degrees.

Scheaffer and Lee (2000) further state the case for undergraduate statistics. Bryce (2005) argues in favor of a strong bachelor's degree in statistics, either as a terminal degree or, perhaps for the top 25% of such students, as a foundation for graduate education in statistics.

A Bachelor's Degree Is Insufficient. Moore (2001), a former President of the ASA, maintains that "no undergraduate program is intended to train professional statisticians. For better or worse, statisticians are defined as having at least a master's degree or equivalent experience. Holders of a bachelor's degree may eventually reach this status via on the job training and practical experience, but their degree does not equip them for professional practice." Ritter et al. (2001) add, "There are very few positions exclusively for BS-statisticians. More common are positions (in statistics) for which BS-statisticians qualify, but for which statistics is only one of several appropriate types of preparation." Also, as we noted in Section 3.7, most statisticians working in the major U.S. federal statistical agencies are hired as mathematical statisticians and are encouraged to get master's degrees to further their advancement.

Our Assessment. There are entry-level job opportunities for those with just bachelor's degrees in statistics (or closely related subjects). The work, at least initially, is likely to be more routine and less challenging than for those with more advanced degrees. There are also fewer opportunities for advancement for individuals who wish to remain principally statisticians (as opposed to moving into an application area or management). Thus, a bachelor's degree may get you into the field, but a higher degree, and certainly the equivalent knowledge, is often needed to advance *within the field.*

7.4.3   Master's Versus Ph.D.?

You have decided to extend your education beyond a bachelor's degree. How far do you need to go?

A permanent faculty position at a 4-year college, and especially at a research university, almost always requires a Ph.D. degree, as further discussed in Chapter 13. However, for most statisticians outside academia, the typical requirement is a master's degree in statistics.

So, if you are not planning a career in academia why *might* you want to go on for a Ph.D.?

Well, for one thing, Ph.D.'s tend to get paid better, especially in an entry-level position. (But they often delay their working careers, thereby reducing their total career incomes, by having to go to school longer.)

A Ph.D. program provides the opportunity for more extensive study, for taking or auditing more advanced courses, and for deeper research. The more you know, the better prepared you generally are. A Ph.D. degree provides more opportunities for internships and to explore academic areas outside statistics. The discipline required to write a scholarly dissertation can be highly beneficial. As a result, Ph.D.'s tend to address problems with greater confidence and are better equipped to handle difficult technical problems—even though many problems often can (and should) be addressed by relatively elementary methods. In organizations that have multiple statisticians, those with Ph.D. degrees are, at least initially, more likely to become involved in more challenging activities than those who do not have such degrees.

A Ph.D. is also highly desirable for certain jobs—the so-called "union card" factor. It may be important, or even a prerequisite, for a position that requires dealing with regulatory or law enforcement agencies, especially in a leadership role. Statisticians working on clinical trials in the pharmaceutical industry, or presenting briefs to the U.S. Food and Drug Administration, are more likely to have Ph.D.

degrees than their counterparts working in an internal quality or reliability assurance operation. Similar considerations apply for statisticians involved in writing proposals to government agencies or giving testimony, both in a legal and in a governmental setting. And if you are an independent statistical consultant, a Ph.D. is likely to be helpful in attracting customers.

A Ph.D. is also useful if you are seeking a job in an organization, such as many research and development laboratories, in which colleagues in other fields have such degrees, for example, in science or engineering. The trend in many places, however, is toward employment based upon what you have to offer in totality, including relevant experience, rather than the degrees that you hold.

In a few organizations, a Ph.D. degree might even be somewhat of a liability. Employers might regard Ph.D.'s as overqualified and too far removed from the real world to be motivated by, or effective at, addressing day-to-day problems. Other employers, especially those with few Ph.D.'s in their organizations, might feel intimidated, or just not willing to pay the higher salary.

In summary, for entry-level positions in applied statistics—especially in areas such as manufacturing, official statistics, and finance—the education provided by a strong master's degree is often sufficient, with much of the rest being learned on the job. A Ph.D., on the other hand, in addition to the further knowledge that comes with it, might lead to greater growth opportunities, especially for those who plan to remain in statistics. And, who knows, perhaps some time down the road, you might want to go into academia—for which a Ph.D. is highly important.

See Crank (2009) for further comments.

7.4.4 Some Alternative Routes

From School to School. Many students start their graduate education immediately after getting their undergraduate degrees. This is the traditional route.

Taking a Breather. Others decide to take a breather, and earn some money, by taking a full-time job before continuing their education, perhaps taking some advanced courses while on the job. Others still might take a break between their master's and Ph.D. degrees. Such an interruption can provide an opportunity for informed reflection in helping you decide whether a career in statistics and/or pursuing a higher degree is really for you. On-the-job experience can also help you target better the educational program that suits you best. Returning full time to school is, however, likely to result in, at least a temporary, reduction in income and may, when the time comes, not be practical for financial, family, or other reasons.

Time Sharing. You might consider getting an advanced degree in statistics by taking courses over a number of years while working full-time (or part-time)—if your job allows. Many employers encourage this practice by paying your tuition fee and, possibly, allowing you some time off to take such courses, especially if you can show their relevance to the job.

Do not, however, underestimate the difficulty of going to school and succeeding on a demanding job at the same time. Trying to do both is likely to be strenuous and can interfere with other life activities. It requires an appropriate school to be geographically close to your home and/or place of employment. Long-distance and web-based learning programs increasingly provide an alternative, but some require time on campus, ranging from a few weekends to several weeks.

Other Possibilities. You might decide to pursue further coursework beyond the master's degree, but not write a dissertation and, thereby, not obtain a formal Ph.D. degree. In this way, you get the benefit of the education, without the prestige and financial advantages that are often associated with having a Ph.D. This path is especially appropriate for those for whom having a degree per se is not that important, but having the technical knowledge is. Others still get a second master's degree, perhaps in an allied field or application area.

Irrespective of the specific path, learning is a lifetime activity and needs to be continued, formally or informally, while on the job. We will return to this theme in Chapter 14.

7.5 SELECTING THE RIGHT SCHOOL AND PROGRAM

7.5.1 A Wide Choice

Before the middle of the past century, the choice of a university to study statistics was relatively simple—there were only a handful of schools that offered degrees, or even formal education, in statistics, and many of these emphasized agricultural applications.

Today, you have a wide choice of schools both at the undergraduate and at the graduate level.

Bachelor's Degree. Bryce (2005) cites a one-time (but not currently accessible) ASA web site, indicating 97 undergraduate degree programs in statistics, some of which might be quite small. Ritter et al. (2001) note that almost half of the bachelor's degrees in statistics granted in 1996–1997 were from 15 cited universities.

Graduate Programs. The ASA informs us that in 2010 there were 95 statistics and biostatistics departments in the United States offering Ph.D. degrees.

The U.S. News & World Report provided a ranking of graduate schools in statistics and biostatistics in the United States in 2010. This ordered 69 schools and listed 6 further unranked schools (some schools with multiple departments, offering, for example, degrees in both statistics and biostatistics, had multiple listings).[7]

A further assessment of Ph.D. programs in the United States is provided periodically by the National Research Council (NRC). Its 2010 study[8] included 61 departments with Ph.D. programs in statistics.

---

[7] http://grad-schools.usnews.rankingsandreviews.com/best-graduate-schools/top-statistics-schools/rankings/.

[8] http://www.nap.edu/rdp/.

Making a Choice. In the balance of this section, we present a variety of factors that we propose you consider in selecting a school. Our comments are especially relevant in selecting a graduate school, but most apply also for undergraduate studies.

There are in addition numerous personal factors that will undoubtedly enter into your choice, such as the school's location and cost (including the school's willingness to provide you a scholarship, assistantship, or other financial support).

7.5.2 School Rankings

The aforementioned school rankings may provide one input to your selection. In using such rankings, you need to be aware of the criteria and weightings employed in arriving at them.

In explaining how the rankings of statistics graduate schools were obtained, the U.S. News & World Report indicated that they were "based solely on the ratings of academic experts. To gather the peer opinion data, we asked deans, program directors, and senior faculty to judge the academic quality of programs in their field on a scale of 1 ('marginal') to 5 ('outstanding')."

Generally speaking, the higher ranked schools tend to be the more prestigious national institutions—and are likely to be the hardest to get in to (and the most demanding once you are there).

In its evaluation of graduate schools, NRC considered 20 factors that include characteristics of the faculty (e.g., number of publications and citations, number of awards) and of the students (e.g., GRE scores, percent fellowships), as well as average time to complete a degree, ethnic and gender diversity, and various other measures. NRC then produced two rankings based on different weighting schemes. The weightings used in arriving at the preceding ratings may or may not correlate well with your own.

Most significantly, you can develop your own weighting scheme using the 20 factors considered by NRC and apply this to the NRC response summaries to arrive at your *personally customized ranking* of schools.

In passing, we note that U.S. News & World Report also ranks undergraduate schools in general categories, such as business schools, engineering schools, and liberal arts programs. These rankings are based on a combination of peer assessments and responses by the individual schools in response to a questionnaire sent to them; see Sidebar 7.1.

SIDEBAR 7.1

CRITERIA AND WEIGHTINGS FOR U.S. NEWS & WORLD REPORT 2010 UNDERGRADUATE SCHOOL RANKINGS

The following are the criteria used by U.S. News & World Report, and the weight assigned to each, in ranking national (as opposed to regional) undergraduate schools:

- Peer assessment (from reviews by college presidents, provosts, deans of admission): 25%.

- Various numbers (based on responses to questionnaires sent to colleges):
  - Graduation and retention rate: 20%.
  - Student selectivity (SAT/ACT scores, high school standing, acceptance rates): 15%.
  - Faculty resources (class size, compensation, top degree, % full-time, student/faculty ratio): 20%.
  - Financial resources: 10%.
  - Alumni giving: 5%.
  - Graduate performance: 5%.

7.5.3 Relative Emphasis on Bachelor's, Master's, and Ph.D. Programs

If a school offers a Ph.D. in statistics, master's degrees are also frequently given, as well as, in some cases, bachelor's degrees. Schools that offer Ph.D.'s in statistics are likely to have greater strength and provide a larger choice of coursework and more flexibility, even for students who do not go on for a Ph.D.

At the same time, schools that offer Ph.D.'s in statistics might give prime emphasis to such programs, possibly at the expense of other degree programs. For example, graduate students might have varying degrees of responsibility for teaching less advanced courses.

If you are going for a bachelor's degree in statistics in an institution that does not offer a higher degree in the subject, you may receive greater personal attention. For example, you may have the opportunity to get involved (sometimes even for pay) in consulting projects in which faculty members are engaged. This is much less likely in an institution in which candidates for master's and Ph.D. degrees are available for such assignments.

7.5.4 Department and School Offering Degrees

There is much variability in the departments that offer degrees in statistics. Many universities have Departments of Statistics or some variation thereof, for example, the Department of Applied Statistics or the Department of Statistical Sciences.

In other universities, the degree in statistics is offered by the Mathematics Department (or the Department of Mathematical Sciences) or by the Mathematics and Statistics Department.

In some schools, the statistics department tends to focus on an education reflected by its name. Most prominent amongst these is the Department of Biostatistics. The Department of Experimental Statistics is a more specialized example. In some cases, the statistics department may be joined with another area, such as the Department of Statistics and Computer Science, the Department of Statistics and Operations Research, the Center for Quality and Applied Statistics, and the Department of Statistics and Actuarial Sciences.

In yet other cases, a statistics degree, or something closely resembling it, is offered directly by a department in an application area. Some examples are the Department of Decision Sciences and Engineering Systems, the Industrial Engineering Department, the Economics Department, the Psychology or Educational Psychology Department, the Department of Management Sciences, the Department of Public Health, and the Department of Information Systems and Quantitative Sciences.

In a few universities, there is more than one department that offers a statistics degree (e.g., a degree in statistics in the college of arts and sciences, and a degree in biostatistics in the school of public health). In some cases, these departments are well integrated and might even offer a joint degree. But in others, the programs operate independent (or almost independent) of each other.

In addition to the department that offers degrees in statistics, you should consider the school within the university in which the department is located. Examples are the School of Liberal Arts and Sciences, the School of Business, and the School of Engineering.

The department and the school offering the degree in statistics are important because they suggest the emphasis in the courses offered. Statistics degrees from the School of Engineering or the School of Business are, for example, likely to stress applications in their respective areas. These departments may also have closer ties to employers of applied statisticians (and provide more internship opportunities) in their respective areas than, say, the Department of Mathematics.

Schools that have their own statistics department vary in the degree of emphasis that they place on theory versus applications. Some departments are highly theoretical (these are often highly ranked in the U.S. News & World Report survey). Others, especially those that are located in an applications-oriented school, such as the school of public health, the business school, or an engineering school, tend to be more applied.

7.5.5 Size of the Program

Statistics programs vary considerably in size. Schools with formal departments of statistics are more likely to have a larger number of faculty and students in statistics than those in which statistics is offered by some other department. Small versus large, however, is a matter of personal choice.

Larger programs are likely to provide both a greater selection of courses and activities, such as seminars, and an exposure to a more varied faculty. They tend to have better name recognition within the statistical community and, possibly, offer more opportunities upon graduation (including more graduates returning to campus on recruiting trips) than more modest-sized programs.

Smaller departments, on the other hand, are likely to provide a greater chance for personal expression, more camaraderie, and, possibly, depending on the student/faculty ratio, a more personalized touch. The degree of cordiality and personal attention accorded to students, however, is not always reflected by the size of the department. Some departments, irrespective of size, have an open-door culture with professors spending considerable time in their offices.

7.5.6 Strength of the Faculty

The strength of the statistics faculty can be an important factor in selecting a school. The professors' reputation and professional accomplishments are generally relatively easy to determine. Some of the factors that may impact you more directly, such as their teaching ability and accessibility, are more difficult to evaluate. Current students and recent graduates may be a useful source of information.

Sometimes, students are attracted to a university by the opportunity to take courses from and perhaps work with one or more renowned professors—both a potentially great experience in itself and a neat thing to have on your résumé. You need to assess, in advance, how realistic this expectation might be. For example, are these professors teaching courses that you are likely to take? How many students are doing research with them and are they open to taking on more students? What are the chances that the professor(s) will leave the school before you complete your studies?

7.5.7 Dissertation Options

Ph.D. degree programs culminate with the writing of a dissertation. Some schools allow students to parse their dissertation into three or four papers (corresponding to chapters) on related topics instead of the traditional one-topic dissertation. This may resemble more closely what students will encounter in the workplace. This practice might also increase the number of publications resulting from your Ph.D. work.

Master's degrees at some schools may require a thesis, have a thesis option, or call for the completion of a substantive project. These are typically less intensive and may be less research-oriented than a Ph.D. dissertation. There may be a trade-off between courses and a thesis/project. Programs that require a substantive project for a master's degree may require fewer formal courses.

We will comment further on the importance of thesis topic and advisor selection in Section 7.9.2.

7.5.8 Strength in Other Areas

In selecting a school, you need to also consider its strength in areas other than statistics, and especially in the one(s) in which you wish, or are likely, to apply statistics upon graduation. If you have not selected such an area, as is frequently the case especially early on, you may prefer a school that is strong in many different areas. Also, you would want to explore the degree of encouragement (and associated credit) that you would receive in venturing beyond the statistics program for some of your coursework.

7.5.9 Internship Opportunities

We discuss the importance of hands-on experience via an internship in Section 7.10. You should try to determine the degree to which a school encourages internships and whether it has an established process to help students set them up. Similarly, you would want to determine whether the school has an active consulting center that provides challenging (including outside campus) opportunities. If you are considering year-round part-time employment, you might favor schools that will provide help in securing such positions and that are located in geographical areas that are likely to provide such opportunities.

7.6 STATISTICAL EDUCATION: SETTING THE FOUNDATIONS

Warning: Doing everything that we suggest in this chapter may not be practical within the confines of one person's college career. You will likely need to select from among our recommendations, or at least prioritize them, based upon your personal interests and aptitudes.

7.6.1 Training in Application Area

As we have maintained from the outset, applied statisticians need to be familiar with the application area in which they are involved. Often, this familiarity is obtained on the job. But it is much easier to gain such understanding if you come equipped with a good foundation. Thus, an advanced degree in statistics is a highly viable option for those with undergraduate degrees in an application area, such as biology, chemistry, physics, engineering, psychology, sociology, or economics, with a minor (or part of a double major) in statistics. This tends to defer intensive study of statistics to graduate school. Another possibility is for a statistics major to get a minor in an application area as an undergraduate. Yet others major or minor in mathematics or computer science as undergraduates.

The earlier quoted statistics showing more than three master's degrees in statistics being awarded by U.S. universities for every bachelor's degree suggest that many graduate students in statistics, indeed, have undergraduate backgrounds in other areas.[9] The common factor is often a strong interest in, and aptitude for, quantitative thinking. For others, the path from application area to statistics is the consequence of an evolution, rather than a deliberate plan—as students get to appreciate the value of statistics through an introductory course and/or work experience.

Finally, those who do not get training in an application area as undergraduates might compensate for this by taking courses in graduate school, or even while on the job, and possibly by self-study.

7.6.2 Mathematical Foundation

> Even though I don't solve differential equations anymore, I use the reasoning process that I learned in the Math Department at Dartmouth every day.
>
> —GE CEO Jeff Immelt

Statistics is based upon mathematics. A good background in mathematics is, therefore, required of those embarking on a major in statistics. It is often ability and

---

[9] This result is also attributable to students with undergraduate degrees from other countries receiving their advanced degrees in the United States.

interest in mathematics that draws students to statistics in the first place. Courses in calculus and linear algebra (including matrix theory) are generally prerequisites for a statistics major. Additional courses in numerical methods, discrete mathematics, differential equations, and real analysis also warrant consideration.

Statisticians, mostly in academia, who are dedicated to extending the frontiers of statistics and to developing new methods need to be especially well trained in mathematics.

7.6.3 Other Fundamentals

As we emphasize in Section 6.3 and throughout this book, excellent communication skills—both written and verbal—are essential for aspiring statisticians. We urge you to focus on improving your ability to communicate effectively, including one-on-one interactions, public speaking, and technical writing. This may be done by taking courses in these areas (and in statistical consulting), seizing opportunities to engage in formal communication (not necessarily on technical topics), and actively seeking criticism of and suggestions on your communication abilities.

If you are planning a career in academia, you should also try to gain some teaching experience while still at school. Finally, if you are aspiring to an eventual management position, you should consider coursework that will help build your organizational skills.

Statisticians also need to be knowledgeable in statistical computing and software (Section 7.8). This knowledge is generally acquired both in courses on statistical computing and in general statistics courses. Becoming computer savvy early on will bring about important payoffs later.

7.7 COURSEWORK IN STATISTICS

Statisticians require an education that instills skill in applying (and sometimes furthering) the concepts and tools of statistics, while still providing a solid theoretical base. Such training needs to lead to an in-depth understanding of

- The mathematical foundations of statistics.
- The underlying concepts of statistics and statistical models, statistical reasoning, and underlying assumptions.
- Basic tools (and software) for data analysis.
- More advanced tools relevant to likely application areas.
- How to rapidly and effectively acquire knowledge in areas that were not studied.
- How to adapt existing methods and develop new ones to address problems.

7.7.1 Required Courses

A major in statistics generally requires, at least, a 1-year sequence in applied statistics (or statistical methods) and a 1-year course in the theory of probability and statistics (sometimes called mathematical statistics). Various additional courses such as statistical computing, general linear models (or regression analysis), sampling, and time series analysis are sometimes also required and highly recommended.

7.7.2 Elective Courses

A typical statistics program, especially in a large statistics department, provides a choice from among a wide variety of courses dealing with advanced methods and theory. Courses targeting specific methodology areas (in addition to those already listed as required in some institutions) such as Bayesian methods, categorical data analysis, design of experiments, multivariate analysis, and nonparametric methods are just a few examples. Additional courses provide training in application areas, such as industrial statistics and quality improvement, and statistical methods for clinical trials. There are also courses in such areas as statistical computing and statistical consulting, and on theoretical topics in mathematical statistics and probability theory.

TABLE 7.1 Authors' Assessment of Importance of Course for Various Application Areas

Course	Product design	Manufacturing	Reliability	Field support	Pharmaceutical products	Financial services	Business processes	Official statistics	Social research	Environment
Analysis of variance										
Bayesian methods										
Categorical data analysis[a]										
Data mining										
Decision theory										
Design of experiments[b]										
Life data analysis										
Multivariate methods										
Nonlinear estimation										
Nonparametric methods										
Operations research										
Probability										
Quality assurance[c]										
Regression analysis[d]										
Simulation										
Spatial statistics										
Statistical computing										
Statistical consulting										
Stochastic processes										
Sampling										
Survival data analysis										
Time series analysis										

[a]Including logistic regression.
[b]Including response surface methods.
[c]Including SPC and acceptance sampling.
[d]Including general linear models.

The color pattern is perceived "relevance score" as follows:

- Essential
- Very useful
- Useful
- Consider (if schedule allows)
- Questionable usefulness

7.7.3 Our Assessment

Of the wide spectrum of courses often available to students, which ones are most useful in planning careers in different application areas? We provide our (highly subjective) perspective, tempered by the inputs of colleagues, in Table 7.1. This tabulation aims to show the importance of various frequently offered courses, from analysis of variance to times series analysis (shown in the rows of the tabulation), in different application areas (shown by columns).[10] For each application area, we placed each of the courses into one of five categories, ranging from questionable usefulness to essential; however, no course was felt to be of questionable usefulness in any area.

The table lists courses included in many programs in applied statistics, although the specific course titles may vary. We omit courses directed at specific application areas, such as biostatistics, business statistics, and engineering statistics. We do, however, include courses that tend to be heavily oriented to a specific area of application, such as quality assurance and survival data analysis. The list is far from complete; universities, especially those with large statistics departments, provide various electives beyond those on our list.

Our scores in many of the application areas are based upon our discussion and examples in Hahn and Doganaksoy (2008). Some of the reasoning that led to our assessment with regard to official statistics is provided in Sidebar 7.2.

SIDEBAR 7.2

TRAINING FOR A CAREER IN OFFICIAL STATISTICS: SOME COMMENTS

Many government statisticians engaged in the collection and reporting of official statistics make extensive use of sampling and need to be highly knowledgeable in that area. It is also useful to have an understanding of the intricacies of such matters as questionnaire construction and handling of nonresponse—topics that are not always taught in courses on sampling and are, instead, frequently learned on the job.

Official statistics are typically collected repetitively over time (e.g., monthly) and describe the past. Predictions of the future are, however, of strong interest as, for example, in population projections by the U.S. Census Bureau, future mortality estimates by the Social Security Administration, and forecasts of energy demand by the U.S. Energy Information Administration (EIA). Thus, many government statisticians need expertise in methodologies, such as regression and time series analysis, for making such predictions and in recognizing their limitations.[11]

An understanding of quality assurance systems is useful for the development and implementation of reliable data gathering systems. And, since the conduct of a survey often requires the optimal allocation of limited resources to achieve high quality within budget, organizational, and other constraints, knowledge of operations research is helpful.

---

[10] Our assessments are based upon our understanding of the topics that are currently included in such courses (rather than what, in some cases, we feel the course should contain).

[11] Such work frequently results in close interactions between statisticians and economists. In fact, the use of statistical and mathematical methods in addressing economic problems has given rise to the field of econometrics.

Like other applied statisticians, those working on government statistics frequently guide colleagues in other fields and, therefore, would benefit from a course on statistical consulting. And, again like their colleagues, they use regression analysis, simulation, and statistical computing in a variety of situations, calling for courses in these areas, if time permits.

We note from Table 7.1 that a few courses are highly relevant to all, or almost all, application areas. These are regression analysis (including general linear models), statistical computing (Section 7.8), and statistical consulting. Other courses receive a high score for some application areas, but a lesser score in other areas, for example, the related areas of survival analysis (most relevant for pharmaceutical product applications) and life data analysis (most pertinent in product design, reliability, and field support applications).

7.7.4 Alternative Assessments

Each experienced statistician would likely score the relevance of courses to different application areas somewhat differently and some may, in fact, strongly disagree with our evaluation. Thus, Table 7.1 should be taken as a starting point for further discussion, rather than as the definitive answer.

Some other assessments are provided in Sidebar 7.3.

SIDEBAR 7.3

OTHER ASSESSMENTS

In discussing (undergraduate) "most desirable coursework" for a position at the EIA, Blumberg (2009) lists

- Mathematical statistics.
- Regression (with a little bit of analysis of variance).
- Data analysis coursework that includes some "dirty" data and multiple analyses.
- Matrix algebra.
- A computing course (either statistical or nonstatistical).
- Economics.

Nancy Buderer (in an unpublished presentation) suggests that for a career in health care statistics, students should have at least an MS in statistics or biostatistics, with courses in mathematics, epidemiology, biology, and computer programming, and advanced coursework in survival data analysis (including Cox proportional hazard models), categorical data analysis (including logistic regression), analysis of variance (including repeated measurement analyses and mixed models), nonparametrics, statistical power analysis, and sample size estimation.

Meeker (2009) lists the following "as tools needed by industrial statisticians": Bayesian statistics, categorical data methods, censored data analysis, design of experiments, graphical methods, image analysis, multivariate analysis, optimization, regression analysis (linear and nonlinear), reliability theory, response surface methods, simulation, spatial statistics, statistical computing and programming, survey sampling, and time series analysis.

TABLE 7.2   Specialized Methods that Have Particular Applicability in Selected Application Areas

Application area	Specialized method(s)
Clinical studies for pharmaceutical products	Planning clinical trials
Developing new food products or chemicals	Mixture experiments
Marketing studies or financial modeling	Multivariate classification, clustering analyses
Product reliability	Censored data analysis

7.7.5   Some Specialized Methods

Individual application areas come with their own technical approaches and methodologies. See Table 7.2 for a few examples.

Some technical challenges that are frequently encountered in working with official statistics, and that those engaged in this area should be especially familiar with, are presented in Sidebar 7.4.

SIDEBAR 7.4

SOME TECHNICAL CHALLENGES IN OFFICIAL STATISTICS

The following are a few of the technical challenges frequently encountered in official statistics:

- Handling missing data.
- "Breaks" or changes in time series data. These may be either due to small changes in the methodology for collecting data or due to major changes in the system. The latter occurred in 1997 when the North American Industrial Classification System replaced the Standard Industrial Classification System (Section 3.4.3).
- The assessment, for example, in much census data, of spatial components in time series data (thus requiring an understanding of geographic information systems).
- Mathematical models based on a combination of sample and administrative data (as, for example, in the U.S. Census Bureau's Small Area Income and Poverty estimates).
- Methods to combine various data sources to produce synthetic data (Section 3.9).

Most of these topics are not sufficient to merit a full course, but are (or should be) topics in various individual courses.

We urge you to determine the specialized methods that you need to know in your likely application area and learn about them (especially prior to relevant job interviews).[12]

---

[12] One of the authors remembers, many years ago, studying up on the then hot, new topic of response surface experiments prior to an interview with a prominent food processing company. (Despite this, he did not get the job.)

7.7.6 Some Further Pointers

Based upon our experiences, and those of colleagues, we feel that there are some important concepts that statisticians need to know on the job that are often not sufficiently emphasized in their education. We urge you to ensure that, in one way or another, you learn from your studies to

- Recognize that getting the right data is frequently a statistician's biggest challenge.
- Always, always, always plot the data in different ways and look at the plots—and do so prior to conducting any formal statistical analyses.
- Appreciate the importance of percentiles.
- Understand the limitations of the normal distribution.
- Appreciate the time dependence of much data.
- Recognize and appropriately handle censored data.
- Not take outliers lightly.
- Beware of extrapolation.
- Appreciate the difference between statistical significance and practical importance.
- Be able to differentiate between confidence, tolerance, and prediction intervals.
- Recognize that statistical precision (in taking a random sample) is determined by the actual sample size (and not by the size of the sample relative to that of the population).
- Be able to respond to the question you are likely to be asked most frequently: "how large a sample do I need?" (and to recognize that usually this is only part of the larger question of planning a study).

To keep the length of this volume under control, we do not discuss the preceding points further—with one exception: our first point. We feel this subject is so important, and often so much neglected in much of the standard education, that we devote all of (lengthy) Chapter 11 to getting good data. We provide further elaboration on the other points stated above on the book's ftp site.

7.8 STATISTICAL COMPUTING AND SOFTWARE

> Anyone who can do solid statistical programming will never miss a meal.
>
> —David Banks (2008)
>
> Computational literacy and programming are as fundamental to statistical practice and research as mathematics.
>
> —Nolan and Lang (2010)[13]

---

[13] Nolan and Lang also advocate a major redesign, broadening (to include advances in modern computing beyond traditional numerical algorithms), and integration of statistical computing in the traditional statistics curriculum to better prepare students to handle very large data files and computationally intensive analyses.

You need to be knowledgeable in statistical computing and software to meet both your customers' and your own computing needs. This requires a good appreciation of the statistical packages that your customers are using or that you might recommend for a particular application. It calls for you to be knowledgeable about the main features, technical correctness, and ease of use of popular offerings. Such knowledge is acquired from various courses in statistical methods in which such products are used, as well as courses in statistical computing, and must be maintained throughout your career as software features change and new products become available.

Software with statistical capabilities can be conveniently categorized as

- General purpose software with statistical features.
- General purpose statistical software.
- Specialized statistical software.
- Statistical programming environments.

This differentiation is sometimes quite blurred—and is continuously becoming more so.

7.8.1 General Purpose Software with Statistical Features

Popular general purpose programs, such as Microsoft Excel™, now come with a variety of statistical features and add-ons (see Berk and Carey, 2009). McKenzie (2008) asserts that "today more statistical analyses are done on Excel than any other software." Unfortunately, some significant technical problems have been found in the statistical features of this program; see Sidebar 7.5.

SIDEBAR 7.5

ADVANTAGES AND LIMITATIONS OF EXCEL FOR STATISTICAL ANALYSES

Although professional statisticians generally favor the more specialized and advanced software offerings described later in this section for their own use, Excel is highly attractive to practitioners who conduct statistical analyses. Most are already familiar with the program's spreadsheet features, have ready access, and generally do not incur added costs to use it. Excel, moreover, greatly facilitates data preprocessing and cleanup. Also, various commercially available packages are designed as add-ons to Excel for specialized analyses.

Various articles in the June 2008 issue (Vol. 52, Issue 10) of *Computational Statistics and Data Analysis* examined the technical adequacy of Microsoft Excel 2007. Some criticisms voiced in these assessments were

- "Excel 2007, like its predecessors, fails a standard set of intermediate level accuracy tests in three areas: statistical distributions, random number generation, and estimation" (McCullough and Heiser, 2008).
- "The accuracy of various statistical functions in Excel 2007 range from unacceptably bad to acceptable" (Yalta, 2008).

- "It's easy to produce Chartjunk using Microsoft Excel 2007 but hard to make good graphs" (Sung, 2008).
- "Persons who wish to conduct statistical analyses should use some other package" (McCullough and Heiser, 2008).

It is possible that these problems will be addressed in future versions of Excel and there has been some progress from earlier versions. (Problems with the statistical features of earlier versions of Excel were noted by McCullough and Wilson (2002).) See McKenzie (2008) for a historical perspective and a summary of some of the remaining deficiencies.

7.8.2 General Purpose Statistical Software

Numerous statistical software packages have been developed over time to permit users to conduct a wide variety of standard analyses (e.g., multiple regression and analysis of variance). Some of these also offer scripting languages that let users add their own code for more specialized analyses.

Many of these packages are directed principally at practitioners and focus on ease of use via point-and-click type interactions. Some packages include more advanced features and often serve the basic needs of nonstatisticians—as well as, on occasion, statisticians—with small to medium size data sets. Students are often introduced to one of these packages in introductory course in statistics. Popular offerings include JMP, Minitab, Statgraphics, Stata, and SYSTAT.

In addition, powerful and versatile packages have been developed for use by statisticians and more sophisticated practitioners who need to do advanced analyses and/or have large data sets. Such packages offer a variety of features, including management of large databases, a highly diverse suite of statistical methods, and data visualization. Students typically become acquainted with such software in more advanced courses in statistics. Popular long-standing offerings include SAS and, especially for social science applications, SPSS. SAS is considered essential for students with terminal master's degrees.

7.8.3 Specialized Statistical Software

Specialized packages have been developed for various application areas, such as the design of experiments and reliability analysis. Such offerings are generally directed at and frequently, but not exclusively, used by practitioners. There are additional packages, used by both practitioners and statisticians, for special statistical evaluations, such as Bayesian analyses, categorical data analysis, classification and regression trees, nonparametric data analyses, resampling, and survey design and analysis.

7.8.4 Statistical Programming Environments

Powerful high-level languages allow users to perform creative computational and graphical analyses that are appropriate for the application at hand, in addition to containing extensive standard preprogrammed features. The open-source R programming environment—based on the S programming language originally developed at Bell Labs—is available for free[14] and has gained much popularity among statisticians. R was initially written by Robert Gentleman and Ross Ihaka of the Statistics Department of the University of Auckland. It continues to evolve as a result of a collaborative effort of a core group of statisticians from all over the world. There is also a commercially supported version of S, called S-Plus, which, in addition to the S-based programming environment, has a point-and-click graphical user interface for many commonly used statistical procedures.

In addition to providing statistical functionality, programming environments let users customize their analyses to the problem at hand and incorporate new methods, thus giving them the freedom to go beyond the capabilities of standard software. For example, SPLIDA and RSPLIDA were developed as add-ons to S-Plus and R, respectively, to address some of the important, but specialized, questions that are encountered in product life data analysis.

Students are frequently required to use one of these languages in more advanced statistical methods courses and in courses in statistical computing. They are often the programming tools of choice by Ph.D. students and professors in their research. Outside academia, such languages are used most frequently by statisticians who have the knowledge and motivation to go beyond conventional data analyses.

7.9 SOME OTHER RECOMMENDATIONS

Once you have selected an institution and developed a proposed list of courses you wish to take, there are still a variety of other decisions that you will need to make—or initiatives that you might want to consider along the way—that can add to the effectiveness of your education. We mention a few of these in this section.

7.9.1 Instructor, Instructor, Instructor

As always, an inspiring instructor makes a vast difference. A marginally important course with a great teacher—and especially one who can provide insights into practical applications and issues—often trumps a possibly more pertinent course with an indifferent instructor. We advise you to explore the track record of instructors of elective courses, through past teacher evaluations and talking to other students, before signing up for courses.

7.9.2 Be Prudent in Dissertation Advisor, Topic, and Committee Selection

The selection of a dissertation/thesis advisor, topic, and committee involves key decisions that most Ph.D. and some master's degree students need to make. These decisions, together with steering the thesis to its successful completion, not only provide significant intellectual challenges but can also test your strategic planning and diplomatic skills. They warrant your careful attention.

[14] http://www.r-project.org/.

Advisor Selection. In selecting an advisor, students need to understand the environment in which professors (and especially those seeking promotion and/or tenure) operate (see Chapter 13). A highly important criterion for faculty advancement in many universities is successful research, as evidenced by publications in scholarly journals. The joint research involved in guiding the dissertations of graduate students often provides a major vehicle for helping professors meet their research aspirations.

Successful research, or promise thereof, is also important in your advisor's securing of funding. Such funding is often required to support your (or the next student's) further research.

The eventual goal of most Ph.D. theses is research that culminates in peer-reviewed publications (typically authored jointly by the student and the professor) in esteemed refereed journals (Section 13.8.4). Such publications enhance students' credentials, as reflected in their résumés—even though, due to the lengthy refereeing process, publication might be after graduation.

Your advisor may select, or at least suggest to you, one or more dissertation topics—more on that shortly—and will be the major person deciding when your research is complete and you can graduate.

You need to assess carefully and realistically how well you will be able to work with a potential advisor. This relationship, in fact, provides you useful experience for working for a future manager or department chair.

The professional reputation of your advisor also warrants consideration. Consider, for example, how your career might have been enhanced if you could have laid claim to having been one of Albert Einstein's students. Keep in mind contacts that your advisor might have that can be helpful to you in seeking an internship and in looking for a job upon graduation. At the same time, you need to check whether a professor—eminent or not—has a reputation for giving students the attention that they seek and require.

To make a good selection, you should get to know your department's faculty members and their research interests. An obvious way to do this, and for potential advisors to get to know you, is to take courses from them. Also, we propose you try to talk to students (some of whom may have graduated) who have worked under potential advisors about their experiences. Were they able to complete their theses in a reasonable time? Did they receive helpful guidance and attention, when needed? Was working with the advisor a pleasant and productive learning experience?

We recommend that you try to find out the likely plans of your advisor for the next few years. More than one promising dissertation has been derailed by an advisor moving to another institution or retiring[15] or significantly delayed by an advisor going on sabbatical.

Finally, matching students with advisors is a two-way street. The advisor also needs to want you. So it is important for you to impress potential advisors favorably. Again, the classroom provides an opportunity to do this.

---

[15] Although many professors continue their work to completion with selected students after moving or retirement.

Dissertation Topic Selection. Professors typically suggest as dissertation topics extension of their previous research and/or that of past students. The proposed research may be part of a research grant or proposal. In this case, you should determine whether the proposal has actually been funded, and if not, the chances that it will be, and how dependent you will be on such funding.

Also consider making your own proposal for a dissertation topic. A problem encountered during, or motivated by, an internship may provide an excellent subject in that it may have greater relevance and impact and you are likely to have a good understanding of what it is all about. Possibly you (and perhaps even your advisor) can receive funding or other support to conduct some or all of the research from the organization from which the problem originated.

A project developed by you might provide you more independence than one "owned" by your advisor, while still allowing you to benefit from your advisor's inputs (although these may be less than those on a subject in which the advisor has been previously directly involved). At the same time, a project selected by you, especially one emanating from an internship, might be more applied than one proposed by your advisor. A possible negative consequence of this, especially from your advisor's perspective, is that the results of such research might have a smaller chance of being accepted for publication in an esteemed journal than would be those from a more scholarly topic.

In any case, you and your advisor need to assess realistically your ability to undertake and complete the required work to your advisor's satisfaction. We suggest that you don't be overly ambitious. Successful completion of the dissertation is a requirement for graduation. Thus, in selecting a dissertation topic, one of your goals should be successful completion within a reasonable time frame compatible with your life schedule. You need to understand—and possibly negotiate with your advisor, as early as possible—what is required of you for successful dissertation completion. Keep in mind that you can always conduct added research after graduation when you are no longer obligated to follow the dictates, and possible whims, of your professors—but might still benefit from their insights—and can do so on your own time schedule. Also, if you take a job in academia, you should be aware that many universities will only count research conducted after completion of your Ph.D. toward promotion and tenure, suggesting that you keep something in reserve.

Finally, keep in mind that the selection of a dissertation topic can have an important bearing on your career. It might determine the application area in which you get your first job—since it is, after all, the area in which you are most knowledgeable and are likely to have most to offer a prospective employer. Writing a dissertation that develops new methodology for analysis and modeling of product degradation data should, for example, improve your chances of obtaining a position with an industrial operation concerned with product reliability—especially if you can successfully relate your work to a likely business concern.

Dissertation Committee Selection. A dissertation committee is typically selected after an advisor and topic have been, at least tentatively, agreed upon. Committee members are picked to provide both added value and ability to evaluate your work. It can be said that the committee is there, in part, to exercise quality control. Typically, but not always, committee members go along with the wishes of the dissertation advisor with regard to successful completion of the work.

Our comments on advisor selection also hold for committee members. It is often advantageous to include at least one knowledgeable member from outside your department, especially if such a person can provide added application area insights. If your topic is based upon a problem generated by an internship, you might want to have your internship mentor or supervisor, or some other person involved, be invited to be a member of your committee.

Overall, you should seek to have your committee consist of individuals who are likely to provide added and constructive insights and ideas, but are not likely to unreasonably obstruct your forward path.

Note on Dissertation Execution. Once your advisor, dissertation, and committee have been selected, you will want to move forward expeditiously in conducting your research. In so doing, you should communicate frequently with your advisor to ensure that you continue to be on track and that you can overcome possible technical or other stumbling blocks in a timely manner.

If you should find that, despite your careful initial evaluation and subsequent work, you have selected a topic that is intractable (or already solved by somebody else) or there has been a significant change in your or your advisor's status, we propose that you face up to this as early as possible and appropriately recalibrate your plans.

7.9.3 Explore Life Outside the Statistics Program

Graduate students in statistics (and often their professors) may be inclined to become immersed exclusively within the small world around them. This does not have to be. We recommend that you learn about and avail yourself of the diverse academic activities on campus.

Attending seminars in other departments, especially ones presented by those from application areas or institutes that might be of future interest to you, can open new horizons—including possible research topics. Such seminars are, sometimes, scantily publicized outside the sponsoring department, and might have to be sought out by you.

Consider collaborating in relevant research with people in other areas, for example, working with chemistry students to quantify sources of variability in their lab trials or assisting in the design of a sampling study for a social science project. This can provide important insights into application areas (and those who work in them) and might enhance your future employment opportunities. It can also help prepare you for the teamwork that will be so important to your future success.

Finally, you might consider participating in the activities, such as meetings and short courses, of a local chapter of the ASA and other professional societies. In addition to being informative, these could result in useful interactions and present important future leads for internships and permanent employment. Many chapters extend membership to students for free or for a nominal fee.

7.10 INTERNSHIPS AND UNIVERSITY CONSULTING

There's no experience like real experience.

—Adaptation of popular show song

7.10.1 Importance

We strongly urge students, even undergraduates, to get professional experience outside the university through an internship (or apprenticeship) or similar program. This should be in a challenging position involving statistical work, preferably with professional statisticians, or individuals knowledgeable in statistics, who can serve as mentors.

An internship allows you to assess firsthand whether the challenges and excitement of a career as an applied statistician are really for you. It may help you select an application area on which you would like to focus your career. It can demonstrate the relevance of your coursework and allow you to tailor better your remaining courses to your needs. And it may provide you ideas for a dissertation topic, and possibly even financial support should you proceed with this topic.

Most importantly, work experience provides students an edge in getting a job upon graduation. Employers prefer students with some experience, both because of the added knowledge this provides and for the interest it portrays. Such experience will allow you to talk at job interviews about real-world problems and how you have helped address them—topics that might interest potential employers more than the details of your dissertation. Finally, your internship employer might even come up with an offer for a full-time job upon graduation. (Identifying promising future hires is often a major reason for an organization having an internship program.)

An internship may delay your graduation, unless you can get course credit for it. This is a worthwhile trade-off for most students, and especially those with no other relevant experience. Income earned from such employment should help offset the additional costs of staying in school longer.

7.10.2 Opportunities

Companies, government agencies, and other institutions offer internship programs to students. Blumberg (2009) reports that the EIA hired 54 interns during the (recession) summer of 2009 with a variety of backgrounds, including six undergraduates with a major interest in statistics.

Each year, the ASA invites organizations with statistical internship opportunities to provide a brief description of these, as well as their location, the number of positions, the type of student desired (e.g., graduate, Ph.D.), the deadline for application, and contact information. These are then published in *Amstat News* (e.g., December 2010, pp. 9–15) and listed online together with subsequent additions.[16]

---

[16] www.amstat.org/education.

The most recent listing (in the December 2010 issue of *Amstat News*) shows 35 employers with a total of more than 330 openings, with the following breakdown:

- Pharmaceuticals and health sciences: 16 employers.
- Business, industry, and research: 5 employers.
- Education research: 4 employers.
- Government: 4 employers.
- Software: 3 employers.
- Other: 3 employers.

See Habiger et al. (2009), Huiras et al. (2009), and Rossetti (2010) for students' descriptions of their statistical consulting experiences.

7.10.3 Implementation

When to Take an Internship. The summer is an obvious time for internships, and this is what is sought in many of the ASA listed opportunities. Employment over multiple summers, with either the same or different employers, is especially desirable.

Some employers prefer around the year part-time internships of, say, 12–20 hours per week because of the continuity they provide. Such internships allow you to witness the unfolding of projects over time and to become more directly involved. You need to ensure, at the same time, that the job does not become so demanding that it detracts you from schoolwork.

To qualify for a statistics internship position you need to have some background in statistics. You might, however, try to get started reasonably early—at a time when your internship experience can have the most impact on your choice of courses and potential dissertation topic. Most internships are for pay. These have the added advantage that your employer is likely to assign you to more meaningful work than if your services were for free. On the other hand, an opportunity might be so enticing, and so much in your long-term benefit, that it warrants your taking it on, at least for a short time, even without financial compensation.

Seeking an Internship. How do you get an internship? Some universities offer these through a formal internship program. In other cases, your involvement may be through a faculty member engaged in external consulting. You may start by working directly for a professor and then transition to employment by the organization—while perhaps still receiving professorial guidance.

In other cases, you might be principally on your own seeking an internship, especially when it involves a summer job. The aforementioned ASA site provides a great starting point for identifying opportunities for students in the United States, as do more general web sites and other methods for seeking jobs discussed in Chapter 8.

Family, friends, and graduates who have held such jobs in the past might provide leads. The contacts you may have established at local chapters of the ASA and other professional societies, or possibly through attendance at national statistical meetings, might prove to be helpful.

Seeking an internship can be a mini-version of searching for a full-time job and can provide good experience for what lies ahead. Just as for a full-time job, it will be important for you to get a good understanding of what the job is all about and what the expectations on you will be—and, more generally, how you can make it a true learning experience and stepping-stone for the future. For example, will you be assigned to a specific project with specific "deliverables" or will you be providing general support as the need arises? What kind of mentoring can you expect? To what degree will you be part of a project team? Will you be responsible for interpreting and reporting the results of your work and, if so, in what manner? How will your performance be assessed and documented?

7.10.4 University Consulting

Many universities have statistical consulting centers. They provide an important service to the university and useful hands-on experience to students.

Such experience, however, may not always be an adequate substitute for an internship within an organization outside academia—because of the differences in environments and in criteria for success—and may not be so regarded by organizations in hiring new employees. As indicated, the ultimate goal of university research typically is to have the resulting work published in a scholarly journal.[17] These objectives are markedly different from those of companies that want to be profitable or government organizations that need to generate useful information for use by lawmakers, regulating agencies, and the general public. An ideal path might be to start off with on-campus consulting and follow this up with an external internship.

The recently developed STATCOM program, offered at a (growing) number of universities, provides an experience somewhere in between university consulting and a formal internship; see Sidebar 7.6.

SIDEBAR 7.6

THE STATCOM PROGRAM

The Statistics in the Community (STATCOM) program is a student-run volunteer effort, started in 2001 at Purdue University and expanded in 2006 with the help of the ASA into a network to include other schools. This program has students provide free statistical consulting to local nonprofit, government, and community service organizations. Ochsenfeld and Olbricht (2009) describe STATCOM and list seven universities that have taken up this program and indicate continuing interest from other institutions.

---

[17] There are exceptions. Some professors and even some students are engaged in research contracts that are tied to specific application deliverables.

Some typical STATCOM projects are

- A survey to solicit residents' opinions on issues (e.g., traffic, pedestrian, developmental, and financing) concerning a stretch of a nearby major local highway.
- Assessment for a local library of the popularity (based on number of times checked out) of different types of loan offerings, as a function of such factors as price, year of issue, and nature of offering.
- Development of methodology to predict future cash flow from donations for a nonprofit organization.

7.11 ENTERING STATISTICS FROM OTHER FIELDS

7.11.1 The Path

Many practitioners have a background in areas other than statistics, but became interested in the subject by discovering its applicability to their work. This route was especially popular before statistics programs and departments became common in universities. It is how such eminent statisticians as Sir Ronald Fisher (former geneticist), Ed Deming (former physicist), and George Box (aspiring chemist) entered the field (as did Leah and Boris in the examples in Section 1.10.1). Deming (1967), describing the statistical quality control pioneer Walter Shewhart, said, "As a statistician, he was, like so many of us, self-taught on a good background of physics and mathematics."

Today, many individuals become involved in statistics as a result of special initiatives within their organizations. For example, Six Sigma black belts or master black belts, across the world, are frequently asked to give statistical guidance— especially in organizations without professional statisticians. They witness the usefulness of statistics firsthand, become engrossed in it, and over time acquire significant statistical knowledge and expertise.

Your transformation from being a chemist, biologist, financial analyst, or social researcher to a statistician is likely to be gradual—and you will continue to benefit from your education and experience in your original discipline.

7.11.2 Education Avenues

If you discovered statistics on the job, you may have received much of your initial statistical education in one of a variety of ways, such as short courses, targeted at making you more effective. Over time, this may suggest new career opportunities— leveraging a combination of your formal education and work experience in another field and your increasing knowledge of statistics.

You might decide to go back to school full-time to get a degree in statistics. More likely, you might get formal education in statistics by taking courses at a local college or university, or even via long-distance or web-based training, while staying on the job. Or you might select a less formal route, getting your education through a combination of disciplined self-study, short courses, and participation at professional society meetings; see our comments in Chapter 14.

7.12 FURTHER RESOURCES

The ASA's web site on education and its monthly *Amstat News* publication provide a wealth of further information on statistical education. The web site currently lists (in addition to the links mentioned earlier)

- Opportunities for undergraduate students.
- Scholarships, awards and competitions, and fellowships.
- Continuing education offerings, including courses at the Joint Statistical Meetings (JSM) and forthcoming web-based lectures.

A typical article in *Amstat News* (Crank, 2010c) discusses government funding opportunities for training programs in statistics (principally in biostatistics) at various levels. Hedin and Vock (2010) provide advice and professional development tips for graduate students.

A detailed discussion identifying obstacles to and strategies for recruiting students of statistics is provided by Landes (2009).

To broaden your horizons, we urge you to consider attending the annual JSM and other national (in addition to local) professional society meetings (Sections 8.3.2 and 14.5). When ready to do so, you might take the opportunity at such a meeting to give a contributed paper on your dissertation or other research. Often, you may be able to use your (or your professor's) research grant to pay for your travel expenses.

7.13 CONCLUDING COMMENT: THE LIMITS OF FORMAL STATISTICAL EDUCATION

> Nearly every problem is different from what you learn in formal statistics courses.
> —Lonnie Vance

We repeat a comment with which we started the preface of our earlier book (Hahn and Doganaksoy, 2008): We asked a recent Ph.D. in statistics what was the biggest surprise in her first year working as an applied statistician. "I was amazed at how little I really knew, even about subjects about which I felt I knew a lot," she answered unhesitatingly.

There is no way you can learn in school all the things that you will need to know to address the problems that you will encounter in the workplace. A goal of your education then is not just the acquisition of knowledge per se, but to help you learn how to work successfully on your own over a lifetime. This will include the ability of gaining an understanding of application areas in which you may become involved, becoming aware of new developments in your own field, learning how to use new software, and adapting what you already know to the problem at hand. We have, during the course of our careers, seen tremendous changes in the way statisticians work. Readers can expect no less during theirs. This will require continued training, either formal or informal—a subject we will return to in Chapter 14.

Society, moreover, tends to assume that you, as a statistician, are knowledgeable in all areas of statistics. You may frequently be asked to respond, both on and off the job, to a wide variety of questions, some of which may be far removed from your current area of application or any that you have worked on in the past. A broad education in statistics should put you in a good position to be responsive in a wide variety of contexts.

The successful completion of even an advanced degree in statistics is only a first—though essential—step in the path to a successful career in statistics. Statistics graduates need to approach the workplace not only with confidence that they are well armed but also with humility, recognizing that they still have much to learn.

7.14   AS OTHERS SEE IT

We present below some of the responses (sometimes in abridged or edited form) pertaining to education to the question that we posed to a number of eminent statisticians: What key advice would you give to an aspiring statistician? As those listed in other chapters, these comments reiterate or amplify points we make within the chapter and add some others.

> Statisticians require strong knowledge of how to use statistical methods in many subject areas and a keen interest in helping others understand how statistical thinking applies to all aspects of an enterprise.
>
> Strive to obtain an internship with a company or with some other institution, such as a government agency, a non-profit organization, or working on an applied university project.
>
> —Thomas J. Boardman, Professor (retired), Colorado State University; Consultant
>
> Take courses in graduate school that give you a broad understanding of statistics rather than just concentrating on the latest hot topic … Getting a broad outlook makes it easier to adapt to different jobs.
>
> —Michael R. Chernick, Director, Biostatistical Services, Lankenau Institute for Medical Research
>
> A solid background in science/engineering is of immeasurable help in understanding the problem of interest in many application areas and buys instant credibility in the eyes of your customer.
>
> —Ronald Iman, President, Southwest Technology Consultants; past President of the American Statistical Association
>
> Be familiar with the techniques used in applied statistics and have a good command of the software necessary to implement them. If specializing in an area, such as the analysis of clinical trials, then be familiar with both successful and unsuccessful case studies in that area. Also, be willing to learn a sufficient amount of the application area to allow you to judge how to apply statistics wisely. Be

willing to keep current on new methodology, such as the bootstrap and Bayesian Markov Chain Monte Carlo methods.

—Ramon Leon, Associate Professor, University of Tennessee; formerly, technical supervisor at Bell Laboratories

For a position in business and industry, get as much training as you can in science and engineering. This training will be invaluable in communicating with your clients. Take lots of applied statistics courses, such as hierarchical Bayesian methods, regression analysis, experimental design, reliability, and quality assurance.

—Harry Martz, Laboratory Associate in Statistics (retired), Los Alamos National Laboratory

Mathematical theorems are true; statistical methods are effective when used with judgment. But you can never know too much mathematics. It's amazing what someone with real modeling and analyzing power can do—if he/she also has the right traits.

Statistics is an art as well as a mathematical science. Arts are learned by experience, preferably apprenticeship.

Be sure that in grad school you learn at least a little about lots of things. If you can recognize a setting, you can learn more as needed: the main purpose of grad school is to teach you how to learn.

—David S. Moore, Shanti S. Gupta Distinguished Professor (retired), Purdue University; past President of the American Statistical Association

As part of your academic training, endeavor to participate in internships or other programs that provide experience in statistical consulting.

—Carolyn B. Morgan, Professor, Hampton University; formerly, Statistician, GE Corporate Research and Development

Get as much statistical consulting experience as possible. Get some teaching experience. If you can teach successfully then you can consult successfully.

—Margaret A. Nemeth, Statistics Center Lead, Regulatory, Monsanto Company

Learn about leadership and about working with other people. If you aim to go into business and industry, learn something about marketing, management and financial analysis. In the area of financial analysis, in particular, learn how to evaluate the financial viability of efforts that have benefit and cost streams. Understand how to do sensitivity analyses on these.

—William C. Parr, Professor, China Europe International Business School; Consultant

It seems that, in many courses, techniques are emphasized over ideas and concepts. It is only as a Bayesian that I have come to understand what lurks behind the techniques used.

—Nozer D. Singpurwalla, Professor, The George Washington University; Consultant

7.15   MAJOR TAKEAWAYS

- The democratization of statistics makes it more important than ever for statisticians to be highly knowledgeable in their field.

- The Advanced Placement course in statistics and similar programs let high school students in the United States get an early taste of the subject.

- A bachelor's degree (alone) in statistics may get you an entry-level position, but may limit your advancement. Most professional statisticians need to have the knowledge (and recognition) that comes with graduate training.

- Most permanent faculty positions in academia require a Ph.D. The typical requirement for most other positions in statistics is a master's degree, although it is important for some to go further. The preponderance of degrees in statistics in the United States is at the master's level.

- Selection of the "right school" for you is important. Considerations include the relative emphasis given to the degree program of interest to you (bachelor's, master's, or Ph.D.), the department and school within the institution in which the degree is offered, the size of the program, the strength of the faculty, the available dissertation options, strength in areas other than statistics, and internship opportunities.

- A good foundation in mathematics is a prerequisite for a career in statistics.

- Some elective courses are highly important in most application areas. Your anticipated application area should be considered in selecting your electives.

- Statisticians need to be computer savvy and have a broad knowledge of the applicability and limitations of general purpose software with statistical features, general purpose statistical software, specialized statistical software, and statistical programming environments.

- Students should give careful consideration to their selection of instructors, dissertation advisors, committee members, and dissertation topic, and are urged to explore life outside the statistics program.

- Job experience while in school, preferably through a well-structured off-campus internship (perhaps preceded by on-campus consulting), can add significantly to your academic experience and can help you select an application area and find a job upon graduation.

- Many professionals have successfully transitioned into statistics after starting their careers in other fields.

- Even the best education will not prepare you fully for a career in statistics. But it should provide you the knowledge to adapt what you know and to learn rapidly needed things that you don't know.

DISCUSSION QUESTIONS

(* indicates question does *not* require any past statistical training)

1. *How far do you feel you should take your formal education to meet your career goals? Explain.

2. *In Section 7.5.2, we state that "Most significantly, you can develop your own weighting scheme using the 20 factors considered by NRC, and apply this to the NRC response summaries to arrive at your *personally customized ranking* of schools." Go to the NRC web site to determine the factors considered by NRC, develop your own weights for each of these factors, and indicate how you would use these weights, together with the NRC data, to arrive at your personally customized ranking of the 61 departments with Ph.D. programs in statistics.

3. *Return to the preceding question. Suggest some other applications for which a personally customized ranking would be informative and describe how you might go about developing such schemes.

4. *Critique the U.S. News & World Report's method for ranking undergraduate schools shown in Sidebar 7.1. How might you improve the approach? What added criteria would you propose?

5. *What opportunities are there for you to get on-the-job experience in your current school? To what degree have you, or are you planning to, avail yourself of these?

6. *Prepare a list of questions that students should ask in evaluating the attractiveness of a school and its statistics program.

7. Identify boxes in Table 7.1 where you (or your professor) disagree with our assessments of the importance of different courses for different application areas.

8. What courses should be added to the list in Table 7.1 and how would you (and your professor) rate their importance for different application areas?

9. Select one (or more) of the challenges stated in Sidebar 7.4 and research and report on methods for addressing it.

10. Select one (or more) of the technical pointers in Section 7.7.6 and elaborate on its meaning and importance. Which of these do you feel have been emphasized insufficiently in your education to date?

11. What is your assessment of the statistical software that you have encountered to date? How would you rate each offering's correctness, completeness of features, documentation, and ease of use for yourself and for practitioners? What changes/improvements do you recommend?

12. Determine and report on the degree to which the technical concerns stated about Excel 2007 in Sidebar 7.5 have been addressed in subsequent versions of the program.

13. *Identify activities at your school outside the statistics program in which it would be useful for you to participate. Have you taken advantage of these? Did

this result in worthwhile experiences and why or why not? What are your plans for the future?

14. Assume that you are involved in one (or more) of the past STATCOM projects briefly mentioned in Sidebar 7.6. What are some of the questions that you would ask to gain an understanding of the problem? What specific data are needed to make the desired assessments and how would you recommend acquiring such data? What analyses might you conduct on the resulting data? How would you expect the results of the study to be used?

15. Describe situations that you (or others) have encountered that require adapting existing methodology to fit the problem at hand. How well were you equipped to do so?

16. What would you add to our recommendations in this chapter to help others make the most of their academic experience?

CHAPTER **8**

GETTING THE RIGHT JOB

8.1 ABOUT THIS CHAPTER

In this chapter, we provide some hints to help you get the right (for you) job as a statistician. We consider

- Defining career goals.
- Identifying opportunities.
- Résumé writing.
- The job interview.
- Follow-up.
- Providing references.
- Some further hints.
- Assessing job offers (hopefully, the fun part).

8.2 DEFINING CAREER GOALS

The first step in seeking a position is having, at least, a general career goal.

Defining your career goal calls for you to decide, and perhaps document, what you wish to accomplish during the course of your professional career. In so doing, you should try to sketch out both your immediate objectives and what you would like to be doing in, say, 5, 10, 25, and 40 years from now. Clearly, the further out you go in time, the vaguer will be your objectives. You may aspire to eventually become (see Chapter 12)

- A renowned statistician in academia, government or industry.
- A corporate vice president with oversight responsibility for activities in statistics and related areas.
- CEO of your own statistical consulting company.

Or you may just not yet know.

Students seeking their first full-time job as statisticians are often unsure even of their immediate objectives (other than gaining experience and making some money)—to say nothing of long-term career goals. For most individuals, career goals

A Career in Statistics: Beyond the Numbers, Gerald J. Hahn and Necip Doganaksoy.
© 2011 John Wiley & Sons, Inc. Published 2011 by John Wiley & Sons, Inc.

evolve over time as they encounter different environments and experiences—and as opportunities arise. The job search itself, and the clarification of available opportunities that comes with it, is often an important catalyst in helping define one's goals.

Formulating career goals nevertheless merits your attention at the start of your job search—if for no other reason than that you will likely be asked about them in your job interviews. You should have at least a first cut career plan—even though this will likely change and be refined over time.

8.2.1 Some Key Considerations

Business and Industry, Government, or Academia? We hope that our discussion in preceding chapters, and especially our description of what statisticians do in different application areas (together with our comments about academia in Chapter 13), is helpful to readers in deciding upon at least their initial career direction.

For those with Ph.D.'s—a requirement for most permanent faculty positions in academia—an important first question is whether to pursue a career in academia or to venture into the outside world. This does not have to be a complete one or the other decision. Industrial and government statisticians frequently serve as adjunct professors, teaching an evening course at a local university (something both authors have done). And many academically based statisticians consult extensively in government and industry. Such cross-fertilization is beneficial to all.

However, (like the majority of statistics graduates) you may not have a Ph.D. in statistics and will be seeking a career outside academia. You will need to decide between business and industry versus government or, possibly, one of the other options described in Chapter 4. Even within these, you might have a favored application area, perhaps a field in which you have minored or taken courses. And if you are planning to go into government, you might have a preference between national and state government.

Your eventual choice will, undoubtedly, be driven by available opportunities; for example, heavy demand for statisticians by the pharmaceutical industry or the fact that in the United States the federal government employs more statisticians than do state governments.

8.2.2 Role Aspirations

You also need to think about the role to which you would like to aspire, at least eventually, within an organization. For example,

- Do you wish to focus on the technical aspects of statistics, eventually becoming a "master statistician"?

- Do you wish to become heavily involved in an application area, such as product design, finance, or marketing—which will call for you to take on a broader role, beyond being a statistician?

- Do you aspire, at least eventually, to take on a managerial or administrative role?

We discuss alternative career paths in Chapter 12.

8.2.3 The Path Forward

Defining your career goals as a statistician will be an iterative process that is constantly refined as you gain more information and experience. An important first step is the realization of the need for having career goals. You will then want to size up your own skills (see Chapter 6) and likes and dislikes, as well as talk with your professors and some that have traveled the road before you.

Some of the suggestions that we have made in Chapter 7, such as taking courses in application areas, exploring life outside the Statistics Department, taking on an external internship, and attending (and possibly participating in) professional meetings, will be helpful in developing your goals. And the job seeking and interviewing process can in itself help you define better what you want to do—one reason for approaching the process with an open mind and closely examining the alternatives open to you.

Armed with, at least a rough, definition of your career goals, you can then proceed to look for a job. In addition to seeking a position that matches your goals and appears to provide you great growth opportunities, there will, undoubtedly, be a variety of other, mostly practical, considerations, such as where the job is located and how well it pays, subjects we will comment on further in Section 8.9.

Whether or not your dream job exists and it will be offered to you depends upon the job market, and on what *you* have to offer—as well as how practical your dreams may be. Most likely, you will have to compromise and at least initially settle for something less than your ideal. Being realistic, however, does not mean that it is not useful to have aspirations.

8.2.4 Opportunity for Change

No career is frozen in stone. As we shall see in Chapter 12, many statisticians, either as part of a preconceived plan or as a consequence of circumstances, change roles during the course of their careers. Some start their careers in industry or government, and then leverage their experience by moving into academia. Others travel the opposite path. Also, it is very common for statisticians to move among application areas during the course of their careers—either because they enjoy diversity and different experiences or because of new opportunities.

8.3 IDENTIFYING OPPORTUNITIES

> Environments that stretch personal development and provide for diverse application opportunities make for satisfying and rewarding careers.
>
> —Edward J. Spar

You are now poised to seek a (first) job as a statistician. We suggest that you explore a variety of opportunities. The interviewing process provides you a unique chance—that you may never again have during the course of your career—to witness different organizations from the inside.

We recommend that you start your job search early and work diligently to seek out opportunities. For federal government jobs, especially, you should start the process as much as 9 months before you graduate. Do not restrict your search to immediate advertised openings. Forward-looking organizations try to keep their pipeline filled and are receptive to initiatives by attractive candidates.

8.3.1   General Job Search Approaches

There are a variety of standard ways in which students, irrespective of their fields, go about identifying opportunities. We suggest that you

- Talk to professors, and especially those who would likely recommend you to colleagues in the workforce, updating them on your interests, aspirations, and accomplishments.

- Sign up for on-campus interviews. You may need to be a bit creative here since some employers will not advertise for statisticians, but the positions they have open can often be filled by someone with a degree in statistics; or the interest you show can lead to an "unadvertised" opportunity.

- Respond to Internet job postings and post your own résumé on applicant job search sites. Online professional networking sites can also be utilized for this purpose.

- Respond to newspaper ads (in the United States in the *Sunday New York Times*, *Washington Post*, and other big city papers, and the *Wall Street Journal*).

- Contact statisticians or managers in organizations of interest to you. (One of the authors got his job interview with GE through a letter to a GE statistician/manager whom he had identified through the membership listing of the American Statistical Association (ASA).)

- Attend professional meetings and get to know people in the field.

- Contact authors of technical papers that deal with application areas that interest you.

- Establish contact with a recruiting agency (sometimes referred to as a "headhunter"). These typically provide free services to applicants, making their revenues from the hiring companies. They might also review your résumé and provide you useful suggestions.

Different organizations use different hiring practices and it is important to understand these differences. Sidebar 8.1 summarizes the process used by the U.S. federal government and by one U.S. state. Those interested in a government position might, in addition, check out the web sites of individual agencies and contact these directly.

SIDEBAR *8.1*

GOVERNMENT HIRING PROCESSES

Blumberg (2009) reports that almost all U.S. federal government agencies now use the USAJobs[1] system; she also provides statisticians helpful hints for its use.

To become part of the system, applicants create an account and enter their résumés for different positions. Once your résumés are created, you can then authorize potential recruiters to contact you.

Applicants are screened and receive a numerical ranking based upon how their knowledge, skills, and abilities seem to match the job requirements. The highest ranked individuals are likely to be contacted. Government agencies are not allowed to contact applicants directly except for those ranked the highest, but you can track your applications via USAJobs. Each vacancy announcement has a contact person listed to whom you can address questions.

You can also search USAJobs for opportunities (via a vacancy announcement) by government agency, occupation, location, and so on, learn about the government hiring process, receive automated job alerts, and apply for a posted job.

New York State, through its Civil Service Department, typically gives examinations once every 4 years for statistician and related positions in state agencies and local governments. In the past, these examinations have been multiple-choice tests covering such topics as knowledge of statistics, research methods, reading comprehension, writing skills, and other subjects based on the skills needed for the position (e.g., administrative skills for senior positions; statistical questions tailored to the position's application area). When an agency has an opening, the state's Civil Service Department ranks the applicants who passed the test for that position and provides this information to the hiring agency. If a suitable candidate for a position is not identified, the agency has the option of hiring anyone meeting the minimum requirements on a provisional basis, subject to that person taking the exam for the position the next time it is given.

8.3.2 Job Search Approaches for Statisticians

Some approaches used specifically by statisticians in the United States are to

- Respond to job listings posted online at ASA's web site and advertised in *Amstat News*.

- Attend the annual Joint Statistical Meetings (JSM, see Section 14.5) and participate in their placement service. Typically, there are more than 50 employer listings. You fill out an information form to briefly describe your qualifications (attaching a résumé). This lets interested employers contact you for an interview at the meetings and permits you to request an interview with any listed employer (who also posts a job description). If you fill out your application in advance, it will be available to employers (who, in turn, might make advance plans to meet you). To find information on the ASA Job

[1] http://www.usajobs.gov/firsttimevisitors.asp.

Placement Service, go to the home page for the next JSM meeting and click on the Placement menu.

- While at the JSM, attend the business meetings and mixers of the ASA sections and special interest groups related to your interest areas and try to talk to attendees about their work and possible job opportunities.

8.4 RÉSUMÉ WRITING

It is essential that you put together an appealing résumé and cover letter. These are often the first things about you that a potential employer sees and on which the decision whether or not to follow up further with you is based. Even in a highly test-based setting, such as New York State's, résumés are reviewed by the hiring manager and provide a critical first impression, as well as information that helps guide the interview.

Throughout the interviewing process, you need to put yourself into the shoes of potential employers and identify those things about you that they might find attractive. We suggest that you desist from stating in your résumé (or in an interview) that your goal is to "gain job experience" or the like. Consider instead something more targeted at your potential employer's interests, such as "leverage my passion and talent for helping address tough technical problems with useful practical solutions and grow on the job." Also, we recommend that you

- For business and industry positions, be brief and to the point (usually one page for a résumé with perhaps a short attachment providing added details, such as papers published or submitted for publication).
- For federal government positions, make sure your résumé is complete and gives detailed descriptions of important relevant experiences.
- Be specific in describing your experience and accomplishments. Do not exaggerate (unduly) or undersell yourself.
- Make sure to describe relevant professional experiences indicating your role and accomplishments. List these in sequence of importance and relevance to an employer, not chronological order, unless the application process specifically requires otherwise.
- Briefly include nonprofessional accomplishments that demonstrate your leadership capabilities and interests, for example, captain of the baseball team, member of the debating society, and senior counselor at summer camp for disadvantaged children.
- State your personal skills, for example, excellent communicator, interested in solving problems, and ability to get along well with others.
- Make it attractive looking and easy to read.
- Be ready in an interview to describe and elaborate on anything that you state in the résumé.
- Modify your résumé and cover letter to address the needs and likely interests of different potential positions and employers.

Sidebar 8.2 shows what one government agency looks for in evaluating résumés.

We have[2] constructed in Sidebar 8.3 an example résumé for an individual with limited experience seeking a position in industry. More experienced people would further highlight such experience and, likely, place it ahead of education.

---

[2] Inspired by actual résumés of recent graduates.

results to customer. Co-taught course on experimental design to company Six Sigma black belts. Consulted with Manufacturing Department on development of multivariate control charts (leading to thesis topic).

Statistical Analysis Intern, Great Company 2, Financial Services, Summer 2008.

Conducted data analysis to develop financial risk model under direction of senior statistician. Received special award for excellence of work.

Statistical Consulting Intern, Graduate University, Department of Statistics, 2009–2011.

Provided guidance to 5 professors and 15 graduate students throughout the university on various research topics.

Publications and Presentations

- Co-author of three papers in refereed journals.
- Book review in Journal of the American Statistical Association.
- Three contributed presentations at national statistical meetings.

Professional Contributions

- President (2010/11) and Vice-President (2009/10), Statistics Club, Graduate University.
- Session Chair at Joint Statistical Meetings (2010).

Student Honors and Scholarships

- Graduate University: Outstanding Graduate Student Award, 2010.
- Student Scholarship to attend ASA/ASQ Fall Technical Conference, 2009.

Skills

- Statistical computing.
- Excellent communicator (oral and written).
- Passion for successfully solving real problems.
- Some proficiency in Chinese.

Personal Accomplishments

- President, Student Council, Undergraduate College.
- Community service volunteer: Two summers in inner city day nursery and in elder care home.

Attachments: List of references and details of publications.

For some further hints to statisticians on writing a résumé and some added references, see O'Brien (2000).

The cover letter, when applicable,[3] should briefly highlight the key points made in the résumé, show your enthusiasm, and interest the reader in you. It should be specifically targeted at the position for which you are applying.

---

[3] Cover letters are generally not allowed for federal government job vacancies, using USAJobs. However, you can insert parts of your cover letter at various points within your résumé(s).

8.5   THE JOB INTERVIEW

OK—you have identified a job that interests you and you have been able to secure an interview. This may be in the form of a preliminary screening interview on the telephone, to be followed, if successful, by an in-person on-site interview.

Your next challenge is to succeed in the interview or round of interviews. Success is measured by getting a job offer, and by the attractiveness of such an offer—in terms of both the nature of the job and the associated compensation and benefits. (The specifics of an offer are sometimes adjusted to meet your qualifications, as well as the importance that the employer places on having you accept the offer.)

8.5.1   Do Your Homework

It is important for you to gain as good an advance understanding as possible of the job for which you are applying, the organization to which it belongs, and the interviewing process. This will help you calibrate your comments, as well as demonstrate your diligence.

The Job and the Organization. Research the organization and the group within the organization, with which you are interviewing. If you are applying to a corporation, you might find reviewing recent annual reports to stockholders instructive.

It is especially helpful to have a good understanding of what an organization is specifically seeking from those that they wish to hire. You might get some idea of this from the job description and get further insight through the process of researching the organization, and possibly by just asking. See Keller and Sargent (2010) for a description of what it takes to be one of the top candidates for one employer (Mayo Clinic).

Publications by members of the group with which you are interviewing provide another useful source of information. Your showing awareness of these will also be flattering to your interviewers (but do not overdo the flattery).

The Interviewing Process. Leading organizations recognize that good people are essential to their long-term success. Hiring "the right people" is, therefore, critical to them. Just as you are eager to get the job that is right for you, they are anxious to hire people that best meet their needs.

Some organizations train their employees in recruiting and interviewing techniques; their human resources group may even have developed a formal interviewing process; see Sidebar 8.4 for one example.

You should try to learn the process used by the organization with which you are interviewing—perhaps, by asking discreetly in advance. Identifying and talking to others who have recently interviewed with the organization, perhaps for different positions, can also be helpful.

SIDEBAR *8.4*

SEEING IT FROM THE OTHER SIDE: INSIGHTS INTO THE RECRUITING PROCESS

The following illustrates one organization's (highly disciplined) interviewing process. The process employed by many other organizations is less formal, and might involve more one-on-one interviews. It is, however, likely to have, at least, some of the features described below. See Tanenbaum (2010a) for some further insights.

Identifying Potential Candidates. Candidates for a job as a statistician are identified by a process similar to that for job seekers (college recruiting, statistical meetings, advertising in the *Amstat News*, write-ins, recruiters, etc.).

The "Preinterview Interview". Before inviting a potential candidate for a formal on-site interview, a preinterview is conducted. This may be through a university recruitment visit, a society meeting, or by telephone, or a combination of these. On-campus interviews are frequently conducted by graduates from the institution, often majoring in some field other than statistics, or by a human resources specialist. Such initial interviews tend to focus on nonstatistical skills, and might be followed by one or more statistically oriented telephone interviews. If a candidate is located nearby, one or a series of preinterview interviews might be conducted on site to screen candidates.

Formal interviews are scheduled with the most promising candidates from such preinterview interviews.

The Formal Interview. Attractive candidates are generally invited on site for one or more interviews at the hiring organization's expense. A typical 1-day interview might proceed as follows:

- Arrival on the evening prior to the interview (for out-of-towners) and, if time allows, informal dinner with a potential future colleague.
- Early breakfast with the hiring manager.
- Applicant gives technical presentation for master's and Ph.D. level positions. The typical time allowed for this is 1 hour, including ample time for questions.
- Technical group interview with two to four potential future colleagues.
- Lunch with one or two potential future colleagues and/or clients.
- Nontechnical group interview with two to four potential future colleagues.
- Meeting with human resources representative.
- Meeting with hiring manager (and, possibly, brief meetings with one or more higher level managers).
- Meeting with a recent hire and/or brief tour of facility.

All those involved will have received a copy of the candidate's résumé and the hiring manager's description of the position to be filled.

The Evaluation. Those who participated in the interviewing process meet to discuss the candidate's strengths and weaknesses and general suitability for the job (perhaps while the candidate is still on site and meeting with a recent hire or touring the facility). All complete a form to provide their impressions, including a ranking, on a scale from 1 to 5, on whether they would recommend an offer be made.

The "Send-Off." The hiring manager takes the candidate to dinner and to the airport or hotel. This provides an opportunity to ask questions raised at the evaluation meeting, to assess a candidate's stamina, and to respond to questions. If it seems likely that an offer is to be made, it also allows the manager to, at least subtly, "sell" the job. This part of the day should be less formal and more relaxed than what preceded it (and might be omitted if the candidate does not look promising).

Checking References. A check of references is conducted, usually by phone. One obvious purpose is to verify the correctness of the factual information provided by the candidate. In addition, it provides an opportunity for the employer to probe in depth how others perceive the applicant and can be more informative than formal letters of recommendation.[4]

Employers are used to getting enthusiastic responses from references. Skillful enquirers make a special effort to seek out and pick up on possible negatives by

- Asking such questions as "what do you view as the candidate's key strengths and limitations?" (and trying to make sure that the response includes both limitations and strengths).
- Listening carefully to identify what is not said, as well as what is said.
- Taking note of any verbal cues, such as hesitation in response to a question.

Making an Offer. Hiring managers in consultation with their management make the final decision on whether or not to make a job offer. They then work with human resources to arrive at the proposed compensation and/or grade level (if not initially specified) and those fringe benefits for which there may be some flexibility (e.g., moving expenses). A formal offer letter is prepared and sent. When the details have been worked out, the hiring manager calls the candidate, offers congratulations, solicits questions, and works on convincing the candidate to accept the offer.

8.5.2 Demonstrate Your Personal Skills

> Interviews are as much about chemistry among people as about completing a technical checklist.
>
> —Gary McDonald

The job interview is your opportunity to sell yourself as a person, as well as technically, and to demonstrate that you possess the characteristics for success described in Chapter 6. The employer will look for these traits in you. Some, such as your listening and communication skills, will be relatively easy for you to exhibit. Demonstrating other skills, such as your ability to serve as a contributing and congenial team member, may require more planning and subtlety on your part.

---

[4] Some employers, at the advice of their legal departments, refuse to provide references, other than verification of employment.

Before the interview, think about experiences that you might use to illustrate your personal skills and strengths as they may relate to the job. Concrete examples that demonstrate your collaborative and leadership abilities are far better than general statements like "I am a committed team player." Think, for example, about volunteer projects in which you have been involved (and might even have led), and your dealings with, perhaps sometimes difficult, fellow students and faculty. Be prepared to provide thoughtful answers to questions about your career goals, short-term objectives, and accomplishments to date. Be ready to respond to questions (raised explicitly or implicitly) such as

- Do you feel you are well qualified for this position and, if so, why?
- What special skills do you have that will allow you to provide important added value?
- What would you enjoy most in this job?
- Provide an example from your previous work or life experiences in which you had to respond to a delicate situation or resolve a conflict and how you handled it.
- Give an example of a team effort in which you were involved and describe your role in the team.

Keep in mind that your interviewers will likely be evaluating you from the perspective of both how well you are suited for the job and how effective and congenial a potential colleague you are likely to be. Be honest, but make sure your positives stand out.

Most importantly, always look at things from your potential employer's perspective. The example in Sidebar 8.5 compares, in a somewhat exaggerated manner, ways in which two candidates might address the same questions in an interview for a job as a statistician. We will leave it to you to determine which candidate received a job offer.

SIDEBAR *8.5*

A COMPARISON OF TWO INTERVIEW RESPONSE STYLES

In the following example, two candidates respond to questions posed by a potential hiring manager.

Manager	Candidate 1	Candidate 2
Good morning, I am delighted that you are interested in working in our group. But what makes you want a job with us in the first place?	Well, I have been going to school for all these years and learning about all the wonderful methods of statistics. Now I would like to learn about how	Well, I have a passion for addressing real problems and working with people. I have learned all this good stuff in school—and now I want to put it to use and help

	they are used in the real world.	make a difference. From what I have heard I am convinced that your company will provide me the opportunity to do just that.
Thanks. OK, what would you say was the most exciting thing you learned in school?	Well, I felt I had a real revelation the day I got to understand the relationship between the Neyman–Pearson theorem and the Cramer–Rao inequality.	I'd say it is appreciating the concept of thinking statistically and understanding the role of uncertainty and variability in all we do.
OK, then, what do you think is the most important thing that statisticians do?	Only we have the expertise for conducting statistical analyses. For example, we alone know how to calculate the expected value of the mean square in an analysis of variance so that we perform the correct significance test.	Well, I would say it is working with a project team helping define the real problem, planning to get the right data, and serving as a resource in helping understand the results and their implications.
Thanks. Now tell me, how did you spend your summers in college?	Well. I stayed in school and took this cool course on measure theory.	I worked as an intern in a company to learn what a job in industry is all about. From this, I decided the challenges and excitement of industry are, indeed, my cup of tea. I also made some money to help me stay in school a little longer to make up for the summers I spent away.
Thanks. I asked you to bring a copy of something that you have done that you were real proud of. What did you bring?	Well, here is a copy of my thesis. It extends my advisor's work on optimal estimation of the kurtosis of the gamma distribution when the data are grouped in unequal intervals. I have proven the asymptotic consistency of this estimator (shows voluminous report).	My thesis was based upon a problem I ran into on one of my summer projects. It deals with ways of reducing impurities in the wastewater recycle stream (shows copy of presentation). I had to come up with a plan to get the right data, and had to do a lot of complex mathematics to perform the correct analyses. But in this presentation I try to explain to a plant manager in 15 minutes why the results merit her attention.

OK. I'd like to end with a technical question. My wife and I plan to buy a new car of a particular model. She is concerned with gasoline consumption. So, before giving me the cash for the car, she asked me to do a statistical study to provide her an estimate with 95% assuredness of the poorest gas consumption that we can expect for the car we are buying. I sample five cars of the same model and from these estimate an average gasoline consumption of 30 miles per gallon, with a standard deviation of 2 miles per gallon. How would you use this data to answer my wife's question?

(Saying to herself: I wonder why he is giving me such an easy question.) I would obtain a lower confidence bound on the mean; that is the mean of 30 minus the 5% t-value with 4 degrees of freedom times the standard deviation of 2 divided by the square root of 5, the sample size. Comes out to around 28 miles per gallon.

Well, I'd first like to ask some questions about how the five cars were selected, and how you measured gas consumption (asks further questions). Your question calls for a prediction interval on a future observation—since I assume you are buying only one car. I bet I can work out the answer and check out my approach in the book on statistical intervals by Hahn and Meeker (1991). My method will also assume a normal distribution for gasoline consumption for cars of the given model. I need to validate this assumption also. I would probably have to do this principally based upon the physical background of the problem and, perhaps, some similar data because of the small sample size in this example.

Thank you very much.

8.5.3 Demonstrate Your Technical Skills

Demonstrating your technical skills is a necessary, though not sufficient, condition for getting a job offer. Your résumé and academic transcripts provide the broad outline. You need to demonstrate understanding of what you have learned and how to apply it successfully. You are likely to be asked technical questions if there are statisticians on your interviewing team. Some may ask direct questions; for example: "How would you develop a fractional factorial experimental design with six variables so that all two-factor interactions can be estimated without any confounding among them or with any main effects and with the minimum number of test runs?" Others may prefer questions that describe a specific scenario and ask you how you might address it (like the example in Sidebar 8.5). Yet other questions may be more open-ended such as "Describe technical work that you have done and that you are especially proud of, indicating your role and contribution."

In general, employers are not seeking complete on-the-spot answers. They are looking more for your ability to grasp the question, show good statistical insights, and communicate effectively, especially to management.

You may be asked to give a seminar on a technical topic. This will provide you an opportunity to demonstrate your technical skills to a possibly mixed-background audience (see Sidebar 8.6).

SIDEBAR 8.6

THE TECHNICAL PRESENTATION

The first step in preparing for a technical presentation is to determine the likely makeup, statistical sophistication, and interests of your audience, and what they aim to get from the presentation. Management might be interested principally in the bottom-line actual or potential impact of your work. Others with a background in statistics may ask you to provide some technical details. All would want to know how much of the work that you describe was conducted by you personally (or under your direction).

You generally have the opportunity to select your presentation topic. If you are an advanced degree candidate, you will most likely think first of basing your presentation on your thesis. This may be a reasonable choice; it is, after all, a topic about which you know much (and your audience might know little). You may have previously prepared material from a presentation to your committee or from a department seminar.

You need to recognize, however, the difference in the background and interests of your audience in contrast to that in school and adapt accordingly. You will need to describe the motivation for and the practical applicability of your research and specifically what you have accomplished and how this has been or might be applied in practice. The fact that you have solved some interesting (to you and your professors) theoretical problem is usually insufficient. You need to convince others of the practical benefits of your work. You may even decide to deemphasize the technical details—making sure, however, that the fact that your work required innovative research and high technical skill is not lost on the audience. Also, an example using real or, possibly, synthetic data is often especially effective.

If you have had past work experience, you may, instead, want to base your presentation on a project in which you have been directly involved. This may especially make sense when you feel that the audience, or an important part of it, has only a limited interest in your thesis topic.[5]

In summary, irrespective of the topic, your presentation should be at a level, and degree of detail, appropriate to your audience, and include

- How you came upon the problem and why it is important.
- The technical and nontechnical challenges that you had to overcome.
- The results and their practical consequences and application.
- Your specific contributions.

Your ability to respond to questions on your work will also be important. Be sure to invite (and leave enough time for) these and to provide clear and meaningful responses. Perhaps even consider holding back some "nuggets" during your presentation for appropriate injection into your answers to questions.

---

[5] In some cases, you may have to change the details, or even the numbers to protect confidentiality, but usually your audience will understand the need for your doing this.

8.5.4 Ask Good Questions

The interview will focus on asking *you* questions. This does not preclude you from asking appropriate questions to the appropriate people at appropriate times. Good questions demonstrate that you have an inquisitive mind and think ahead.

Your questions should be principally job-related, rather than dealing with, say, housing or the community. The focus of the interview is to assess your suitability for the job and your main goal is to sell yourself as a potential employee. It is likely that questions that are important to you—especially those that are not job related—will not be answered during the interview. But make clear that you will need such information before making a decision if a job offer is to be made. We will return to this subject shortly.

8.5.5 Some Less Structured Interviewing Modes

The process that we have described, or some variation thereof, is most common for larger organizations in business and industry. For most government groups and many smaller organizations, the approach might be much less structured and the formal interview might be as short as 1 to 3 hours, with all interested parties participating. In this case, the major focus will likely be on describing the job, assessing your suitability for it based on your past experience and responses, and evaluating the chemistry between you and the hiring manager. However, the general guidelines that we have proposed still hold.

8.6 FOLLOW-UP

If the job appeals to you, be sure to communicate your enthusiasm and express your thanks for the interview by a written note shortly after your return home.[6] If you expect an offer to be forthcoming, you need to be careful not to overdo this, so as not to leave the impression that you would accept any offer, thereby possibly compromising the terms. (During the interview, you might also subtly suggest that you are considering other offers, should this be the case.)

If you have promised to provide further information on a topic, supply this rapidly—or indicate when your response will be forthcoming—and make sure that you meet this commitment. If you feel that there is something important that you failed to convey during the interview, your follow-up provides an opportunity to do so. You may also try to amplify, perhaps based on further research, your responses to questions that you feel you could have answered better. This can demonstrate your diligence and persistence in addressing problems—but needs to be balanced against drawing undue attention to things that you may not have gotten quite right.

---

[6] It is also a good idea to write thank you notes for positions that you do not find attractive in the recognition that these might interest you some time in the future.

8.7 PROVIDING REFERENCES

Employers may check out your references carefully and, on a close call, might base their decision as to whether to make an offer on these—especially if one or more of the responses is less than enthusiastic. You will want to identify individuals who are well acquainted with you and (except for personal references) your work and who are likely to speak highly about you. People of stature in the profession are especially desirable as references. Individuals who already know the person who will be contacting them can be especially convincing and candid references.

Graduate students would most likely use their advisors as references (and need to keep this in mind from the outset in their dealings with them). Past or present employers and clients are also obvious references. A prospective employer might be suspicious if you excluded these from your list.

Be sure to contact your references in advance to get their permission to serve as references and perhaps to get a better handle on how enthusiastic their recommendation might be. Apprise them of your interviewing plans. Provide them with your résumé and perhaps subtly make them aware of relevant positives about you, such as difficulties that you might have overcome, of which they may not be aware. Also, update them promptly after your interview and share with them your impressions of the job—keeping in mind that they might relay these to your potential employer when they are contacted.

8.8 SOME FURTHER HINTS

Here are some added suggestions that may help you in seeking the right job:

- Try to use your initial interviews to build up your interviewing skills and avoid, if possible, having (what you regard to be) your most important interview first. You need to balance this against the following dilemma. You receive a job offer from one of your early interviews and are asked to make a decision by a specified date. But you have not yet interviewed (or heard from) your "first choice" potential employer. One way around this problem might be to apprise all, up front, of the time frame in which you will be making your decision.

- Dress appropriately. Even though your interviewers might be dressed casually, it is good practice for you to be dressed smartly and be well groomed.

- Keep in mind the need to be on your toes throughout what might be a trying and lengthy interview. Make sure you are well rested and interview-focused—avoid scheduling your interview immediately after taking (and cramming for) an important exam.

- Do not let modesty get in the way of talking about your accomplishments; however, avoid appearing brash or to appear to know it all.

- If you have changed jobs in the past, or are now planning to do so, try to provide a good explanation for this and to dispel any incorrect suspicions that the change might be due to less than outstanding performance on the job (or that you are a job-hopper who uses one job merely as a stepping-stone to the next).

- Maintain good spirits and your sense of humor—after all, it is only an interview.

As in other areas, the *Amstat News* provides further, and current, perspectives on how statisticians proceed in searching for a job; see, for example, Gauvin (2007).

8.9 ASSESSING JOB OFFERS

You have followed all our advice and succeeded in lining up a number of attractive job offers. Congratulations!

Now you have the pleasant, but difficult, task of deciding which one to accept. If you have well laid-out career plans one of the first things you will undoubtedly do is assess how well each offer addresses these. In so doing, there are some specifics about the job that you will want to consider. There are also a variety of other considerations that you will want to explore in making your decision.

8.9.1 Job Growth Opportunities

Clearly, a major criterion for your first full-time job will be the degree to which it provides learning opportunities that will expedite your growth and advancement. You will want to consider

- The nature and breadth of the job. Is it likely to provide you wide exposure to different problems or is it a "cubbyhole" position, such as performing strictly supervised computer analyses in a single application area? Will you be included in key meetings and have the chance to witness how experienced professionals operate?
- Your level of responsibility. Is it commensurate with your abilities?
- Opportunities to learn from others and potential mentors. Will you be working with statisticians from whom you can learn? To what degree are there others who would be willing to serve as mentors, either informally or formally, and how helpful is their guidance likely to be?
- Opportunity for further formal learning. Will the organization encourage and support (both time-wise and financially) you in taking courses at a local university or remotely, or attending intensive short courses and professional society meetings?
- What will be the mechanism and timing for formal assessment of your performance and what will be the nature of the feedback you will receive?
- The path forward: Is there a clear path for continuing to learn and advance on the job? Will your responsibilities increase as you gain increased ability and confidence and demonstrate your worth? Is there a defined or likely stream of advancement opportunities with increasing challenges and responsibilities?

8.9.2 Some Other Considerations

> In every job I've had, I've gone in thinking I was there to work and then realized it is impossible to separate work from the people with whom you work.
>
> —J. Bose (2010)

You will also want to try to assess how a specific job offer stacks up on

- Personal chemistry. How do you think you will hit it off with your future coworkers and, especially, your future manager?

- Management commitment. Does your potential manager have a realistic appreciation of your capabilities and aspirations and is he/she committed to advancing these?

- Stability. How secure do the position and the company appear to be? Is the job in a growing field? To what degree will it hinge on one or two individuals—and how will you be impacted if they change jobs? How likely does this seem to be?

- Opportunities for and recognition given to statistical research and professional engagement (Section 12.6).

Reiterating some of our earlier points, you will also want to consider the

- Areas of activity in which the organization is engaged. If you are considering a position in business and industry, you might want to be in finance because of the high potential monetary impact of (and rewards for) your work. Or you might want to be in pharmaceuticals because it is a large, well-established, and well-defined opportunity area for statisticians or because of the opportunities it provides to help improve people's health. Or you might select an emerging area, such as genetics or nanotechnology, because of its excitement and growth potential.

- Function within an organization. Different operations within a company, such as research, engineering, manufacturing, and marketing, present different challenges to applied statisticians.

- Size of organization. A large organization might, arguably, provide more opportunities than a small one. A small organization might offer more of a personal touch and offer broader immediate involvement.

- Travel. Is there much travel associated with the job and, if so, where to and for how long?

- Working hours. To what degree is there flexibility in working hours? How frequently do crash projects and demands call for what is euphemistically referred to as "casual overtime?"

- The general morale of the "troops." More on that shortly.

Different individuals will weight the preceding criteria differently. Some might regard the ability to do statistical research highly; others might not. And some will find the opportunity to travel invigorating; others will regard it as a time-absorbing burden that intrudes on family and personal life.

See Nyberg (2010) for some further comments concerning finding your fit.

8.9.3 Geographical Considerations

A new job often calls for moving to a new community. This affects not only you but also your entire family, and can profoundly impact your entire life. This is especially so if you are moving to a new and, perhaps, unfamiliar, part of the country or even the world. Thus, where the job is located is a highly important personal and family consideration.

Small communities typically offer fewer job alternatives (and fewer educational opportunities) for statisticians—as well as their family members—than large metropolitan areas. This means that changing jobs in the future will likely require a further move for you and your family. Also, your children are less likely to stay in the community when they grow up. On the other hand, smaller communities might provide greater intimacy for a newcomer and be easier to settle into than larger ones.

We will comment further shortly on getting needed information about a new community.

8.9.4 The Money Factor

Last but not least, there is remuneration in the form of salary (taking into consideration local living costs) and other benefits (health plan, vacation, payment of moving expenses, etc.). Important as these are (and anxious, as you may be to pay off student loans), we advise that you do not focus on this unduly on your first job. Instead, we propose you give heavy weight to growth (including potential monetary growth) and other important professional and personal considerations.[7]

8.9.5 Getting the Needed Information

As already suggested, the prime purpose of the interview is for the prospective employer to evaluate your suitability for the job and for you to sell yourself. You are, therefore, often limited in asking questions, especially about topics not directly related to the job.

Once an offer has been extended, however, you become the boss. If you have the opportunity to decide among multiple offers, it is especially important for you to have a good understanding about each—and everything that comes with it. After all, it is your future that is at stake.

Your decision needs to be based upon a realistic assessment, rather than wishful thinking. You will never have perfect information to make your decision, but you want to have as full and accurate answers to your questions and concerns as possible. Or, statistically speaking, you want to have as much relevant data as possible in making your decision.

---

[7] One of the authors (GH) accepted his first full-time job as a statistician in 1955 paying $4800 yearly over a $6300 offer with another organization (in locations with comparable living costs), perceiving, correctly, that the first offer provided great growth opportunities (as indeed it did and which—in a reasonably short time—was also reflected in his paycheck).

The Reverse Interview. Ideally, your potential employer will invite you back for a reverse on-site interview. Such reverse interviews are not uncommon in business and industry. If not extended, do not hesitate to request such an interview. (Should your request not be granted this would speak for itself.) However, in academia, government, and nonprofit organizations, such reverse interviews are quite rare. Reverse interviews also need to be differentiated from invitations *after* accepting the job offer to return to a new community for house hunting.

Your return visit provides you the opportunity to learn about the details of the job and to talk further with your potential manager and future colleagues. This should help you better assess the chemistry factor (are these people that *you* would enjoy working for and with?) and the likelihood of mentor support. More informal meetings also help you assess the morale of the organization. Are your potential colleagues enthusiastic about their jobs or do they seem to be holding something back? If you have not met colleagues with whom you are likely to interact closely, request to be able to do so.

Assessing a New Community. If the job involves moving to a new geographical location, hopefully, your partner will be invited to join you on the reverse interview and have the chance to look over the community and, if relevant, assess work opportunities, while you are further exploring the job.

There are some obvious things to consider, such as cost of housing, quality of the school system, and educational, recreational, cultural, and religious activities. Meeting with real estate agents might be a good idea even if you expect to rent initially. In addition to providing housing information that might be of long-term interest, they can inform you about schools and local activities. You may also want to explore some less tangible factors, such as the degree to which newcomers are welcomed into the community and the apparent conduciveness for building social relationships. The hiring organization's human resources group might also help in this exploration.

You can learn much about a community by searching the Internet. You may want to subscribe to the local newspaper[8] and seek out information about local attractions and issues.

Other Information. Try to identify people who previously worked for the organization, and especially in the group in which you would be working— one of whom you may even be replacing—and, if possible, talk with them directly. They may be more likely than current employees to give you frank assessments. A key question is why they left the job. Some may have been promoted within the company. Others might have sought greater opportunity elsewhere. Yet others might have been asked to leave. What did each like and dislike about the job?

---

[8] Some companies automatically provide subscriptions to those to whom they make offers.

8.9.6 Making the Decision

Once you have the needed information, you will have to sort it out and come to a decision. How well does each offer match up to your dream job? What are the pros and cons associated with each offer? In making a decision, you need to assess to what degree you would consider each position as a potential long-term career opportunity versus a short-term stepping-stone, and the relative importance of these to you.

Don't regard offers, and especially those in business and industry, as nonnegotiable. Employers who make you an offer want you to accept. Say you like the job with Company A, but Company B offered you more money. Then don't hesitate to tell Company A about your alternatives, and respectfully ask them to reconsider salary. You might not succeed (and still take the job)—but you will have made your point.[9] Also, your discussion might lead to some other improvement, such as a more generous package for reimbursement of moving expenses or more vacation time—especially if you make clear that these are important to you. There is generally less flexibility for government jobs, in light of the grading system and fixed rules on vacation time, among other factors.

If possible, do not allow yourself to be rushed. Offers come with a specified decision date. But you can politely request a short extension if you need more time to evaluate alternatives.

8.10 MAJOR TAKEAWAYS

- Strive to define your career goals and your immediate objectives before embarking on your job search. Refine these as you gain further information.
- Start your job search early and avail yourself of different channels for identifying job opportunities.
- Develop an informative and appealing résumé and tailor it to each job for which you are applying.
- Get good advance understanding of the job for which you are interviewing and of the interviewing process. Prepare carefully for the interview and work toward selling yourself both technically and as an effective and congenial colleague who possesses the needed personal characteristics for success.
- Follow up all interviews with thank you letters and promptly provide further promised information.
- Pay careful attention to who you give as references and keep these individuals updated.
- In selecting among job offers, closely assess the pros and cons associated with each. A reverse interview can be especially useful.

---

[9] Your coauthor failed in a request to improve on the $4800 offer, but was promised reconsideration after 6 months, and received a $500 raise then.

DISCUSSION QUESTIONS

(* indicates question does *not* require any past statistical training)

1. *In Section 7.1, we noted the importance of planning ahead, while still a student, to acquire skills that will be attractive to a future employer. What actions might you want to take to potentially add to your résumé and improve your credentials in a future job search?

2. State your current long-term career goals and short-term aspirations.

3. *What is commendable about the résumé for Jimmy Applicant, shown in Sidebar 8.3? Is there anything that you recommend Jimmy do to further improve it?

4. *Develop your own one-page résumé. Exchange this with other class members. Critique their résumés (while they critique yours) from the perspective of a potential employer, and suggest improvements. Then improve your own résumé.

5. What are some specific organizations and positions within organizations that especially appeal to you for your first job? Why?

6. How might you tailor your résumé to make your credentials maximally attractive to the needs of the organizations that you identified in response to the last question?

7. You now have been invited to interview for a position with the organization that you identified in response to Question 5, or some other specific organization. Study the organization, and develop some questions that you might want to ask prior to and during your interview. Suggest one or more seminar topics that you might present that you expect to be of strong interest to the organization.

8. Assume again that you have been invited for a job interview with the organization that you identified in response to Question 5, or some other specific organization. How would you respond to the questions raised in Section 8.5.2?

9. *Critique the responses to the interview questions of the two candidates shown in Sidebar 8.5. What are the key points that we are trying to make in this example?

10. *Assume that you have received a job offer from a number of organizations of interest to you and have been granted a reverse interview. What are some of the questions you would ask?

11. *How would you rate, on a scale from 1 to 5, the importance to you of the various considerations for assessing a job offer that we discussed in Section 8.9? What other factors would you want to consider and how would you weight these?

PART III

BUILDING A SUCCESSFUL CAREER AS A STATISTICIAN

THE **PURPOSE** of this part of the book is to provide hints to help statisticians succeed on the job.

We propose strategies dealing with project initiation and execution (Chapter 9) and with communication, publicizing, and ethical considerations (Chapter 10). We discuss the statistician's all-important, and frequently underappreciated, role in getting good data (Chapter 11). We then present alternative career paths that statisticians follow (Chapter 12).

CHAPTER 9

ON-THE-JOB STRATEGIES: PROJECT INITIATION AND EXECUTION

9.1 ABOUT THIS CHAPTER

> Nothing is less common than common sense.
>
> —George Box

In our earlier chapters, especially those on the environment in which statisticians work (Chapter 5) and the characteristics of successful statisticians (Chapter 6), we have tried to convey some tips to success for statisticians. In this and the next chapter, we present what we believe to be some useful on-the-job strategies. We comment on project initiation and execution in this chapter and on communication, publicizing statistics and statisticians, and ethical considerations in Chapter 10.

9.2 PROJECT INITIATION

Early in your career, your manager may assign you to specific projects. As you advance, you will likely have more choice—both through your own initiative and, hopefully, because of the excellent reputation that you have established. Potential customers may, in fact, be approaching you directly. As a result, you will have a greater say about the work in which you become engaged.

The specific way in which you become involved in projects depends upon many factors, including the nature of the problem, the culture of the operation, management acceptance, organizational and geographical considerations, and, often, your own initiative. In a reactive environment, statisticians usually enter a project after the data have been gathered and help is sought for analysis. In the more proactive environment that we advocate throughout this book—and that we believe is becoming increasingly more prevalent—you may be recruited initially as part of the project team and even asked to propose projects.

A Career in Statistics: Beyond the Numbers, Gerald J. Hahn and Necip Doganaksoy.
© 2011 John Wiley & Sons, Inc. Published 2011 by John Wiley & Sons, Inc.

9.2.1 Statisticians Recruited During the Course of a Project

As indicated in Section 6.4, statisticians are often contacted during the course of a project to respond to narrowly stated technical questions, sometimes in a vacuum and with limited description of the real problem. Often data have been collected and those responsible realize that they do not have the expertise to conduct a proficient analysis. In its extreme form, this is the "if all fails call in the statistician" scenario.

Entering a project in midstream puts you at a disadvantage. You need to play catch-up to understand the project goal and its general context. The available data may be inadequate, incomplete, or ambiguous and there may be insufficient time and budget for additional data acquisition.

Excellence in your reactive work, to the degree possible, and some prodding can lead others to appreciate that, to really contribute effectively, you need to have a fuller understanding. This may result in requests for you to review the project in its entirety, to propose data acquisition and data analysis schemes, and, perhaps eventually, to your direct involvement as a member of the project team.

Example. One of us was contacted by a biologist who had been working on the development of materials used in the manufacture of surgical tools. He presented us with data on "counts" from a trial consisting of a sequence of readings on 16 repeat samples from each of 10 materials; see Table 9.1. He requested a formula to test the

TABLE 9.1 Counts on Repeat Samples from Each of 10 Materials

Position	Materials									
	A	B	C	D	E	F	G	H	I	J
1	4191	5409	3077	5457	3935	5229	5394	5122	2641	6266
2	5429	5187	1875	4522	5311	7001	3930	4906	3147	6139
3	4467	4727	2399	5710	4827	5540	3952	4354	2702	6182
4	4973	5752	2803	5143	4868	6128	4880	5536	2082	6635
5	4553	4355	3469	5483	4767	5684	4530	5916	1941	5817
6	4499	3355	501	3448	2913	3606	2772	3709	131	4979
7	3356	3109	1033	4351	4464	4175	3612	3809	1351	5167
8	2885	3323	353	3784	4140	5007	4474	3567	1148	5603
9	3045	3585	1175	4637	3740	3604	2926	3304	282	5054
10	4499	5504	3786	4953	4561	4788	4750	6432	1765	6303
11	5258	5131	3734	4285	4928	5943	3336	5251	2175	6044
12	4665	4374	2815	6041	4,453	5389	5471	4868	2301	6612
13	3544	5374	3912	4556	3720	4926	3277	5402	2151	6003
14	5005	4866	3342	4447	5182	5881	5283	4724	1224	6942
15	4507	6565	1796	4561	5281	6080	4977	4927	638	5554
16	4454	4995	3187	4079	6768	4358	5225	5350	1620	6736
Mean	4333	4726	2454	4716	4616	5209	4299	4824	1706	6002
Std. Dev.	755	981	1189	706	869	946	909	879	866	602

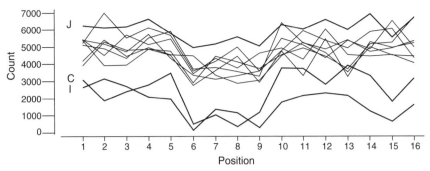

Figure 9.1   Plot of counts against sample position for each of 10 materials.

statistical equivalence of the sample means for the counts of the 10 materials. He was planning to continue this type of testing on additional materials.

Based on limited exposure to the problem, one may be tempted to suggest one of the several well-established statistical tests that respond to the biologist's direct question. After a quick glance at the tabulations—which also identified the position of each sample—we decided to first plot the readings for each sample against its position number, connecting the sample values from each material; see Figure 9.1.

This plot clearly revealed differences between materials, with materials C and I yielding consistently lower counts than the other materials and material J having typically above average counts.

In addition, the plots showed distinct differences among positions. The readings in Position 6, for example, were lower than those for Position 5 for each material.

This discovery came as a surprise to the biologist and changed the tone of our discussions. While some variability was anticipated, there was no reason to expect a systematic difference among the positions. We began to talk about what the data actually represented and how the data had been gathered.

It soon became evident that a better understanding of experimental and measurement variability was needed. We were asked to make recommendations to plan a further study to obtain data to help make such evaluations. This, in turn, led to an invitation to serve as a mentor to the project and to participate in the design and analyses of subsequent experiments.

9.2.2 Statisticians Recruited Up Front as Part of Project Team

In some application areas, statisticians are routinely sought and included as part of the project team from the outset. This is often the case in situations that involve a formal regulatory process, such as in obtaining approval from a regulatory agency for market release of a pharmaceutical product. In other cases, you might have contributed significantly to past projects, and therefore are a natural candidate for participation in subsequent ones.

Up-front involvement, starting at the project proposal stage, carries with it some distinct advantages. It provides you a say in the essential tasks of developing a clear and quantitative problem definition and in getting the best possible data.

Example. We had a long-standing relationship with a team developing new chemicals used in electronics applications. Along the way, we gained good understanding of the business, ranging from appreciation of the jargon to knowledge of the potential business applications and issues. When a key team member was promoted to lead supply chain management for the business, she decided to undertake a comprehensive study of the current supply chain so as to understand potential problem areas for improvement. From our past involvement, she recognized that we could be important contributors to the project. She invited us to participate in the kickoff meeting, even though our specific role was not yet known. Being involved from the outset allowed us to provide help in defining the problem quantitatively, bringing to bear statistical thinking and, most significantly, contributing to the study planning and data acquisition.

9.2.3   Projects Initiated by Statisticians

> The job of a statistician is to find problems that other people cannot be expected to perceive.
>
> —W. Edwards Deming

As statisticians become more proactive, they seek to identify opportunities and initiate projects that they believe to be of critical business importance and to which they think they can make an important contribution. Such projects are often best driven in collaboration with working partners from the application area, as in the following two examples.

Example 1. A commercial lending business provided loans involving tens of millions of dollars. An accurate assessment of potential risk for each loan was, therefore, critical to the overall success of the business. Statisticians who had intimate knowledge of the business believed that a statistically based algorithm, constructed from data on past loan performance, could be used to predict, based on various loan default predictors, the potential vulnerability of each loan currently on the books. The statisticians partnered with colleagues from the operation to prepare a proposal (Section 9.2.5) for the development of a system to routinely assess and update the probability of default of individual loans. The proposal was funded and the resulting work and its implementation led to a significant savings to the business (see Neagu and Hoerl, 2005).

Example 2. We had been involved for some time in helping plastics development engineers, in different locations around the world, plan designed experiments to determine the product formulation that best met individual customers' product performance specifications.

All concerned realized that conducting isolated experiments was repetitive and inefficient. Instead, one should be able to leverage information on the relationship between process settings and product performance, based on past test data collected throughout the business. We, therefore, proposed a web-based system with searching capabilities that would permit sharing the large amounts of previously obtained data, and, for each new order, supplement such data, with additional runs, as needed, to fill in holes in the available information and to provide validation information (such data would, in turn, be fed back into the system); see Sidebar 2.5 of Hahn and Doganaksoy (2008) for further details. This idea was then formalized in a work proposal (Section 9.2.5). The effort was funded and led to the successful development and implementation of the proposed system.

9.2.4 Project Selection

Avoid doomed projects.
—David Banks (2008)
(Our addition: Unless you are confident that you can save them.)

Successful statisticians often find there are more opportunities than time to pursue them. Even though at the outset of your career you may have limited choice in picking the projects that you work on, we recommend that you start early in developing a system that will, at least eventually, help you to be selective.

So how might you go about deciding whether or not to pursue a project? We suggest four major criteria:

- *Importance of the Work*: This might be measured by the potential benefits or "payoff" from successful execution of the project (Section 9.2.5). Important work is often readily recognized. But occasionally the likely benefits need to be sold to management and/or the appropriate funding agency. When the payoff is mainly long term, some short-term benefits should also be identified, if at all possible.

- *Likelihood of Success of Your Work*: This involves both technical considerations, such as the difficulty of the task, and nontechnical ones, such as timing restrictions and the opportunities for getting needed data. You also need to assess the ability, interest, and commitment of your likely colleagues on the project and whether you will be comfortable working with them (and they with you). Can you expect free and constructive flow of information? Are the expectations placed upon you realistic? How will success be measured?

- *Likelihood of Success of the Overall Project*: Doing your part successfully may be of limited benefit—and receive little recognition—if the overall project fails or the results are not successfully implemented.

- *Project Appeal*: Does the project have potential for "opening the door" to other desirable work? Will it provide opportunities for learning and advancement? Will it be fun?

Going a step further, you may wish, for a proposed project, to score each of the preceding criteria on a scale from 1 to 10, and then multiply the resulting four scores (possibly assigning different weights to each criterion) to arrive at an overall attractiveness index for the project. You can then use the resulting attractiveness indexes to compare and rank proposed projects.

Admittedly, it might be difficult for you to assess each of the criteria in advance (even without quantifying them) for a proposed project—and especially the likelihood of overall project success. The closer you are to the action, the better are the chances that you will be able to make reasonable assessments. But even if you do not go through the preceding process formally, it will be helpful for you to acquire the line of thinking it suggests.

The actual selection of a "portfolio" of projects will depend, in addition, upon such factors as funding, timing, and your availability—and, of course, your management's wishes. If a project seems worthwhile, but does not make your personal cutoff, you might try to provide a qualified colleague the opportunity to become involved.

For a related discussion, see Kenett et al. (2003), who propose assessment of "practical statistical efficiency" in project selection.

9.2.5 The Proposal Process

Requirements. A new project, or even the renewal of an existing one, generally requires the development of a proposal to do the work for approval by management and/or a funding agency. This may be the case even if the work has been requested by others. Statisticians in academia, research organizations, and consulting companies typically spend substantial time in developing detailed formal proposals for submittal to government or other agencies often closely following the guidelines of a RFP (Request for Proposal).

Key prerequisites for a successful proposal are a good idea that addresses an important need and an agency or organization that is likely to be interested and has the resources to provide the needed funding. A good understanding of how the proposal process works for the agency to which it is being submitted and of the submittal process and timeline is also required. Typically, the proposal includes a statement of the reason for the work, the specific work to be performed, potential benefits that would result, the resources required for successful implementation, milestones along the way, the planned work completion date, and, of course, the cost. The specific requirements vary among application areas and organizations; see Hulley et al. (2006) for a discussion of writing and funding research proposals for clinical research.

Effective selling of a work proposal through formal as well as informal means is frequently required. This typically calls for eliciting support from those directly involved and presenting the idea to key stakeholders and management. A long-standing working relationship and an appreciation of organizational issues are important advantages.

Short Proposals. Work proposals within an organization tend to be more informal and briefer than those submitted to an external party. They may, therefore, be less burdensome—or at least less bureaucratic (since the basic ideas still need to be

TABLE 9.2 Short Proposal 1

Proposal: **Early warning system to predict loan defaults for commercial lending operation**

Objective	*Completion date*
Develop a system to assess the probability of individual loan default	• 1 year effort
	Required resources (in person-years)
Work to be performed	• 1.5 statisticians
Build and validate an early warning system for loan default prediction based on past loan performance	• 1 system architect (25%)
	• 1 software developer
	• 1 portfolio manager (50%)
(Scope: North American, public, nonfinancial companies)	*Funding request*
	$800K (including overhead)
Benefits	*Project leaders*
• Estimated reduction of 15–25% in current yearly losses of $625MM by appropriately acting on likely defaults	C. Diverse, Portfolio Management
	A. Skew, Statistics Group
• Greater focus by narrowing the "watch list"	
• Improved consistency and objectivity in decision making	

developed)—than those required in other areas. In some cases, the written documentation in support of the project may be limited to a one-page standard proposal sheet.

Two such short proposals are shown in Tables 9.2 and 9.3, based on the examples dealing with loan default prediction and custom design of plastics products, described in Section 9.2.3.

TABLE 9.3 Short Proposal 2

Proposal: System to expedite custom design of plastics products	
Objective	*Completion date*
Develop a web-based system to expedite the custom design of plastic products that meet customer specifications of physical properties	• 1.5 years effort
Work to be performed	*Required resources (in person-years)*
Use data based on past experiments throughout business to establish relationship between process settings and product performance; utilize these relationships to establish process setting for new design and determine needed added experimentation, as necessary	• 2 statisticians
	• 1 system architect (30%)
	• 2 web developers
	• 1 product designer (25%)
Benefits	*Funding request*
• Reduce development cycle time for custom products by an average of 3 weeks; cut costs for such development by 25%	$900K (including overhead)
• Create capacity for $20MM/year new product sales	*Project leader*
	G. Kurtosis, Statistics Group

TABLE 9.4 Deliverables Timetable for Early Warning System to Predict Loan Defaults

End of Month 1	*End of Month 6*
• Define data requirements and data gathering process	• Complete validation of statistical model
• Begin data gathering	• Begin development of implementation software
End of Month 3	*End of Month 9*
• Complete data gathering and data cleaning	• Complete development of implementation software
• Begin development of statistical model for loan performance	• Begin software testing
• Define software requirements for implementation	• Begin user training
	End of Month 12
End of Month 5	• Complete software testing and user training
• Complete development of statistical model	• Begin routine system implementation

A more detailed short proposal might include a planned "deliverables" time-table; see Table 9.4 for an example.

Initial Project Benefits Estimation. An important element in preparing a proposal is making a first cut, preferably quantified, estimate of the benefits that will result from successful project conduct and implementation. The magnitude and credibility of such claimed benefits are important elements in determining whether a proposed project is to be funded.

In Sidebar 9.1, we provide an example dealing with estimating the potential savings associated with developing a system that can successfully distinguish real helicopter malfunctions from false alarms.

We will discuss the challenge of quantifying the benefits of our work in greater detail in Section 9.3.7.

SIDEBAR *9.1*

QUANTIFYING THE EXPECTED SAVINGS FROM DEVELOPMENT OF A SENSOR WARNING VERIFICATION SYSTEM

The sensing equipment for a fleet of newly designed helicopters recorded about 30 suspected engine malfunctions during the first 3 months of operation. Each time this happened, the helicopter's mission had to be aborted and the engine inspected at an average cost of $25,000 per incident. On each such occasion, the "problem" turned out to be a false alarm. Due to the serious consequences of missing a true engine malfunction, such alarms could not be ignored. They were expected to increase rapidly, due to new additions to the fleet, at least until the difficult long-term task of designing, building, and installing a more precise sensor could be successfully completed. Thus, the development of a statistically based algorithm that could consistently distinguish between sensor warnings signaling real engine mal-functions and false alarms was proposed.

In quantifying the benefits associated with successful implementation of this project, the expected savings needed to be assessed. This included estimating the expected cost of *not* proceeding with such development (i.e., the cost of the resulting unnecessary mission abortions), the likelihood of project success (i.e., developing an algorithm with an extremely high probability of clearly distinguishing engine malfunctions from false alarms and its unequivocal validation), and the anticipated timing of both the implementation of the new algorithm and the successful development and installation of a new sensor to replace the faulty one.

Project Cost Estimation. A further task in preparing a proposal is to provide a cost estimate for the work. This is a task that building contractors are skilled at doing, but is generally more difficult for many statisticians. Unlike building contractors, statisticians do not have a blueprint of the job to be done. Instead, they are often faced with a less than well-defined problem and a limited understanding of the perplexities that may lie ahead. Moreover, in addition to their own costs, they may have to estimate those of others, such as subcontractors, that the job might require.

A frequent mistake is to assume everything goes right, resulting, when it does not, in an underestimate of the actual cost and in missed timelines. (Something that is also known to happen to building contractors.) As a result, some add a percentage, ranging from 10% to 30%, to the initial cost estimate to take into account unforeseen happenings and higher than anticipated inflation. Increasing the cost estimate in this manner, however, decreases the chances of the proposal being accepted.

Sometimes it may be possible to conduct an initial scoping study to hone in on cost estimates; see Sidebar 9.2.

SIDEBAR 9.2

SCOPING STUDIES

One way to help address the uncertainties of up-front project cost estimates that is sometimes used in business and industry is to propose and conduct a preliminary scoping study. The goals of such a study are generally to

- Define the problem as precisely and quantitatively as possible.
- Develop an initial detailed plan for the gathering and analysis of the needed data.
- Describe, in some detail, the specific work that remains to be done, the approaches to be used, and the associated costs and timing.

The scoping study might include a preliminary analysis of currently available data.

In a typical application, you may suggest a 10-day effort over a 1-month time span for such a scoping study. In so doing, you might also provide an initial "ballpark" range for the entire project cost—to be refined as part of the scoping study.

Prospective customers might be willing to fund a scoping study—if there are no severe time pressures—because it provides an opportunity to get a better understanding of you and the proposed work.

Do not take the task of developing cost estimates lightly. Strive to base your estimate on as much relevant information as possible. And bear in mind that you may have to live with the consequences.

9.3 PROJECT EXECUTION

> Consider yourself a business partner—put your skin in the game.
> —David Coleman

9.3.1 Use a Disciplined Approach

> It takes some art to tease a clear statement of the goals of their studies from clients.
> —Peter Wludyka

Disciplined use of project management and tracking tools, such as Six Sigma (Section 5.2.2), is key to successful project execution. A disciplined approach starts with developing a clearly stated, and generally agreed upon, definition of the problem or purpose of the project, the objectives of the work to be done, and what will constitute successful conclusion of the effort. An important next step is to establish appropriate metrics and getting the right data—a subject to which we devote Chapter 11.

It is often helpful to break a project down into small, realizable parts or building blocks, and address these separately—while still keeping the whole in mind. This can make seemingly daunting projects seem more doable. For example, a project aimed at reducing the time to make loan approval decisions was "piloted" at one of the eight divisions of a financial services company. Once shown to be successful there, it became easier to convince other divisions to adopt the approach.

9.3.2 Understand the Process or Product

> The statistician cannot evade the responsibility for understanding the process he applies or recommends.
> —Sir R.A. Fisher

An essential first step in addressing a problem is to gain a good appreciation of the process or product with which you are dealing. For example,

- *New Product Development*: Get to know the fundamentals of how the product works, to whom it is to be sold, and how it is to be used.
- *Development of a Quality Assurance System*: Gain an appreciation of what constitutes a defect, how defects arise, and how they are found. This should include a tour of the production facility, paying particular attention to the measurement process, and talking to those who do the measuring about possible problems in the process (and carefully listening to their responses).

- *Development of a System to Minimize Unscheduled Field Maintenance*: Learn how malfunctions occur, how they are detected, and how they are addressed. This might call for accompanying a member of the maintenance team on field trips.

- *New Drug Approval*: Learn the basic biology underlying the disease or condition being treated and the underlying chemistry of the drug that has been proposed to address it. This might require reading up and/or taking a (short) course(s) on these subjects.

- *Development of a System that Automates Approval of Credit Applications*: Get to know how the current system works and how it can be improved. This might involve visits to local branch offices to observe operations.

- *Planning a Sampling Study*: Seek an understanding of the procedure for conducting the study, the population to be sampled, the sampling frame, and how similar surveys have been conducted in the past (and might be improved).

9.3.3 Gain a Clear Understanding of the Problem and Its Significance and Dynamics

> A problem well defined is a problem half solved.
>
> —Popular saying

We discussed in Section 6.4 the importance of statisticians being able to size up problems and see the big picture. These skills need to be applied to a project from its outset to ensure a clear understanding of the real problem and the associated nuances (and the personalities and idiosyncrasies of those involved).

You need to gain a good holistic appreciation of the business/organizational implications and of the management expectations from your work. For example,

- How does the problem relate to the overall business/organizational strategy?
- How does it tie in with other parts or aspects of the operation?
- What will be the impact of your work, if successful, and how will the results be implemented and measured?
- Who are the key players and what are their roles?

You also need to recognize, as discussed in Section 5.8.2, that the focus of your work and the associated expectations are liable to change over time. You need to keep apace with the dynamics of the environment in which you are working. All of this requires your continued immersion, or at least careful attention.

Gaining and maintaining the required understanding is easier if you are physically and organizationally colocated with your working partners and, in general, close to the scene of the action; see Sidebar 9.3.

SIDEBAR *9.3*

ON-SITE AND OFF-SITE INVOLVEMENT

Statisticians learn faster when they get their hands dirty.

—Martha Gardner

On-site presence at the facility that is the focal point of your work enhances your effectiveness by allowing you to interact continuously with others involved in the project and to witness the actual operation with which you are dealing (sometimes referred to as "kicking the tires") and to witness its evolution over time.

Statisticians working in a pharmaceutical company are often located at the site at which the test of a new drug is being conducted. They can take advantage of this presence by stepping out of their offices to witness the process firsthand and interact with all involved. Such direct participation was also easy for traditional plant statisticians since they were generally integral members of the operation where the work was being conducted.

In contrast, in today's global operations, there is often not a single facility at which the work is focused. Different factories, sometimes located long distances apart, might be responsible for different stages of the development and manufacture of a product. Similarly, users of a system that automates and expedites credit approval are likely to be spread out in branch offices throughout the world. Official statistics (e.g., gathering of agricultural data) might be sought nation-wide and their collection administered at various local offices. In each of these situations, it is useful to meet personally early in the project with the key individuals from different locations with whom you are working. This allows you to size them up, both technically and personally, and to establish good working relationships. You can then establish a system for continuous two-way flow of information, perhaps via routinely scheduled video/teleconferences, with only occasional direct meetings.

9.3.4 Be Strategically Proactive

We have emphasized throughout this book, the importance of today's statisticians having a proactive mindset and being on the lookout for new opportunities that will best serve their employers and society. We add a few practical hints here.

Be Realistic. Put yourself into the proverbial shoes of your management and customers and look at things from their perspectives. The problem of projects with long-term payoffs often is this. The costs for undertaking the work are incurred immediately, and get charged directly against this year's profits. The benefits, in the form of increased revenues or future cost avoidance, will, though possibly substantial, be realized only a few years hence. The manager who is to fund the project hopes to be promoted shortly in light of excellent current performance—as reflected by *this year's* balance sheet. In short, the manager will bear the costs of the proposed work, but a successor will be the direct beneficiary. Understanding this situation will lead you to think of ways of making your proposal more immediately attractive.

Watch Your Timing. Timing might make the difference between success and failure in being strategically proactive. Slow periods (e.g., annual scheduled manufacturing shutdowns) may provide an opportunity to conduct investigations that might not be feasible at other times. Or it might be possible to "piggyback" an important special feature on to a study that is scheduled to be conducted anyway. The Current Population Survey used to track the unemployment rate in the United States includes, most months, at least one supplement[1] on a topic of special interest.

Leverage Reactive Successes. Circumstances and management requests sometimes dictate involvement in reactive efforts. The contacts and goodwill gained from succeeding in such work provide you the opportunity to promote more proactive approaches in support of the long-range goals of the organization. Immediately after helping identify and remove a serious existing problem is an ideal time to suggest the implementation of processes to avoid future problems. The projects summarized in Tables 9.2 and 9.3 were both outgrowths of helping resolve narrower more immediately focused issues. The personal relationships established and the insights gained into core operational issues through past activities paved the way for us to take a lead role in proposing and working on the subsequent projects.

9.3.5 Stay Focused

You need to stay focused on the immediate requirements of the project and the long-term goals of the organization, and not let marginal issues divert you. Cost and practical considerations dictate how far to take a project. Carry your work forward only as far as it is of practical usefulness to your client—keeping in mind that often 80% of the value of the work may be gained by 20% of the effort (known as the Pareto rule). That last 20% often may not be warranted pursuing, based on economic considerations; see Section 6.8 for an example. Of course, an important requirement in all of this is to understand what the 80% point represents and to know when you have reached it. More fundamentally, the issue is determining the point of diminishing returns—it might be the 90% or 70% or even the 50%, rather than the 80% point—and then being able to recognize when you are there.

9.3.6 Perform

> Take responsibility and deliver ... do what you say you will.
> —Martha Gardner

All that we have said so far is prologue. What you will eventually be judged on is the perceived quality, effectiveness, and timeliness of your work. This means providing results that successfully address your clients' needs (and, perhaps, go beyond).

---

[1] For example, the Annual Social and Economic Supplement obtains added information on income, health insurance coverage, and education, every March.

Based on his extensive life experiences Starbuck (2009) urges you to

- Keep commitments. Nothing turns people off more than you agreeing to do something and then not delivering.
- Ask for feedback from colleagues and for help from people inside and outside your circle or comfort zone.
- Deliver value that far exceeds your cost of employment.
- Make yourself invaluable to a few good influential people.

It is especially important that, during the course of a project, you have a realistic perception of how well you are doing. You need to recognize rapidly if and when you are not meeting your customers' highest expectations—and, if so, why not. Such understanding allows you to correct course expeditiously if needed.

All of this requires ongoing communication—including keeping your customers abreast of your progress (and possible setbacks) and striving to obtain, and then listen carefully to, their feedback. Such communication also provides you an opportunity to bring to the forefront any potential obstacles, as a first step to resolving them.

9.3.7   Quantify Benefits

> Make your boss aware of your accomplishments.
> —Nancy Geller

Applied statisticians are evaluated by the contributions that they make. Estimates of the benefits associated with the successful execution of a project are required both at the project proposal stage and upon completion of the work.

Business and Industry. In business and industry, a statistician's contributions are typically stated in terms of their payoff or bottom-line impact, as measured by the resulting increase in income, market share, or asset growth or by the resulting cost reduction.

In many, mostly reactive, projects aimed at addressing existing problems, the benefits of successful project execution can be readily quantified—at least if one is willing to make some assumptions. The savings from work aimed at reducing scrap and rework or the added income from increasing market penetration can often be readily measured.

Quantifying benefits is also frequently possible in financial services applications since these generally deal directly with money. In the proposal for developing an early warning system to predict loan defaults (Table 9.2), savings were estimated from

- Predictions of defaults under the existing system.
- An assessment of the effectiveness of the proposed system in averting such defaults.
- An evaluation of the revenues that might be gained or lost by the specific actions that are taken as a consequence of replacing the existing system with the proposed new one.

Government. Government projects have bottom-line payoffs too, but these are often more subtle and can only occasionally be readily expressed in monetary terms, such as savings to the taxpayer. They may result in faster execution of a survey or more informative and/or precise estimates of an important statistic, such as the national unemployment rate. These, in turn, may lead to better decisions, improved drafting of legislation, and a better informed public. Translating such benefits into meaningful monetary terms is often not possible.

Need for Early Planning. To quantify the benefits gained from a project, one needs to put in place, as early as possible, a mechanism for measuring such benefits, as precisely, honestly, and convincingly as possible.

In many applications, the system originally developed to obtain data to diagnose the problem may be adapted to quantify savings. This holds true for projects ranging from reducing scrap and rework to minimizing waiting time in a doctor's office.

Challenges. For proactively oriented projects, the biggest gains often come from cost avoidance such as *averted* field repairs and recalls and from greater customer satisfaction. The resulting savings from such work are more speculative than those from most reactive projects, take more time to accrue, and, sometimes, are unknown or unknowable.

Consider a project to improve a product feature. Successful execution will likely result in increased sales. But sales volume is a function of many factors, and it is difficult or impossible to ferret out the impact of any particular one of them.

Measuring project impact becomes especially difficult in trying to quantify product reliability improvement for two major reasons. It often takes considerable time for the impact of the work, such as increased product life, to be realized. In addition, the consequence frequently is the avoidance of a serious problem, such as premature field failures. But how can you quantify the impact of avoiding a problem that, due to your (and others') efforts, did not happen?

Similarly, as we have already suggested, improved official statistics—such as the development of a more accurate consumer price index—carries with it many benefits to society that do not lend themselves to precise measurement.

Surmounting the Challenges. Difficulties in quantifying the impact of your contributions should not dissuade you from striving to develop some credible (even if approximate) assessment of the payoffs to your client, employer, or society of your work—even though such assessments may not always be monetary, or even quantitative.

Consider again the problem of measuring the payoff from a project to help redesign a product to avoid a previously observed field failure mode. Data on past failures might be used to predict the expected number of failures had the redesign not been implemented. One can then compare these with the number of failures that actually occurred (or are predicted to occur) for the new design.

The process of estimating expected benefits accruing from successful execution of a proposed effort can in itself be instructive in gaining an improved understanding of the underlying dynamics. In order to help ensure objectivity and improve credibility, it is, moreover, useful to involve knowledgeable outsiders, such as auditors, in helping develop measures of improvement.

Many of the projects in which statisticians participate—such as product redesign—are team efforts. In such cases, there is often no direct way of isolating the statistician's direct contributions, nor should there be. But statisticians are in a good position to help quantify the team's overall contributions to help ensure that all contributing team members will be able to share in the glory (and rewards) of project success.

9.3.8 Report

> No one will know what you have done unless you tell/show them.
> —Martha Gardner

Your work, including its quantification, is of little value to others if you don't effectively communicate your findings, especially to your clients and management—both during the course of a project and at its conclusion. We will discuss the challenges of communicating effectively in the next chapter.

9.3.9 Help Implement

To help make sure that your work comes to its full fruition, you often need to participate in its implementation. Guiding implementation may be a formal part of the project itself, such as in developing a quality assurance system or in conducting a survey. In other cases, implementation might be the major responsibility of others. It might take place after you have presented your findings, but may still require your involvement.

See Snee (1999) for some examples of projects—ranging from product development to reducing newspaper copy errors—for which, to achieve success, statisticians had to go beyond developing solutions and needed to participate in implementation.

9.4 AS OTHERS SEE IT

We present below some of the responses (sometimes, in slightly abridged or edited form) that pertain to on-the-job strategies to our question, posed to a number of eminent statisticians: What key advice would you give to an aspiring statistician planning or embarking on a career? These comments, like those provided in other chapters, reiterate or amplify points we make in the current chapter and add some further perspectives.

> Listen first, then communicate well and be creative enough to find the tool that gives roughly the right answer. There are always practical considerations of time pressure and available funds that affect what is the "best" solution. Be tactful, explain your reasons in a soft-spoken unemotional manner and be patient. Compromise is sometimes necessary. But firmness is important too.

—Michael R. Chernick, Director, Biostatistical Services, Lankenau Institute for Medical Research

Strive to become an expert on something of importance to your organization.

—Martha Gardner, Six Sigma Leader, GE Global Research Center

Statisticians have no authority of their own but should position themselves as the advocates of the client.

—Lynne Hare, Director (retired), Applied Statistics, Kraft Foods

Early on identify one or more mentors who can advise you and support your development. Don't try to do it all on your own savvy. Then, first and foremost, focus on the customer and his/her needs by delivering results. Use the 80/20 Pareto rule. Don't get enamored by the beauty of the methodology and end up spending a lot of customer-funded time in overdoing what is needed to deliver the results. Meet with the customers in their workplaces and don't expect them to come to you. Communicate frequently so the customer and your supervisor know where you are on a project. Listen to their inputs. Don't assume they will be patient and wait. Also, share your success and recognition with others that were instrumental in the project.

—William Hill, Six Sigma Leader (retired), Center for Applied Mathematics, Honeywell Company

Re-think everything that you have learned about statistics and adapt it to the specific environment in which you work. There is a good chance that some of what you have learned is irrelevant. Identify a killer application where good statistical thinking is under-utilized, and then focus on making a large impact, which is visible to upper management. Remain ethical in all that you do.

—Bruce Hoadley, Research Statistician (retired), Fair, Isaac and Co. Inc. (FICO)

Don't allow the label "statistician" to limit your thinking about how you can contribute and add value to your organization. Be prepared to "de-program" yourself from some of what you have been taught (common assumptions of normality, independence, random data, etc.). In the real world, the "best business solution" is more important than the "best statistical solution"; know the difference. Always question the data (How were the data collected and by whom? How were measurements taken? Were there any data you couldn't obtain, or that were not included? Can I speak to those involved?) and assume data are "guilty" until proven "innocent." Always plot the data; over time if possible.

—Roger Hoerl, Manager, Applied Statistics Laboratory, GE Global Research Center

When I describe what statisticians and actuaries do, I emphasize the need to be to the point and brief (KISS—Keep It Simple Statistician) and to avoid being boring!

—Robert Hogg, Professor (retired), University of Iowa; past President of the American Statistical Association

Always remember that the goal is to show the customer the impact of statistics on the company's bottom line—not how much you know.

—Ronald Iman, President, Southwest Technology Consultants; past President of the American Statistical Association

Listen, learn, reflect (i.e., confirm understanding by restating in your own words), interact (get intimate with the business problem, not just "arms length"). Fine-tune the ability to assess the practical problem (not just the statistical problem) and drive to the practical solution (not just the statistical solution). Employ the principle of parsimony to gauge the level of technical sophistication required ... technical marvels are not a requirement. Strive for enterprise-wide statistical thinking over the long haul.

—Tim Keyes, Manager, Strategic Portfolio Decisioning, GE Corporate Financial Services

Be willing to really listen and be quick at extracting the essence of a problem. Have the courage to take on a problem, before knowing how to solve it, and then the courage to meet a deadline, before having solved the problem completely. It's easier to be successful if you have a network, and it's easier to build a network if you maintain a reputation for being open to listening to problems and to treating people fairly and openly.

—Diane Lambert, Research Scientist, Google

Learn to "see" statistical problems from a more macroscopic, as opposed to microscopic, point-of-view. Always look for the "bigger picture" in which smaller statistical problems are imbedded.

—Harry Martz, Laboratory Associate in Statistics (retired), Los Alamos National Laboratory

Be willing to ask questions and listen rather than leap to conclusions. (The worst mistakes I've made came when I thought I knew what the client was saying, but didn't. I should have started at the very beginning.) Use flexibility and judgment in choosing appropriate methods that fit the design, the problem, and the client's level of sophistication. Simpler is better unless complexity is truly demanded.

—David S. Moore, Shanti S. Gupta Distinguished Professor (retired), Purdue University; past President of the American Statistical Association

Avoid appearing to tell clients how to do their jobs. Recognize, for instance, that some engineers feel that they were hired to plan and carry out experiments—and not to go to someone to tell them how to do this.

—Lloyd S. Nelson, Statistical Leader (retired), Nashua Corporation

Learn the regulatory, policy, economic, legal, technical, and enforcement ramifications of the problem and

1. Realize you have to get out in the field to see how your survey or application is working. You probably can't see it from the office.
2. Carefully hone your communication skills. You have to explain and convince someone about your approach, analysis, conclusions, and recommendations

3. Get proactive. In applications, frequently the statistician is the last to be included in the team. If data have already been collected, "Better late than never may not be good enough."

4. Learn to explain you work SIMPLY. See if the other guy can "play it back to you." Keep your eye on the prize; not the maximum likelihood estimator of the fourth moment.

—Barry D. Nussbaum, Chief Statistician, Environmental Protection Agency

Understand that companies, as opposed to statistics research institutions, view statistics as a means to an end and not the end itself. Strong impact requires developing and honing your soft skills and learning about the products and processes of your clients. While a sound background and competency in statistics is needed and advantageous, your advancement will be judged by the business impact of your work.

Corporate environments that stretch personal development and provide for diverse application opportunities make for satisfying and rewarding careers.

—Charles G. Pfeifer, Principal Consultant, DuPont Quality Management and Technology

Read the *Wall Street Journal*, *Fortune* and/or *Harvard Business Review* to find out what direction managers might turn to next; take some business courses to find out how managers think. Learn how to explain to managers how they could more effectively interpret performance indicators (profit, costs, hours of overtime, cost per unit, etc.); i.e., explain that a period-to-period increase or decrease is not necessarily indicative of a change. There is enormous waste in business organizations due to demands for explanations of why a particular measure went up or down this period compared to last period.

—Gipsie Ranney, Statistical Consultant

Learn to: Listen, coach (without seeming like a coach), present in front of groups, communicate in as jargon-free manner as possible, hold a strongly felt position in a graceful manner.

—Charles B. Sampson, Director of Statistics (retired), Eli Lilly & Company

Learn the nomenclature of the business world and/or government. Don't get pigeon-holed.

Learn what product or service you are dealing with and see statistics within the context of that entity. Do not be afraid to ask questions! Theory is fine, but the realities of what you have to work with take precedence. Accept that the statistical issues are only one part of a set of complex decisions by "management." Finally, be prepared to make a definitive statement. Management has no use for equivocation. Understanding that there's statistical error is important but eventually, to be useful, the applied statistician has to reach conclusions.

—Edward J. Spar, Executive Director, Council of Professional Association on Federal Statistics

9.5 MAJOR TAKEAWAYS

- To be most effective, statisticians should be part of the project team from the outset. Sometimes you need to work to make this happen. Eventually, you may propose important work yourself.

- As you advance, you will have greater opportunity to select the specific work in which to become involved. We propose the following criteria for project selection: the importance of the work, the likelihood of success of both your work and of the project overall, and the project's general appeal.

- Most projects require an up-front proposal for approval by management and/or a funding agency. This can involve a lengthy and highly formal process or a shorter, more informal one. It calls for skill in proposal writing, in estimating the cost to conduct the project, and in stating the benefits that would result from the work.

- Successful project execution requires you to use a disciplined approach; understand the underlying process or product; gain a clear understanding of the problem and its significance and dynamics; be strategically proactive; stay focused on the work; (most importantly) effectively perform the promised tasks (and maybe more) in a timely manner; quantify the benefits (monetarily if possible); clearly report your findings and recommendations; and, often, participate in their implementation.

DISCUSSION QUESTIONS

(* indicates that question does *not* require any past statistical training)

1. Describe a situation in which you have been asked a narrow technical question that led to a deeper engagement. How did you handle it?

2. Study and report on the approach used by Neagu and Hoerl (2005) in developing a system to predict the vulnerability to default of existing loans for a commercial lending operation (Section 9.2.3).

3. *How would you weight each of the four criteria suggested for consideration in project selection in Section 9.2.4? What further criteria would you add? How might the relative weighting of the criteria change at different stages of your career?

4. *Consider one of the projects mentioned in this chapter, or a project that you are familiar with. How might you go about assessing its importance?

5. Consider the quantification of expected savings from the development of the sensor warning verification system, described in Sidebar 9.1. Elaborate on how to get the information needed to quantify the benefits of developing such a system, both at the proposal stage and after the project has been completed.

6. *Think of a study that you feel warrants conducting. Identify a potential funding agency (e.g., parents, school administrators, government). Prepare a short proposal to request approval and funding of the study.

7. In the last bullet point in Section 9.3.2, we enumerated some of the steps required in planning a sampling study. Assume that you want to survey students' satisfaction with the education that they receive at your school. Describe how you would apply the stated steps (and possibly others) for such a study.

8. *We have tried to emphasize that performance is the key to (though not the sole requirement for) success. Discuss a project in which you have been involved and what you did to deliver on what was expected from you (and more).

9. *Describe a situation in business, government, or public life in which a particular course of action might detract from the measured performance of current management, but enhance that of their successors. How might this be rectified?

10. *In Section 9.3.7, we ask "How can you quantify the impact of avoiding a problem that, due to your (and others') efforts, did not happen?" Consider a specific example and discuss the difficulties in quantifying the impact of your work on this problem and how you might try to overcome such difficulties.

CHAPTER *10*

ON-THE-JOB STRATEGIES: COMMUNICATION, PUBLICIZING, AND ETHICS

10.1 ABOUT THIS CHAPTER

I want to talk to an expert—not a know-it-all.

—Charles Schwab advertisement

In this chapter, we discuss important on-the-job strategies for statisticians that pertain to

- Communication.
- (Subtly) publicizing statistics and statisticians.
- Ethical considerations.

10.2 COMMUNICATION, COMMUNICATION, COMMUNICATION

The ability to communicate effectively at all levels is essential to a statistician's success, as we have tried to make clear throughout this book, and especially in Section 6.3. In this section, we provide some hints for effective communication.

10.2.1 Adjust to the Environment

Statisticians can become enamored with statistics.

—Scott Pardo

Most students of statistics are surrounded by other students and professors who talk the same technical language and think along similar quantitatively oriented lines as themselves. Moving to an environment where statistical jargon is an unfamiliar language, and statistics itself may be viewed as a boring discipline, often comes as an unpleasant surprise.

A Career in Statistics: Beyond the Numbers, Gerald J. Hahn and Necip Doganaksoy.
© 2011 John Wiley & Sons, Inc. Published 2011 by John Wiley & Sons, Inc.

There is much variability in the level of interest and understanding of statistics among those with whom we interact. Some rely heavily on gut feeling assessments and have limited knowledge of—or interest in—statistical concepts. In contrast, others, such as many quantitatively trained financial analysts, engineers, and scientists, have a good appreciation of and a high interest in using a statistical approach.

You need to gauge the technical sophistication and level of interest in statistics of those with whom you are working. Often, this can be gained by informal conversation. A good strategy may be to assume that others have minimal statistical knowledge and gradually become more technical in what you say—to the point that it continues to be well received (and is needed). We also suggest that you ask to be interrupted when you use terminology or discuss concepts with which those whom you are addressing are unfamiliar (but don't rely on this happening).

Effective communication is a particular challenge in working in a global work environment; see Sidebar 10.1.

SIDEBAR 10.1

EFFECTIVE COMMUNICATION IN A GLOBAL WORK ENVIRONMENT

Sometimes those with whom you are communicating have a different native language from your own. This is becoming increasingly frequent as a consequence of globalization.

English is the generally accepted international language for communication. Fluency in English is an essential skill; professional meetings are typically conducted in English. In fact, you will often find participants from different non-English-speaking countries conversing with each other in English! The problem of communicating comprehensively and effectively is especially acute for those for whom English is not a native tongue, but who are working in an English-speaking country.

At the same time, being able to speak with others and perhaps even address a foreign audience effectively in their native tongue (when this is not English) can gain you appreciation and respect—and perhaps enhance understanding.

We have the following suggestions for individuals who, such as one of the authors (ND), need or want to communicate in a language different from their native tongue (whether that native tongue be English or some other language):

- Try to obtain relevant written background documentation; this is often easier for you to follow and can aid subsequent oral communication.

- In giving a formal presentation, provide sufficient detail in your slides to make them as self-explanatory as possible.

- Speak slowly and clearly.

- Check to ensure that you are being understood (recognizing that others are often reluctant to tell you if you are not).

- Get a mentor with whom you can rehearse your comments in advance, who can point out difficulties and help make your written communication more understandable.

- In transmitting your talk (or providing handouts), add further explanatory comments next to your presentation slides.

10.2.2 The Art of Listening

The first step in effective communication is to listen carefully to others, understand their priorities, and elicit the information that you need to be most effective.

Addressing a problem requires a solid understanding of your customers' perception of the situation and their expectations concerning your contributions. Problems are often not clearly stated, and may be understood differently by different people. As indicated earlier (Section 9.3.3), helping define the problem and doing so as quantitatively as possible is a key part of what statisticians do. This is often an iterative process requiring skillful probing, asking the "right" questions—at well-selected times—and carefully listening to the answers, including all accompanying nuances.

Different people often have different perceptions of the goals of a project and of what has to be done. You need to understand such differences, and sort them out in your own mind—as well as help narrow possible gaps in perception. It is often wise at the beginning of an interaction—after you have established your credibility—to focus on listening rather than talking.

10.2.3 Opportunities for Communication

> Statisticians should be heard as well as seen.
>
> —Gerald Van Belle (2002)

Project Meetings. Project meetings offer an opportunity for exchanging ideas, providing updates, jointly addressing issues, and in general fostering progress. They also allow you to explain important concepts, such as the need for obtaining well-targeted data. When the participants are at the same location, such meetings allow you to observe others' body language.

Meetings need to be well planned in advance and well managed to avoid becoming a bureaucratic drag; see Sidebar 10.2.

SIDEBAR *10.2*

TIPS FOR EFFECTIVE MEETINGS

Statisticians are sometimes responsible for organizing and leading project meetings. The following are some simple guidelines for conducting meetings effectively:

- Invite the key participants—balancing the desirability of including all involved against the need to conserve people's time and the recognition that the larger the meeting, the harder it might be to get things done.
- Schedule the meeting sufficiently in advance to avoid conflicts and whenever possible negotiate the date, time, and place with those whose participation is most important.
- Have a meeting goal; this might be as general as "Project Update."

- Prepare an agenda for advance distribution, indicating allocated time to each subject.
- Consider requiring brief written reports in advance.
- Set ground rules for the conduct of the meeting (e.g., no speeches, one person speaks at a time) and enforce them.
- In conducting the meeting, be fair, but insist on brevity and keeping to the point. Also, keep an eye on the clock and strive to enforce the time allocated to each subject (while allowing some flexibility when this seems appropriate). Use judgment as to what can be addressed constructively at this meeting, and what cannot.
- For larger meetings, consider splitting into working groups, possibly after some initial discussion involving everybody. Then reassemble to allow each group to report its conclusions and recommendations and to discuss these briefly.
- Focus on action and what is to happen next.
- Conclude with a clear statement of the agreed-upon actions and who will follow up on each.
- Document major results and action items and distribute them to attendees (and others involved) in a timely manner.

Project Progress Reports. Project progress reports provide important vehicles for communicating the current status of the work and for updating all involved. They can build enthusiasm, elicit feedback, and lead to corrections if some aspects of the work may have gone offtrack. They also provide an opportunity to identify, articulate, and respond to problems that might arise. Most importantly, such progress reports serve to keep everybody updated and interested. They can be helpful, irrespective of whether or not they are a formal requirement.

Quarterly, monthly, weekly, or even daily progress reports have been employed historically. Online postings, when important results become available, make this process more dynamic. E-mail can apprise all or a targeted group of important developments, but see Sidebar 10.3 for a note of caution.

SIDEBAR *10.3*

A NOTE OF CAUTION ON ELECTRONIC COMMUNICATION

Electronic communication tools can simplify transfer and sharing of information with others anywhere in the world. Instant messaging enables rapid access to those whose inputs are needed to speedily address a question. E-mail is minimally intrusive and is especially useful for transmitting (or soliciting) factual, and often routine, information or when a considered, but not necessarily immediate, response is desired.

These tools are, however, not well suited for give-and-take discussion and should not replace more direct interactions, especially on controversial issues. The speed with which messages can be composed and transmitted can lead to your using wording or sending messages that you might later regret. Whenever possible, compose important, and especially controversial, messages in draft form and reexamine them after a "cooling off" period before transmitting them. In fact, some advise against engaging in *any* disagreement electronically.

We also suggest moving to a more direct form of communication if an issue has not been resolved after two or three e-mail exchanges.

Compared to in-person and oral communication, electronic communication lacks the ability to easily and clearly get across nuances of meaning and emotions. When these are important, try, if possible, not to use electronic or any other written form of communication.

Final Reports. Traditionally, statisticians wrote formal and detailed project reports documenting their findings at the conclusion of a study. This is still done today in some areas—such as dossiers prepared for regulatory agencies by drug companies. In many other areas, the emphasis on speed and simplicity has led to the formal report being supplemented with or even replaced by a verbal presentation. The slides prepared for this presentation (to be discussed shortly) may then serve as the formal project documentation. This is especially the case in areas in which formal written documentation is not mandated.

When writing a project report, keep foremost in mind the needs and level of sophistication of those to whom the report is being addressed, recognizing that this is often a diverse audience. Stress the bottom-line findings up front (in perhaps an executive summary) and generally relegate the technical details to a later section or an appendix.

Especially avoid statistical jargon that might be unfamiliar to your readers. Strive for completeness and precision in what you state. If possible, give your mentor or possibly your manager the opportunity to review your report in draft form.

For some further advice to statisticians on preparing written communication, see Fenn-Buderer (2000).

Informal Opportunities. Communication can also be less formal, such as luncheon meetings or just dropping by a colleague's office. Some important encounters arise by chance—and you need be prepared for these; see Sidebar 10.4.

SIDEBAR *10.4*

THE ELEVATOR SPEECH

The best spontaneous remarks are practiced ahead of time.

—Bill Cheetham

You find yourself going up an elevator with your CEO or division chief—whose office is typically on the top floor—and who, noting your presence, asks "So what is new in the world of statistics?" This and similar occasions are your chance to convey concisely what you are doing and its significance.[1]

[1] If your office is on the second floor, we propose that you go along for the ride rather than truncate your comments.

Do not take the question literally. Unless you are close to retirement or have lined up your next job, we suggest that you refrain from responding, for example, "I have an elegant new proof for the independence of the sample mean and standard deviation from a normal distribution." Also, you might, not necessarily, want to respond with what happens to be on the top of your mind (most likely, today's project) and certainly not with "nothing much."

Instead, select a subject of interest and concern to your CEO or division chief in which you played an important role—although your participation might come as a surprise. Your comments should describe some significant—maybe even exciting—recent results, subtly suggesting the part that you played together with others, such as

- Working with Joe Smith, we have conducted a market study that suggests that the proposed overhaul of our XYZ product is expected to result in a 15% increase in market share.

- Following up on our presentation to you a few weeks ago, we have been working with design engineering to assess the damage that a humid environment might do to our new superwidget. The news is "so far, so good." Our accelerated test of 10 widgets has shown no appreciable degradation after the equivalent of 3 years in the toughest environment that we can expect in the field.

- Initial evaluations of our new drug suggest that its cure rate might be between 75% and 150% higher than that of a placebo, with no evidence of adverse reactions during its first year of application.

We propose that you always have an elevator speech in mind for use when appropriate—and periodically update it. If it presents an unexpected finding, all the better.

10.2.4 Giving Effective Presentations

Statisticians, like many other professionals, are called upon frequently to give presentations. These might be in the form of a project proposal, a progress update during the course of the project, or a final briefing at the end of a project. There are also other situations that require presentations, for example, as part of a job interview or at a technical professional society meeting.

The presentation may be directly to a "live" audience or it may be to a remote audience— possibly involving participants throughout the world—or some combination of the two.

We briefly discuss some concepts that we have found helpful in giving effective presentations. Courses on this subject provide further suggestions and practice. Joining a local chapter of Toastmasters or a similar group that provides speaking practice can prove to be helpful for those with little experience in giving presentations.

Tailor Comments to Audience. Statisticians, like doctors, lawyers, and many other professionals, face the challenge of explaining complex technical concepts to others in simple terms. You need to talk in your customers' language, instead of your own jargon—and do so in a nonconfrontational and informal manner without seeming to lecture. In structuring what you say and how you say it, always use your assessment of the knowledge and interests of your audience. A compliment sometimes paid to statisticians who have mastered the ability to communicate is that they do not sound like statisticians!

Your contact with higher levels of management is especially important. You need to be crisp and to the point—as in the elevator speech. One of us (GH) was required to present yearly to his Vice President the highlights of the past year's work of 16 statistics professionals—and had 20 minutes (possibly cut to 10 or 15 minutes if the program was running late) to do so!

Preparing Presentation Charts. Formal presentations are typically built around initially prepared charts/graphics viewed on a screen while you give the verbal presentation. These charts allow you to emphasize key points. They also provide cues to you as you give the presentation, and sometimes serve as a summary to those attending and to others who may not be there.

Preparing the presentation charts and graphics is often tantamount to preparing the presentation itself. We recommend that you

- Initially state the objective of the presentation and outline what you are planning to present.
- Include graphics to enliven the presentation. Well-selected and correctly executed graphics can significantly enhance communication from the early stages of discovery to the presentation of the final results of a project.
- Do not cram too much material on a single page and make sure that your presentation is readable. This might require your checking in advance on projection capabilities.
- Consider including some of the more technical material as an appendix for subsequent reading or as backup.
- Don't let background graphics drown out your presentation.
- Conclude with a summary of the key points that you want your audience to remember.

Sidebar 10.5 addresses the question of how much detail to include in your charts.

SIDEBAR *10.5*

HOW DETAILED SHOULD YOUR PRESENTATION CHARTS BE?

Make slides that reinforce your words rather than repeat them.
—S. Godin

An initial inclination is to have your presentation charts contain essentially your entire verbal presentation. But keep in mind that your charts are competing with your oral comments for attention; the audience may be split between listening to you and reading your charts. It is usually advisable to keep the presentation charts succinct to emphasize key points in abbreviated form—keeping separate notes as cues for your own reference. Your presentation will then elaborate on your charts and not make them (or you) seem redundant.

A possible exception, calling for more detailed presentation charts, is the situation in which it may be difficult for some to understand you, such as when, as previously mentioned, you are giving a presentation to people whose native language differs from the one you are using.

You might also strive to include more detail in the slides if the presentation is to serve as a final report of project findings, if the presentation is to be read by important people who were unable to attend the presentation, or if the presentation is likely to be referred to in the future. In these cases, a better alternative might be to make up two sets of slides—one for presentation and one for reference. Another possibility is to use the notes section accompanying your slides to include additional information.

Giving the Presentation. Having good presentation charts is a necessary but not sufficient condition for giving a good presentation. The focus needs to be on you; your charts are only a vehicle for making your comments more effective. We propose that you

- Start off with a light initial comment—referred to as the "ho-hum crasher" in effective presentation courses. (This also gives your audience a chance to adjust to your speech if your native language is different from the one in which you are giving the presentation.)
- Decide in advance how much emphasis to give to each point in your presentation.
- Depending upon your style and spontaneity, and the importance of the presentation, consider preparing a written version of your comments. Avoid reading this—or, at least, making it seem that is what you are doing.[2]
- Similarly, do not read your presentation charts—your audience can generally do so faster than you—instead amplify on the charts (another reason for keeping them concise).
- Keep eye contact with your audience—not your charts. This will help you maintain interest and let you gauge reaction to what you are saying.
- Pace yourself to meet time requirements—but be ready to adjust, based upon what you observe to be your audience's interest. Keep track of your remaining time and retain ample time for discussion.
- In addressing an audience whose native language is different from the one you are using, speak especially clearly and more slowly than you might otherwise. Avoid examples to which other cultures might not relate, for example, unfamiliar baseball or cricket slang.[3]
- Consider inviting audience questions as you go along (another voice can enliven things). Answering questions also provides you an opportunity to amplify

---

[2] Politicians are skilled in reading prepared remarks from a teleprompter, often without the audience noticing.

[3] One of us (GH) experienced some blank looks in talking about the reliability of an electric dishwasher in a presentation in a developing country.

points that might not be clear to all (or you might have forgotten to make). However, feel free to delay (perhaps for a subsequent one-on-one discussion) responding to questions that are likely to interest only the person asking the question—unless that person happens to be your manager, CEO, division chief, or a key customer.

- For an important presentation, rehearse your comments with knowledgeable and critical colleagues who can put themselves into your audience's shoes. This also provides a good check on your timing.

10.2.5 Further Reading

Numerous articles have been written in *Amstat News* and other publications to guide statisticians to be effective communicators; see, for example, Tanenbaum (2010b) and Younger (2009).

10.3 PUBLICIZING STATISTICS (AND STATISTICIANS)

Statisticians need to publicize what statistics (and they themselves) have to offer. Your best advertisement, unquestionably, is a record of past accomplishments. Once you have proven your worth, your services are likely to be in high demand. The hardest task is, therefore, often that faced by newly hired statisticians on their first jobs. In this section, we suggest a few subtle approaches.

10.3.1 Short Courses and Workshops

Teaching short (and sometimes not so short) courses within your organization on topics in statistics or on the use of software gives you a chance to demonstrate the excitement and relevance of the field and to present yourself as an accessible and easy to communicate with source of expertise. Such courses are often offered on special subjects that are not typically covered in introductory courses, but are important to the organization; for example, teaching design and analysis of accelerated life tests to design engineers or mixture experiments to chemical engineers involved in product development.

You might also have the opportunity to participate in more general courses taught by others. Our company for many years offered a 3-week Modern Engineering Course. The purpose was to provide senior professionals and aspiring managers with insights into important developments in science and engineering, focusing on subjects that have come to the forefront in the years since many of the course participants went to school. Typical topics were modern physics and computer science, often taught by college professors. We were invited to give two 3 hour presentations on the use of statistical methods to address important company problems. We tried to demystify the subject and provided introductions to such topics as product reliability, planning investigations, and simulation, and illustrated these with company application. The course gave us excellent exposure to future company leaders, who often contacted us, years later, when a problem arose in their operations.[4] Six Sigma (Section 5.2.2) training programs provide similar opportunities for teaching statistical concepts and tools.

10.3.2 Newsletters

For many years our organization published what we called *Statogram Newsletters*, distributed widely across the company, providing information on statistical concepts (e.g., Don't Let Statistical Significance Fool You!; Confidence Intervals, Tolerance Intervals and Prediction Intervals—Vive La Difference; the Ubiquitous Mr. Markov and His Remarkable Chain; and Removing Measurement Error in Assessing Conformance) or on recent successful applications (e.g., Scrap Reduced 77% in 6 Weeks; Complete Reliability Data System Installed; and A System for Proposal Risk Analysis). Some of these were subsequently published, over a period of 15 years, starting in 1973, in a series of articles entitled "Random Samples" in *Chemtech* magazine. This followed an approach pioneered by Jack Youden in *Industrial and Engineering Chemistry* from 1954 to 1959.[5]

Today's statistics newsletters have gone electronic and can be posted on web sites (and announced by e-mail notifications). Contributing author Leonard Gaines frequently posts briefs on his New York State agency's site based on recently released data, highlighting findings of interest to members of his agency. These typically consist of up to one page of bulleted news items, graphs, maps, and tables.

Like company courses, newsletters and bulletins can inform potential clients of the power of statistics and point to you as a source of knowledge.

10.3.3 Building on Casual Questions

Timely and effective responses to casual consulting requests provide you the opportunity to broaden the discussion of what statistics really has to offer and also open the door for your more extensive involvement. A casual technical question about software to do curve fitting might, for example, lead to your participation in addressing a much bigger problem (Section 9.2.1).

10.3.4 Further Informal Contacts

Informal conversations, such as casual chats with office neighbors and interactions at receptions, provide further opportunities for telling others about statistics and

---

[4] Many remembered best our demonstration of the Poisson distribution by counting the number of chips on the faces of a randomly selected sample of chocolate chip cookies.

[5] The American Society for Quality's publication *Quality Progress* currently carries, most months, a similar column entitled "Statistics Roundtable" written by various statisticians (including authors of this book who contribute a yearly article, together with Bill Meeker, on product reliability improvement using statistical methods).

establishing potentially useful contacts. You might develop a repertoire of interesting and informative stories (similar to the elevator speech) for this purpose. Some of our favorites are

- An experimental program to assess the claim that birds mistake jet engine noise for mating calls (and, as a result, are ingested into the engines).
- A study to assess dogs' preferences among various brands of dog food.
- The ability of public opinion polls, especially prior to elections, to accurately gauge public sentiments. (How can a random sample of 1200 provide valid information about a potential electorate of about 27 million in California?)
- The statistical validity of a study recently highlighted in the news (e.g., children with big feet read better, tall people earn more, married people live longer).
- The estimation of the number of wild animals in a forest through capture–recapture studies (Section 6.13).

10.4 ETHICAL CONSIDERATIONS

> Statistical tools and methods, as with many other technologies, can be employed either for social good or evil . . . all practitioners of statistics . . . have social obligations to perform their work in a professional, competent, and ethical manner.
>
> —ASA Ethical Guidelines for Statistical Practice

10.4.1 The Challenge

Your employer and/or customer often have a vested interest in your findings. Typical examples, some of which we have discussed in earlier chapters, are

- Disputes between manufacturers or service providers and their customers on whether a product or service meets claimed or agreed-upon specifications; see Sidebar 10.6.
- Reports by a government agency, in response to a request by a Congressional committee, of the anticipated costs and consequences of proposed legislation.
- Demonstration to regulators by a financial services company of compliance with international standards of capital requirements for risk management.
- Assessment of the performance of a company's product, perhaps in comparison with that of competitors, in making an advertising claim and the appropriate wording of the claim.
- Planning and evaluation of clinical trials to demonstrate the effectiveness (and assess the side effects) of a new drug and to obtain regulatory permission to market the drug.

- Assessment of claims concerning illegal discrimination in an organization's hiring, retention, or compensation practices.
- Evaluation of the environmental impact of a product or undertaking.

In some such situations, you may be asked to examine (or refute) another analyst's statistical findings. Often you may be working in support of a lawyer, on either side of an issue. You may on such occasions feel pressured, either directly or indirectly, to skew your findings in a certain direction or simply to bless somebody else's findings.

SIDEBAR *10.6*

ADDRESSING PRODUCT DISPUTES

It is not uncommon for a customer to claim that a product does not meet specifications and for the product manufacturer to dispute the claim. Resolution of such disputes frequently hinges on an analysis of all the available data and/or the procurement of new data. In Section 5.8.5, we briefly described a situation in which a statistician was called upon to serve as a technical arbitrator. Often, however, the statistician does not represent a neutral party, but is in the employ of one of the participants to the dispute.

Statisticians need in such situations to have a thorough understanding of both the manufacturer's and the customer's data. Contributing factors for the differences in conclusions—in addition to statistical variation and differences in analysis methods—might be the manufacturer and customer interpreting the specifications differently, different product subpopulations, different measurement techniques or calibration methods, or different test environments.

For example, both the manufacturer and the customer might have data on product life. The manufacturer's data might be based on in-house testing that the customer claims does not adequately represent the field environment. The customer's data might be from the field and the manufacturer might claim that the observed failures were due to improper use of the product. To complicate matters further, the manufacturer might have tested units from an earlier production period than that for most of the units shipped to the customer.

As in most other situations in which statisticians become involved, resolution of this problem requires knowledge of both statistics and various other disciplines. One of the statistician's responsibilities is to ensure that statistics is used properly and that the principles laid out in this chapter are strictly followed.

It is highly desirable in such cases for the statisticians from the opposing sides to get together to review the existing data and, most likely, propose the judicious gathering and analysis of further data. This has, in the past, helped diffuse some highly contentious disputes and forge a common approach that was then jointly presented to, and eventually accepted by, both sides.

In this section, we discuss general ethical issues that govern your work as a statistician, and then elaborate on their relevance to some specific application areas.

10.4.2 Statisticians' Responsibilities and Guidelines

The statistician's job frequently resembles that of a lawyer hired to support and often argue a particular case—recognizing that the other side is likely engaging their own statistician. Your employer and clients merit strong support. How can you represent them most effectively—and still do so with the highest ethical standards?

To respond to this and other questions, the American Statistical Association (ASA) in 1999 developed Ethical Guidelines for Statistical Practice[6] to help statisticians make and communicate ethical decisions. These guidelines comprehensively address eight areas:

- Professionalism in our work.
- Responsibilities to funders, clients, and employers.
- Responsibilities in publications and testimony.
- Responsibilities to research subjects.
- Responsibilities to research team colleagues.
- Responsibilities to other statisticians or statistical practitioners.
- Responsibilities regarding allegations of misconduct.
- Responsibilities of employers, including organizations, individuals, attorneys, or other clients employing statistical practitioners.

A further (and more compact) code of ethical statistical practice was developed in 2004 by the Statistical Society of Canada.[7] It deals with statisticians' responsibilities to society, their employers, and clients, and to other statistical practitioners and with professionalism. Other statistical societies around the world have similar guidelines.

The International Statistical Institute (2010) recently revised and updated its *Declaration of Professional Ethics*. It considers "the obligations and responsibilities of—as well as the resulting conflicts faced by—statisticians to forces and pressures outside of their own performance, namely to and from society; employers, clients and funders; colleagues; and subjects." This document enumerates ethical principles in 12 areas, such as guarding privileged information, avoiding preempted outcomes, and protecting the interests of human and animal subjects.

A key statement in the ASA guidelines is "All statistical practitioners are obliged to conduct their professional activities with responsible attention to the avoidance of any tendency to slant statistical work toward predetermined outcomes. It is acceptable to advocate a position; it is not acceptable to misapply statistical methods in doing so."

The various guidelines have numerous practical implications, some of which we briefly consider below.

Up-Front Avoidance. Ethical problems can often be avoided by early judicious action. Before accepting an assignment, strive to determine whether it might create potential ethical conflicts. Make your stance clear to your customer (or potential employer) at the outset. In the words of the ASA guidelines, "Do not join a research project unless you can expect to achieve valid results and unless you are confident that your name will not be associated with project or resulting publications without your

[6] www.amstat.org/about/ethicalguidelines.cfm.
[7] http://www.ssc.ca/en/webfm_send/3.

explicit consent." Also, make sure your manager is aware of potential ethical conflicts. If necessary, consult with your organization's legal staff.[8]

Getting the Data. One of your first responsibilities is to help ensure that the data that are to be obtained will lead to as complete and fair an assessment as possible. A sampling study of a human population, for example, requires careful consideration of such issues as sample size, selection of participants (and sampling frame), and how questions are worded. In some other cases, assessment of the possible bias and precision of measurements and when and where to conduct the study is highly important.

We discuss getting the right data and data quality further in Chapter 11.

Performing the Analyses. Data analysis requires you to make some subjective decisions, not only with regard to the prime analysis methods themselves but also with regard to such issues as the handling of missing data and outliers and distributional model assumptions. You need to ensure that the data are analyzed completely and objectively using the most appropriate and up-to-date statistical methodology. Your analysis must be directed at obtaining the most meaningful information possible, irrespective of how the chips might fall. You need to ensure that the analysis does not "cherry-pick" data or analysis methods to slant the results. On a more technical level, you should strive to remove the impact of possibly confounding variables (Section 11.5.2) and clearly state which confounding variables were addressed. And you need to document and justify any subjective decisions that you have made.

In some situations—such as the analysis of clinical trials—it is appropriate, and even sometimes required, for you to decide upon and document the analysis methods in advance of obtaining the data. Unforeseen circumstances, as well as examination of the data themselves, occasionally impact and change these initial plans. In such cases, you again need to carefully document such changes and their justification.

Presenting the Results. Just as you need to avoid selecting your analysis methods to further a desired conclusion, you need to avoid cherry-picking from your results to slant the reported outcome.

Your presentation of findings, both verbal and written, need to be easily understandable to your targeted audience at their level of statistical sophistication. Refrain from "snowing" others with statistical jargon and technicalities beyond their grasp. At the same time, convey in a clearly understandable manner the key assumptions that underlie your analyses and the consequences thereof—as well as how the data were obtained.

Do not expect nonstatisticians to appreciate the statistical limitations of a study or to understand important technical subtleties—without these being made explicit. For example,

---

[8] Most large organizations have one or more lawyers on their staff who are experts both in ethical issues and in related legal issues.

- Make clear that the establishment of a statistical relationship in an observational study *cannot* be interpreted per se as proof of a causal relationship (Section 11.5.1).

- If you use statistical significance tests, explain what such tests convey. Most nonstatisticians do not understand the difference between statistical significance and practical importance and the impact of sample size on the results of such tests; see the book's ftp site for further discussion and Hahn and Doganaksoy (2008, Section 9.8.3) for an example dealing with formulating an advertising claim.

- If you are conducting multiple significance tests, make clear how this increases the likelihood that one or more tests might show statistical significance just due to chance.

Occasionally, there is a need to convey unfavorable results. This presents additional issues; see Sidebar 10.7.

SIDEBAR *10.7*

CONVEYING UNFAVORABLE RESULTS: DON'T SHOOT THE MESSENGER

Statisticians need at times to be the bearers of unfavorable or unexpected results. During up-front evaluations, such as in product design or the initial stages of planning a survey, clients often accept potential setbacks or bad news gracefully and are anxious to rectify the underlying problems. Matters often become more complex in dealing with a finished product or a long-established procedure that might be difficult, expensive, time-consuming, or embarrassing to change. This may especially be the case in situations that involve a customer or a regulatory agency. Statisticians need to convey the findings in a timely, clear, and objective manner, possibly providing suggestions on how to proceed from a technical perspective.

Make the Details Available. You need to make available the details of your analyses, including the raw data or links thereto, subject to confidentiality and proprietary restrictions. This information should be in a sufficiently specific form to allow other statisticians to evaluate, and check, your work. You also need to make clear the source(s) of your data, how you went about "cleaning" the data, and how you handled missing data and outliers. Such information is often best provided in the form of a technical appendix to your reported findings.

10.4.3 Other Considerations

We emphasize in this brief discussion ethical considerations that have particular relevance to statisticians. In addition, you need to be governed by considerations that apply to all professionals, such as revealing possible conflicts of interest, avoiding plagiarism, striving to ensure that your name is not being used without your full involvement and consent, protecting confidentiality and classified information (see Sidebar 10.8), and giving appropriate credit to others.

SIDEBAR *10.8*

PROTECTING CONFIDENTIALITY AND CLASSIFIED INFORMATION

Statisticians in business and industry often work with information that is company confidential, possibly to maintain a competitive advantage. In addition, those working on defense-related projects are often required to work with government classified information (after obtaining security clearance). In dealing with official government statistics (discussed further in Section 10.4.4), the raw data often need to be considered as highly confidential with severe penalties, including prison sentences for violating such confidentiality. In all situations, it is your responsibility to know what information is confidential and to fully protect such data's confidentiality.

10.4.4 Official Government Statistics

Government statisticians provide the facts upon which many political claims and decisions are based. As a result, government statisticians, especially at a senior level, may be especially subject to political or other pressures to skew official statistics (and other findings) to support a particular cause or candidate. Like their colleagues elsewhere, they must have the highest level of integrity and backbone to resist such pressures.[9] Janet Norwood (former Commissioner of the U.S. Bureau of Labor Statistics) offers the following advice (in a 2006 unpublished presentation), based on her firsthand high-level experience: "The best defense against the politicization of data has to be the professional quality and the integrity of those who compile our statistical information. . . . The professionals who compile the nation's statistics must be courageous enough to insist their own work remains free of political influence; they must refrain from participating in political discussion, and they must insist on following careful, objective procedures for the manner in which their work in compiling the data is carried out."[10]

Who can view the data before it is released to the public, and when, is of special significance. In part, to prevent pressuring an agency to change the data for political purposes, it is important that nobody—even the President and members of

---

[9] Senior statisticians in the federal government have resigned from their positions to avoid having to change statistical procedures in a way that could have politicized the data.

[10] Len Cook, former National Statistician of the United Kingdom, adds (in a personal communication): "However, based on my experience, the protection of the independence and integrity of official statistics needs also to rest on the articulation of good practice, in legislation, protocols or codes, so that it is not simply the judgment of the statistician that is seen to be tested at any time of challenge, but the accumulated wisdom of legislators, statistical users and leaders collectively."

Congress—receives the information until shortly before it is officially reported. Statistical agencies in the United States have strict rules to accomplish this, based on a policy developed by the Office of Management and Budget. For further details, see this book's ftp site.

Other major ethical and legal issues in gathering and reporting official statistics include ensuring confidentiality, privacy, and security. Sidebar 10.9 describes the constraints imposed on government statisticians in dealing with these important matters in gathering and reporting official statistics and the general manner in which such issues are addressed.

SIDEBAR *10.9*

ENSURING CONFIDENTIALITY, PRIVACY, AND SECURITY IN GATHERING AND REPORTING OFFICIAL STATISTICS[11]

Government statisticians collect and have access to extremely sensitive individual and business data in many of the studies that they conduct, such as the decennial census. Nearly all of the data must be protected to ensure that the source of individual observations cannot be determined when released to the public. Government agencies are usually successful in collecting sensitive data, even in voluntary surveys, because respondents are assured and understand that their information will be used only for summary purposes and is protected from individual release.

Some federal statistical agencies such as the Bureau of the Census and the National Agricultural Statistics Service have long operated under legislated privacy authorizations. These allow the agencies to withstand requests for release of identifiable data under the Freedom of Information Act, court subpoenas, and Congressional committee requests. The 2002 Confidential Information Protection and Statistical Efficiency Act extended such confidentiality requirements to all statistical agencies. Snider (2009) describes how the U.S. Census Bureau protects the confidentiality of respondents' information while releasing as much high-quality data as possible. This requires modification of the data in a way that provides reliable aggregate information, while ensuring that individual responses are not identifiable. Many states have similar regulations covering the data collected by state agencies.

A second privacy aspect is prevention of inadvertent release of an individual's or a company's data in summary statistics. Most agencies, for that reason, use a "zero tolerance" approach in publishing information on data cells that provide absolute or approximate size information for any entity or organization. Thus, they go to great lengths to identify and suppress summary data cells that are based on data reported by a very few entities or that are so dominated by one entity that a knowledgeable party could closely surmise the data for that entity. See the book's ftp site for further details.

A complete issue (Volume 27, Number 1–2) of the *Statistical Journal of the International Association of Official Statistics* (2011) was dedicated to the topic of "Statistical Ethics and Official Statistics."

---

[11] This sidebar is based upon material prepared by Rich Allen.

10.4.5   Clinical Trials

Ethical considerations are of particular concern to statisticians working for pharmaceutical companies preparing dossiers for approval of a drug by a regulatory agency based upon the results of clinical trials. This is due to the critical human benefits and risks often associated with such trials and the fact that humans, as well as animals, are often the direct subjects of the study. Such concerns are a key reason for requiring statisticians to document the proposed analysis methods *prior* to the conduct of a clinical trial. In addition, important ethical questions arise in the selection of human participants in such trials, in the incentives offered for participation, and in the information that such subjects receive prior to, during, and at the conclusion of such trials.

Whether or not to terminate a clinical trial in response to early findings is especially wrought with ethical concerns. If a treatment appears to be beneficial, members of the placebo group should reap the benefit as early as possible, thus generally ending the study. If the treatment appears to be ineffective, and especially if it has potentially harmful side effects, it should be discontinued.

Unfortunately, the right decision is often far from clear at a given point in time. A treatment might have a mild bad effect initially but could be beneficial in the long run, at least for some patient groups. Early termination of a trial results in the loss of data concerning the impact of an extended treatment and can result in what would eventually be regarded as a faulty decision if all were known.

A report by the Pharmaceutical Research and Manufacturers of America (2009) discusses this difficult topic and deals with

- Protecting research participants.
- Conduct of clinical trials.
- Ensuring objectivity in research.
- Disclosure of clinical trial results.

This report calls for timely communication of study results, regardless of the outcome, and that these be reported in an objective, accurate, balanced, and complete manner, including a discussion of the limitations of the study. Where differences of opinion or data exist, the parties are asked to try to resolve disputes through honest scientific debates. These specifications are especially cogent in light of the fact that statisticians in the pharmaceutical industry are typically called upon to assess their own company's products.

For further discussion of ethics and stopping rules in clinical trials and other technical details, see Nguyen and Fan (2009). For some more general discussions, see Chapter 14 (on addressing ethical issues in designing clinical research studies) of Hulley et al. (2006) and Chapter 17 (on ethics in research involving human populations) of Aschengrau and Seage III (2008).

10.4.6   Serving as an Expert Witness

Statisticians are often called upon and paid by one of the parties to be expert witnesses at—or to prepare information for—courtroom trials. As Kadane (2008) makes clear, such witnesses still have the responsibility "to tell the truth, and nothing but the truth" and not just "those truths that help my client."

In such situations, you must first determine if there is any possible conflict of interest in your serving as an expert witness for the client due, for example, to a prior association with, or employment by, the other party. Absent a conflict of interest, you need to have a frank exchange with your client both preceding your involvement and after you have closely examined all the relevant data (as well as at various times along the way). At the first of these discussions, you need to make clear your professional ethical obligations and be sure, before signing on, that your client understands and is agreeable to them. Kadane (2008) cites an example of a situation in which he was expected (and refused) to testify that sampling could not possibly be relied upon. In the second discussion, you need to clearly lay out your technical assessment of the issues, including pointing out weaknesses of your client's, as well as the opponent's, case from a statistical perspective.

The climax of your involvement in a courtroom situation will likely occur when—after preparing a written report of your findings (which in a judicial court or quasi-judicial proceeding in the United States would be made available to the other side), testifying, and being asked friendly questions by your side's attorney—you will be vigorously cross-examined by the opposing attorneys. It is at that stage that you will be especially challenged to support your client's case in a manner that remains fully ethical and not lose "your cool" in the process. In so doing, note Easton's (2006) comments and recommendations concerning ethical issues associated with giving testimony:

- Be truthful.
- Answer questions asked.
- (You have) no duty to answer questions not asked.
- Be willing to testify and base overall opinions on all analyses done.
- What analysis you choose *not* to do is an important issue.
- (You have) no duty to represent the other side's case—that is their responsibility.

Fienberg (1997) comments on "how the statistician can attempt to protect against the most pernicious aspects of the game associated with the use of expert witnesses in the adversarial process."

10.4.7 Further Reading

Deming (1972) was among the first to highlight the importance of statistical ethics. Subsequent discussions include articles by Lesser (2001) and Vardeman and Morris (2004), an invited address by Bacon (2000), and chapters in the books by Utts (2005, Chapter 26) and Spurrier (2000, Chapter 15).

The ASA's Committee on Professional Ethics continues to address issues pertinent to statisticians, including the posting of case studies, general comments on statistical ethics, and links to numerous relevant articles and presentations; see *Amstat News* (2009c) and the committee's web site.[12]

10.5 MAJOR TAKEAWAYS

- The ability to communicate at all levels is an essential requirement for success for applied statisticians. To contribute effectively, you need to gain a thorough understanding of, and adjust to, the application environment. You must listen carefully, and skillfully avail yourself of all opportunities for communication, including project meetings, written reports, and informal contacts. Also make sure that your clients and your management are, and remain, fully informed of your progress and accomplishments.

- You need to build your skills in communicating orally and to tailor your comments to your audience. In a formal presentation, you need to develop well-targeted charts that support, rather than compete with, your comments, and acquire an effective style of verbal presentation.

- You can publicize statistics (and make others aware of you) by teaching short courses and workshops, writing and transmitting newsletters, and building on casual questions and informal contacts.

- Like other professionals, statisticians need to have the highest ethical standards. This requires refusing to be pressured by employers or customers and following practices that apply to all professionals, such as ensuring privacy and proper handling of confidential and classified information. The American Statistical Association, the International Statistical Institute, and various other professional groups have developed formal guidelines of ethical conduct for statisticians. Special challenges are presented in the gathering and reporting of official government statistics, in dealing with clinical trials, and in serving as an expert witness.

DISCUSSION QUESTIONS

(* indicates that question does *not* require any past statistical training)

1. Prepare and deliver to your class, or other designated audience, a presentation on a technical topic (possibly your thesis subject), in a manner that is understandable and motivational to an audience that has little training and/ or interest in statistics, and is concerned only with the practical significance of your findings.

2. Consider the same scenario as in the preceding question, but now prepare a written report.

---

[12] http://www.amstat.org/committees/ethics/index.html.

3. *Describe specific (different) examples of situations in which it would seem appropriate and inappropriate to communicate electronically.

4. Consider a person whom you would like to impress with your professional credentials and whom you may encounter on brief occasions. Develop your elevator speech— addressed to that person (and similar persons).

5. Imagine that you are at a party and are introduced to a leader of a company with which you might like to interview. The conversation, perhaps with your help, turns to your career goals and your relevant training, experience, and personal qualifications. What would you say?

6. Prepare some examples that you might use at a party to demonstrate the vitality, excitement, and value of statistics.

7. Propose an experimental study to assess the claim that birds mistake jet engine noise for mating calls or to assess dogs' preferences among various brands of dog foods.

8. Explain how a random sample of 1200 can provide valid information about a potential electorate of about 27 million people in California.

9. Your company would like to claim in their advertising that their washing machine removes $x\%$ more stains than that of a competitor. How would you go about getting valid data to potentially support such a claim? What analyses would you conduct on the resulting data? What issues might you expect to encounter and how might these be addressed?

10. In connection with the preceding example, assume you have collected and analyzed the appropriate data. How might you use different assumed results to word potential advertising claims?

11. *You have been asked by a lawyer to be an expert witness in a case involving alleged discriminatory practices. We suggest assessment and discussion of potential ethical issues *before* agreeing to pursue a project. How might you go about making such an assessment in this situation and what are some of the points/questions that you might raise in your preliminary discussion?

12. *Seek a recent example in the news in which a decision was made, or might have been made, concerning the early termination of a clinical trial. What decision was made (or might have been made) and what was its justification?

CHAPTER 11

GETTING GOOD DATA: A KEY CHALLENGE

11.1 ABOUT THIS CHAPTER[1]

> The most important information about any statistical study is how the data were produced.
>
> —David Moore

Rational decisions require transforming data into useful information. Your analyses, however, can be only as good as the data upon which they are based. Given good data, it is frequently possible to extract much meaning from graphical displays and simple analyses. But even the world's most sophisticated statistical analysis cannot compensate for or rescue inadequate data.

Statisticians expend much effort, often with limited payoff, in trying to understand and to compensate for poor data. DeVeaux and Hand (2005) provide a detailed discussion of different types of "bad data," including numerous examples, and claim that common wisdom puts the extent of the total project effort spent in cleaning the data before doing any analysis to be as high as 60–95%.

Examples of the undesired consequences of inadequate data abound. Sometimes, these are evident, such as a recent report that Japanese official records show numerous 150 year-olds and one 200 year-old! At other times, faulty data are far from obvious, especially to the casual observer.

Having the right, or at least the best possible, data is critical to the successful use of statistics. Bad data are often a consequence of the practice, discussed earlier, of statisticians being called in only *after* the data have been gathered—instead of involving them in the planning of the study and the data gathering. One reason for this unfortunate practice is that training in statistics has traditionally focused principally on the analysis of data with little attention given to their effective procurement and

---

[1] This chapter is in places more detailed and technical than the other chapters of this book. It deals with a fundamental topic that is not sufficiently emphasized in most curricula and one to which even beginners should be exposed. It also provides a further taste of some practical problems in which statisticians are engaged.

A Career in Statistics: Beyond the Numbers, Gerald J. Hahn and Necip Doganaksoy.

handling.[2] Typically, it is assumed that the right data are just there. This, sadly, is hardly ever the case. As a statistician, one of your major responsibilities will be to ensure that the best possible data are obtained—subject to practical constraints—to address the problem at hand.

Different situations call for different kinds of investigations. The type of investigation depends on the goal of the study and the nature of the underlying processes and/or populations. In addition, the choice depends on a variety of practical factors—such as economic and timing limitations and, often, ethical and legal considerations.

We briefly describe four different types of investigations that involve the gathering of data:

- Designed experiments.
- Census and random sampling studies.
- Systems development studies.
- Observational studies.

We focus especially on observational studies and on systems development studies since these are not discussed prominently in many introductory courses, yet arise frequently in practice. We then turn to a major theme of this chapter—hints for getting the right data. We conclude by proposing a disciplined step-by-step process for data gathering, illustrating this with a study to assess the reliability of a newly designed washing machine.

We should emphasize at the outset that investigations typically involve an iterative learning process (see Box et al., 2005) and therefore are multiphased. One might, for example, start with an observational study (using readily accessible existing data) and use the results to subsequently conduct a planned experiment or frequently a series of such experiments.

11.2   DESIGNED EXPERIMENTS

A designed experiment is a planned study in which one employs randomization methods to systematically assign various treatments or conditions to a group of objects or participants so as to evaluate the impact of potential explanatory variables on one or more measured response variables.

A well-known early application of a designed experiment, stemming from the 1920s, by Sir Ronald Fisher dealt with determining whether a lady drinking tea could distinguish between the milk being placed in the cup before as opposed to after the tea is poured; see Salsburg (2001).

Some more recent examples of designed experiments are to

- Determine the effects of such factors as construction methods, material properties, usage and environment, and levels of maintenance on pavement durability.

---

[2] The planning of investigations is addressed in specialized courses on topics such as the design of experiments, sampling studies, and the planning of clinical trials. These, however, are typically advanced elective courses and frequently devote considerable time to the associated data analyses (such as analysis of variance for designed experiments), as opposed to the practical details of planning the investigation to get the best possible data.

- Compare the effectiveness of different "skip tracing" strategies for locating a missing debtor.
- Evaluate the impact of product strength, bottle size, and variability between batches on drug stability.
- Assess the effectiveness of different layouts and content alternatives on hits registered for a web page advertisement.

The design of an experiment involves specifying for each planned observation the experimental conditions to be applied, as well as a variety of other details, such as material or subject selection, testing sequence, and the measurement of response variables.

Some designed experiments involve comparative investigations. For example, pharmaceutical studies often call for evaluating the performance and safety of a new drug versus that of a placebo, or some other appropriate reference (such as an existing drug). Participants in the experiment are randomly assigned to receive either the new drug or the placebo/reference drug. Such studies are regarded as the gold standard in evaluating a new drug or medical device.

Because designed experiments are controlled, they can generally be set up so as to avoid biased results. It should be noted, however, that even controlled studies are not immune to misleading conclusions. One reason for this is the inability of such studies to always provide a realistic representation of the real-life situation of interest.[3]

There are numerous texts on design of experiments (often abbreviated as DOE). Popular books, tending to focus on industrial applications, include Box et al. (2005), Mason (2003), and Montgomery (2009). Various other books address the use of designed experiments in the social sciences (e.g., Maxwell and Delaney, 2004; Kirk, 1995) and in other areas such as marketing and service applications (Ledolter and Swersey, 2007). Articles by Coleman and Montgomery (1993) and Hahn (1977) focus on the steps involved in designing an industrial experiment.

11.3 CENSUS AND RANDOM SAMPLING STUDIES OF HUMAN AND OTHER POPULATIONS

Designed experiments deal principally with the study of processes. Other applications (and ones upon which much traditional work in statistics is based) deal with characterizing a finite identifiable collection of units or population[4] with regard to one or more variables of interest.

---

[3] Meeker and Escobar (1998b) provide examples dealing with accelerated stress life testing of products (a specialized type of controlled study). In one example, the results led the experimenters to predict a high 5-year product reliability for an automobile air conditioning system. Yet the subsequent field experience resulted in a substantial fraction of units failing within 2 years. Physical evaluations showed that the field failures were caused by the drying out of materials during the seasons when the air conditioner was not in use. This long-term dormant condition was excluded from the continuously conducted accelerated stress life test, either because the experimenters did not recognize it or because of time limitations in conducting the study.

[4] Deming (1975) emphasized the difference between enumerative studies, which deal with characterizing a population, or finite identifiable collection of units, and analytic studies, in which one typically uses the results from a current process and environment to draw inferences about a future process and environment. Estimating the proportion of defective units in a production lot, based on a sample from that lot, is an example of an enumerative study. The washing machine example, presented later in this chapter, illustrates an analytic study.

Populations of interest are often specified groups of humans, such as all residents aged 18–45 of the U.S. State of Nebraska on July 15, 2011. Other populations may consist of objects, such as all the cars of a particular model that came off a specified assembly line on a specified day.

Further examples of studies of nonhuman populations are

- Estimating the percentage of defective units in a manufacturing lot.
- Estimating the error rate in bills sent to customers.
- Evaluating the delay time in processing orders.
- Conducting surveys on agricultural output, air traffic volume, or new housing starts.

Studies might entail a census of all members of a specified population or, more often, an appropriately selected random sample that is to be used to draw inferences about a population (sometimes referred to as "survey sampling").

11.3.1 Census Studies

A census involves obtaining (or attempting to obtain) information on one or more characteristics for *every member* of a specified population.

In some applications, it is easy to conduct a census. This may often be the case if the population is small. Sometimes it is important to obtain a complete enumeration of a population. We generally would not want to limit measurements of a potentially life-threatening property of an aircraft engine to a sample of engines. In other cases, a census might be mandated by law, such as the census of the U.S. population conducted every 10 years.

11.3.2 Random Sampling Studies

> If you don't like sampling, next time you have a blood test, tell them to take it all.
>
> —CBS News (quoted by Katherine Cramer Walsh)

In many situations, it is impossible, impractical, or uneconomical to conduct a census study. Measuring the size of movie and television audiences provides a comparison of a sampling and a census study—see Sidebar 11.1.

SIDEBAR 11.1

MEASURING AUDIENCE SIZES: MOVIES VERSUS TV

Surveys of movie and television viewing in the United States are similar in that weekly results of the most viewed productions are identified and widely publicized in the media. The types of data upon which these results are based are, however, quite different.

For movies, the results generally come from box office receipts from theaters nationwide, and these are used to infer audience size. In this case, we are dealing with a census study since the entire population is presumably being surveyed. Because no (deliberate) sampling is involved, it would be wrong to construct a statistical confidence interval to contain the true, or actual, number of viewers in the population. These are known exactly—at least if all movie theaters responded and did so correctly (if they did not, it would be highly questionable to treat those that responded as a random sample). A census study is appropriate in this example since information has already been acquired about every member of the population for other purposes (i.e., sales of movie tickets).

In contrast, in television viewership surveys, the results may be based on a random sample of say 7000 households from the population of approximately 110 million households with television sets in the United States. The results from this sample are then used to draw inferences about the number of viewers of different programs in the population from which the sample has been selected. The random sampling process introduces statistical uncertainty that is typically quantified by a statistical confidence interval.

Statisticians have developed scientific methods for conducting sampling studies of defined populations using random sampling. These are directed at providing unbiased estimates of quantities of interest, together with their estimated margins of statistical error (often expressed in the form of confidence intervals). Public opinion polls are sampling studies that generally use some form of random sampling (Section 4.6.1).

Texts on how to conduct random sampling studies and the analysis of the resulting data include Kish (1995), Lohr (2010), Groves et al. (2009), and Scheaffer et al. (2005). These books describe both different random sampling methods (e.g., simple random sampling, stratified sampling, and cluster sampling) and common pitfalls to avoid in conducting such studies (also see Sidebar 11.2). Asher (2010b) considers some of the challenges in collecting data in sampling studies in general and in the underdeveloped world in particular. Lampone (2009) discusses ensuring high-quality data from online studies.

SIDEBAR *11.2*

SOME NONSAMPLING PITFALLS IN STUDIES OF HUMAN POPULATIONS

The following are some frequently encountered flaws that need to be addressed in many census and sampling studies:

- "Loaded' wording of questions either purposefully or inadvertently; see Churchill and Iacobucci (2010, Chapter 8).
- Nonresponse bias—those who choose to participate differ in their answers from those who do not.
- Differences between the sampling frame and the target population, as in the *Literary Digest* example (Section 4.6.1).
- Changes over time, as in the 1948 Presidential preelection poll (Section 4.6.1).

Such nonsampling flaws add, usually nonmeasurable, uncertainty to the results of both census and random sampling studies. For random sampling studies, this further uncertainty is in addition to the statistical uncertainty due to sampling variability. Helping to avoid or minimize the impact of nonsampling errors is one of the statistician's major (but not always recognized) responsibilities in planning census or sampling studies.

11.4 SYSTEMS DEVELOPMENT STUDIES

Statisticians, especially those in business and industry, frequently become involved in investigations that cannot be neatly classified as designed experiments, census or sampling studies, or observational studies (to be discussed in the next section)—or at least are not given much consideration under these headings—but still require the judicious gathering of data.

For lack of a better title, we will refer to these as "systems development studies" since they typically lead to the development or validation of a system. They are often action-oriented in that they result in a specific action or decision that relies heavily upon analysis of the gathered data. Such studies frequently allow at least some degree of control over the data acquisition process.

The following are some examples of systems development studies for which the gathering of data is required:

- Estimating the reliability of a newly designed product. Frequently, testing is conducted on a new product, prior to going into high-volume manufacture, to demonstrate that the product meets customer reliability requirements. We provide an illustration dealing with a washing machine in Section 11.7.2.

- Many web sites utilize automated advisory systems that recommend selected products and/or services to users. Such systems are generally built from appropriately gathered data about the user's recent browsing patterns and purchases, or from purchases by other users with "similar" profiles.

- Many modern products (e.g., automobiles, medical scanners, locomotives, aircraft engines, and power generation equipment) are equipped with sensors that enable real-time data gathering on the state of health and the operating environment of the product. Such information is used for proactive servicing (e.g., maintenance scheduling, parts replacement, etc.) and automated monitoring for impending failures. Appropriate data need to be gathered to establish (and validate) the rules that are used in these systems.

- Control charts signal when a measured product property is no longer in statistical control and corrective action is needed.[5] Appropriate data on past product performance are required to establish and validate the system.

In Section 11.7.1, we propose a five-step data gathering process that has particular applicability to systems development studies.

---

[5] Control charts had their origin in industry, but today are used in more general applications, such as tracking response rates to a weekly government survey or studying the monthly variation in retail prices of a product.

11.5 OBSERVATIONAL STUDIES[6]

> The combination of some data and an aching desire for an answer does not ensure
> that a reasonable answer can be extracted from a given body of data.
>
> — John Tukey

Designed experiments and random sampling studies are characterized by randomization in the selection of test subjects or objects. Such studies also aim to control identified potentially important explanatory variables other than those under investigation. They are highly desirable if the goal is to establish unambiguous cause and effect relationships—as is often the case.

In many applications, it is, however, impossible, impractical, or even unethical to conduct a randomized study. Obvious examples are assessing the effect of cigarette smoking or of alcohol abuse on one's health, or evaluating the impact of alternative forms of punishment for certain crimes on the incidence of future crimes. In developing a credit score to differentiate between desirable and undesirable future applicants for credit based upon the payment performance of past credit applicants, it would be informative to grant credit to some randomly selected poor-risk applicants—so as to obtain a sample that is representative of all future applicants. However, the costs due to defaults from such poor-risk applicants make it unlikely for a business to be willing to include them in the study.

Because randomized studies are often not feasible, a frequent alternative is to conduct a so-called "observational study." Rosenbaum (2002) defines an observational study as "an empirical investigation of the effects caused by a treatment, policy, or intervention in which it is not possible to assign subjects at random to treatment or control, as would be done in a controlled experiment." Such studies often involve the analysis of data that were obtained principally for purposes *other than* to conduct analyses thereon. Flanagan-Hyde (2006) provides a brief overview of observational studies and Hulley et al. (2006) describe their application to clinical studies.[7]

Advances in computer technology and in the instrumentation for taking measurements have significantly enhanced the ability to generate massive databases—and, therefore, the opportunity and urge to conduct observational studies. This has led to the emergence of the field of data mining. As its name suggests, data mining typically involves tapping into existing databases to gain understanding or, as a minimum, to establish clues that can be pursued by further research.

As we will try to show, improper analyses of data from observational studies can lead to erroneous conclusions. But observational studies, in addition to being feasible in many situations for which randomized studies are not, are often also less expensive to conduct than randomized trials. Furthermore, observational studies when analyzed

---

[6] This section benefited greatly from inputs provided by contributing author Carol Joyce Blumberg and by Judy Hahn.

[7] http://www.medpagetoday.com/Medpage-Guide-to-Biostatistics.pdf provides a comprehensive summary, directed at workers in the field, of how research is classified in clinical studies and the applicable terminology.

appropriately—and their findings reported responsibly—can lead to useful (even if often not definitive) findings and open the door to important further investigation, sometimes using designed experiments or random sampling studies. To achieve success, it is again essential to pay careful and early attention to the data gathering process—even though what one can do in an observational study is more limited and passive than in a controlled study.

11.5.1 Limitations of Observational Studies

Evidence of statistical correlation in an observational study is NOT per se indicative of a cause and effect relationship.

The limitations of observational studies in establishing cause and effect relationships are illustrated by the examples in Sidebars 11.3 and 11.4. The first of these examples deals with a study resulting in an ambiguous finding. It did, however, succeed in narrowing the focus of subsequent investigations. The second example illustrates a situation in which the analysis of data from an observational study led to incorrect conclusions.

SIDEBAR *11.3*

PINPOINTING THE SOURCE OF COMPRESSOR FAILURES

Compressors for a new line of residential air conditioners experienced an excessive number of failures in the first year following the introduction of a new product line by its manufacturer. There was an urgent need to understand the root cause of the problem so that steps could be taken to mitigate its adverse effects and, ultimately, to eliminate the failure mode altogether. The available data revealed that nearly all failures occurred in one particular region of the country, which was characterized by high humidity and high air conditioner usage. Most of the units installed in this region were manufactured at the same plant (one of three plants that built the product). It was not clear from this observational study of the available data to what degree the source of the problem was heavy usage of the air conditioners in a high-humidity environment or the manufacturing plant at which the units were built. This question was not fully resolved until a randomly selected sample of units from each of the plants was subjected to an accelerated life test under various operating regimes and then closely examined. The resulting analyses indicated that the major culprit was the more severe operating conditions. This led to a redesign that made the product more robust to heavy usage in a high-humidity environment.

SIDEBAR *11.4*

THE IMPACT OF HORMONE RELACEMENT THERAPY

The Nurses Health Study was an observational study, started in 1976, with a questionnaire to 170,000 married registered nurses in 11 states of the United States; approximately 122,000 nurses responded. Follow-up questionnaires were subsequently mailed to these study participants every 2 years. This provided extensive information on incidence of diseases and health-related issues, including hormone use.

Grodstein et al. (1997), on the basis of analysis of data from 1976 to 1994 from this observational study, noted that "on the average, mortality among women who use post-menopausal hormones is lower than among nonusers." Hormone replacement therapy (HRT), in general, appeared to have a number of beneficial effects, including helping prevent heart attacks and strokes. This, at that time, led to a large number of women to turn to such therapy.

Subsequently, a statistically controlled study involving the random assignment of either hormone replacement therapy or a placebo to participants was conducted. This resulted in findings contrary to those of the earlier study with regard to the ability of hormone therapy to combat heart attacks, strokes, and other diseases.

The earlier incorrect conclusions were attributed to the fact that, unlike the subsequent randomized study, the comparison of results from the observational study was between women who *voluntarily chose* to go on hormone replacement therapy versus those who did not. It seems, in retrospect, that those women who chose the therapy might also have been healthier and led a better lifestyle than those who did not.[8] Irrespective of the specific reason, the inherent differences between the two groups of women, rather than their use of hormone therapy, led to what is now felt to be the initially incorrect conclusions in the interpretation of the results of this observational study.

11.5.2 Lurking Variables: A Key Culprit

The HRT example illustrates the important point that evidence of statistical association in an observational study should not be mistaken as evidence of a cause and effect relationship between the explanatory variables (X) and the response variables (Y). In fact, Y might be the cause of X, rather than the reverse. Or, frequently, one or more intermediate, or lurking, variable(s) (Z) that are themselves related to *both X* and *Y* are the true causal variables, but have been excluded from the analysis—often because their values have not been recorded and might not be known. Sometimes, Z might not even be suspected to impact Y. In that case, X serves merely as a surrogate for the actual causal variable(s) Z. In the HRT example, Y was the incidence of heart attacks and strokes, X was the use of HRT, and Z might be measurements of the individual's general health and lifestyle. Using common statistical terminology, the variables X and Z, in this situation, are said to be "confounded."

A simpler example is the association between a child's foot size (X) and reading ability (Y). In that case, the lurking variable (Z) is likely to be the child's age. Some further examples are provided in Sidebar 11.5.

---

[8] Another study (Herman et al, 2008), using methodology aimed at analyzing observational studies as if they had been conducted in the same manner as randomized experiments, attributed the discrepancy in conclusions between those obtained by Goldstein et al (1967) and those in the subsequent statistically controlled study to be due to differences between the two groups of women (i.e., the women who used hormone replacement therapy and the women who did not use such therapy) in the distribution of the time since menopause and in the distribution of the length of follow-up.

SIDEBAR *11.5*

CONFUSING CORRELATION WITH CAUSE AND EFFECT: SOME FURTHER EXAMPLES

Schield (2006) noted that, according to UN estimates for 2005–2010, the death rate in the United States (8.4 per 1000) was 80% higher than that in Mexico (4.7 per 1000). The United States fared even worse when compared with Ecuador (4.3 per 1000) and Saudi Arabia (2.7 per 1000). So what do these statistics mean? They should not necessarily be interpreted as establishing that the health care system in the United States is inferior. One important lurking variable in these comparisons is the age distributions within each country, with the United States having an appreciably older population. (This lurking variable might be addressed by making separate comparisons over different age groups, assuming the availability of such data; there might, however, be added lurking variables.)

Some further examples of (the many) media headline-grabbing stories that tend to suggest cause and effect relationships based on observational studies are

- Married people live longer.
- People who spend considerable time watching TV weigh more.
- Tall people earn more money.

See Joiner (1981) and Andrade (2007) for further discussion and additional examples and Weisberg (2010) for some added thoughts on bias and causation.

The cause and effect relationship that is frequently concluded or implied by such examples may, of course, be correct—and there may not be any important lurking variable. The problem is that often the potential lurking variables are not identified, and when they are, have not been measured.[9]

11.5.3 Combating Lurking Variables

As suggested in the example in Sidebar 11.5, the impact of lurking variables can sometimes be neutralized by statistical analysis. Thus, if *both* the cause variables (X) and the true lurking variables (Z) were included as explanatory variables for the response variable (Y) in, say, a statistical regression analysis, an unbiased estimate of the impact of X on Y is obtained. This requires identifying lurking variables, measuring them, and including them in the analysis as covariates. The last step—the statistical analysis—is usually the easiest to implement. The major challenge is generally in identifying and measuring lurking variables. And even if you feel that you have been able to do so, you can never be quite sure that there are not some added lurking variables that you may have missed.

---

[9] One might also suspect that some reporters who believe they have a hot story may be less than highly motivated to dig deeper and uncover what might be a less interesting explanation.

11.5.4 Some Other Limitations of Observational Studies

Limited Variability. Another problem that is sometimes encountered in observational studies, especially in manufacturing, is that there is limited variability (sometimes referred to as restriction of range) in the explanatory variables (i.e., the X's) in the data, thereby impeding the ability to establish meaningful relationships.

Say, for example, that you are using data from a manufacturing process to relate in-process variables (e.g., raw material properties, processing conditions, ambient temperature, and humidity) to end-of-line product performance. Specification limits may exist on each of these process variables that restrict their variability. This makes it difficult to establish statistically significant relationships, especially from small samples, and might lead some to incorrectly conclude that such variables are unimportant—even over a broader range—in impacting performance. In a somewhat similar vein, humans at one time believed that the earth was flat.

Inadequate Understanding. Since those who conduct observational studies are often not directly involved in the process that they are studying, they may not fully understand it. This can lead to faulty conclusions; see Sidebar 11.6.

SIDEBAR *11.6*

AN EXAMPLE FROM INDUSTRY

An analyst desired to quantify the relationship between the amount of catalyst added to a chemical reaction and the viscosity of the manufactured material. Theory suggested that increasing the amount of catalyst reduces the viscosity of the material. The analyst used available data on past manufactured batches to construct a scatter plot of the two variables. The plot (see Figure 11.1) suggested, surprisingly, that there was little association between the two factors.

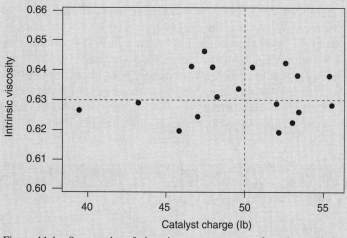

Figure 11.1   Scatter plot of viscosity versus amount of catalyst added.

Further discussion revealed that the process engineers had adjusted the amount of catalyst added to a manufacturing batch based on early readings from a sample taken from that batch, so as to achieve the target viscosity. These adjustments compensated for various other variables that affected viscosity (e.g., incoming raw material quality, reactor temperature, etc.) and succeeded in keeping the final viscosity relatively constant from batch to batch. This masked the true relationship between catalyst addition and viscosity. Only the final data (i.e., after adjustments had been made) on viscosity and catalyst addition were available; the initial measurements and subsequent adjustments were not recorded (and, therefore, could not be included in further statistical analysis).

Overall Quality of Data. The data that are often available from observational studies—at least without proper advance planning—are sometimes deficient in various other ways. Typical problems are inconsistent, inaccurate, and incomplete data recording and storage; see DeVeaux and Hand (2005). Of course, this can also happen in designed experiments, when those involved in the data collection and storage have not been well trained or fail to follow instructions.

11.5.5 Making the Best of Observational Studies

Observational Studies for Prediction. Much of the preceding discussion dealt with the sometimes questionable use of the results of observational studies to explain how some explanatory variables impact one or more response variables, that is, to establish "cause and effect."

There are other situations for which the major goal is to *predict* future performance, rather than to establish cause. The use of credit scoring models constructed from past data to determine to which future applicants to grant credit and how much credit to extend, using legally permissible predictor variables, provides an example. Another illustration is that of college admissions officers using observational data to relate students' high school grades and test results to college grades in order to predict performance at their institutions, and then using the resulting models to make admissions decisions on future applicants.

Models constructed from observational data can be effective in providing useful predictions as long as the structure of past relationships holds into the future, irrespective of whether or not the true causal variables have been included. This requires that the population or process for which predictions are to be made and the underlying environment remain unchanged from—or, in many practical situations, stay reasonably close to—those of the observational study; see Sidebar 11.7.

SIDEBAR 11.7

THE PROBLEM OF DIFFERING POPULATIONS

A frequently encountered problem in using data from observational studies to make future predictions is that the population or process concerning which one wishes to make predictions differs in some important ways from the one from which data are available. In the credit scoring example, the population of applicants for whom you are likely to have data—namely, those past applicants who were previously accepted—is, as already suggested, different from that of the future "walk in the door" applicants for which the model is to be used (see Hand, 2001).

Similarly, the past data on college grades available to the college admissions officer is limited to those students who were accepted by the college and chose to attend. This is clearly a different population from that of all future applicants for which the model was developed.[10]

Observational Studies as Research Catalysts. Well-conducted observational studies often *suggest* patterns or provide clues that—when followed up by more rigorous investigations—can lead to important findings. An early example is an investigation by Dr. John Snow that identified the Broad Street pump as the source of the cholera epidemic in London in the 1850s, resulting in action by the authorities to end the outbreak (Tufte, 1997, Chapter 2).

A later example is the research in the 1950s and 1960s (itself conducted, based upon some earlier limited study) that linked smoking to lung cancer; see Sidebar 11.8.

SIDEBAR *11.8*

SMOKING AND LUNG CANCER

Various observational studies during the middle of the past century suggested a strong relationship between cigarette smoking and lung cancer. These studies did not show definitively that smoking "caused" lung cancer. One could always argue that smokers engage in less healthy behavior and/or are exposed to more dangerous environments than nonsmokers and that these rather than smoking are the true "cause" factors. In fact, Sir Ronald Fisher argued that those with a genetic predisposition to lung cancer could also have a genetic predisposition to smoking (Salsburg, 2001).

The findings from these observational studies, however, spurred on scientists to conduct more definitive research, including both physical evaluations and animal studies.[11] Based on such research, the causal link between smoking and lung cancer (as well as various other diseases) is now widely accepted; see Freedman (1999) for further discussion and references.

Leveraging Observational Studies. Observational studies might not provide ideal data, but, as we have seen, they are often all we can get, especially early in an

---

[10] Similar problems can arise even in randomized experiments on humans aimed, for example, at evaluating the effectiveness of a proposed new drug. Those willing to participate in such studies and adhere to their regimens over time may be a select group that differs from the general population for which the drug is intended.

[11] It also led to additional observational studies to assess Fisher's argument further, including studies on monozygotic twins, only one of which smoked.

investigation, and can provide important clues. Also, statisticians have developed schemes for leveraging such studies to provide the best possible analyses. This has resulted in various approaches, such as case–control and cohort studies, that are used extensively especially in medical and social science applications. Detailed discussion of these methods is beyond the scope of this book. However, we provide a brief description in Sidebar 11.9.

SIDEBAR 11.9

CASE–CONTROL AND COHORT STUDIES

Case–Control Studies. In a case–control study, a group of individuals, the case group, with certain characteristics (e.g., diseased) is compared with a control group that does not possess such characteristics (e.g., not diseased) but is otherwise essentially equivalent in all other ways that can be measured. Then it is determined whether the exposure of interest, for example, some element of the environment, is more common in the case group than the control group. The impact of other explanatory variables that can be identified and measured, for example, age and gender, is presumably removed by comparing individuals from the diseased and nondiseased groups that are similar with regard to such factors.

Getting a control group that is "otherwise equivalent" to the case group can be difficult. It may, however, in some cases, such as the study of some rare diseases with long latency periods, be the best way to proceed.

Cohort Studies. In a cohort study, an initially "good" sample of individuals or objects is selected and tracked over time with regard to both its "deterioration" and characteristics that might be related to that deterioration. Thus, in a medical study, initially disease-free individuals are observed over a reasonably long period of time. Measurements are taken both on the degree to which such individuals develop the disease under study and on characteristics of the individuals and of the environments in which they operate. These characteristics are then related statistically to disease onset and/or severity. A disadvantage of cohort studies is the long time that they typically take to conduct and the associated expense.

Focus on Data Gathering. The essential key to conducting the preceding studies successfully, as well as for correcting for identified and measured lurking variables, is again early and careful planning of the data gathering process. We cannot adjust for a lurking variable that has not been measured. And we cannot compare groups that are equivalent in all respects except for the characteristic under study (as in a case–control study) if we have not taken the measurements that allow us to establish such equivalent groupings.

Much progress has been made, over recent decades, in gaining recognition of the importance of conducting investigations in a statistically sound manner. DOE has become a buzzword in business and industry; approval of a drug requires a statistically based comparative study; and statistically based public opinion polls are the norm.

In contrast, the need for statistically directed advance planning for observational studies has traditionally received less attention, with some notable exceptions, such as in much medical research. Flanagan-Hyde (2006) characterized observational studies as "the neglected stepchild in the family of data gathering." One reason for this is that, as already noted, obtaining data so as to subsequently conduct analyses that might provide insights is generally *not* considered as one of the prime purposes of the underlying process or operation.

In summary, in observational studies, as much as in randomized trials, up-front development of a sound data gathering process—keeping in mind the analyses that may be conducted on the resulting data—is imperative. We provide some specific hints for doing this in the next section.

11.6 HINTS FOR GETTING THE RIGHT DATA

Some have characterized the statistician's role as that of a "data detective." When the needed data are not there, the detective is operating without any clues.

This is especially the case for some observational studies. Often analysis of existing data is the consequence of some unanticipated, and frequently undesirable, event—such as inconsistent or poor product performance, an environmental issue, or an unanticipated epidemic. Because the underlying databases were often constructed with other goals in mind, such as to meet financial, legal, or management reporting requirements, they often fail to contain much of the information needed to address the specific questions of interest in the study.

In this section, we make some specific suggestions for improved data gathering. Our comments are especially applicable, but not limited, to observational studies in business and industry. Sidebar 11.10 indicates some guidelines for government agencies.

SIDEBAR *11.10*

U.S. GOVERNMENT AGENCY GUIDELINES FOR ENSURING HIGH-QUALITY INFORMATION

In 2002 (based upon a law passed the preceding year), the U.S. Office of Management and Budget (OMB) was asked to provide policy and procedural guidelines to federal agencies for ensuring and maximizing the quality, objectivity, utility, and integrity of information (including statistical information) disseminated by such agencies. At the same time, individual agencies were directed to issue their own guidelines for implementation. The procedures that were developed have been posted by OMB[12] and on the web sites of individual agencies, such as the Environmental Protection Agency and the National Institutes of Health.

---

[12] http://www.whitehouse.gov/omb/fedreg_final_information_quality_guidelines.

11.6.1 Strive to Obtain Information on Key Variables and Events

Obtaining information on key variables and events that may impact the process under study is essential to successful data analysis. Recording information on potential lurking variables in observational studies will allow these to be included in the analysis and improve the chances of establishing meaningful relationships, as in the examples that follow.

Determining the Root Causes of Field Failures. When unanticipated product failures occur in the field, the manufacturer wants both to address the immediate problem (by, for example, selective recall) and to gain a rapid understanding of the cause(s) of the problem to avert its future occurrence. As suggested in Section 2.2.4, this frequently calls for analyses of the available data. To maximize the likelihood that such analyses will succeed, provisions need to be made to routinely and consistently obtain and record data on such factors as

- Parts identification and manufacturing history.
- Processing conditions during manufacture.
- Product usage.
- Product performance measurements over time.
- Specifics of product failures.

The gathering of such information can be considered as insurance. You hope that you will not need to use it—but it is prudent to obtain it in case you do.

Developing a Credit Scoring Model for Consumer Loan Approval. Your ability to develop a useful credit scoring model for future consumer loan approval, based upon the performance of past loans and their characteristics, requires the gathering of information on variables that may be predictive of loan repayment performance. This calls for identifying potential predictive variables—such as past payment history, applicant family earnings, and other existing debts—and then developing and implementing a system for consistently recording such information from applicants.

Assessing the Impact of Processing Conditions on Product Performance. One often wants to use data from a manufacturing process to relate potentially important in-process variables or ambient conditions to end-of-line product performance. This requires recording information on such variables as humidity, operator, and raw material lot.

Many unplanned events typically take place during product manufacture. Machines may be upgraded, test methods are changed, voltage surges occur, labor strikes take place, and so on. Such events may impact the performance and quality of the resulting product. Their occurrence needs to be conscientiously recorded, together with relevant qualitative and quantitative information on them.

Note on Controlled Studies. In controlled studies, the attained values of the controlled variables, such as a pressure setting, often differ somewhat from those that were planned. The actual values should be recorded and used in the analysis. In addition, there may be important uncontrolled variables, such as temperature and other environmental conditions, which also impact the results but had not been included formally as variables in the study. Advance plans should be made for these to be measured and recorded so that they can also be included in the analyses.

11.6.2 Strive to Obtain Continuous Data

A fundamental and frequent issue is to obtain continuous, as opposed to attribute or go/no-go, measurements. Measurements on a continuous scale have greater informational value than their attribute counterparts; see Sidebar 11.11.

SIDEBAR *11.11*

FLAMMABILITY TEST STUDY

Plastic material used to make parts for automobiles, computers, and other products must pass a flammability test prior to acceptance. The standard test for this situation is to burn five specially produced bars from each manufactured lot. If the flame on at least four of these five bars extinguishes in less than 5 seconds, the manufactured lot is deemed to meet flammability requirements.

This test was adequate for a materials acceptance test for individual lots. However, it was severely limited in providing information for new product development, such as in evaluating the flammability characteristics of alternative materials and for comparing processing methods. The simple pass/fail results on five bars often were insufficient to detect small, but potentially important, differences. To obtain improved precision, it was initially suggested that the number of bars to be tested be increased to 30. This would have been quite expensive and would have resulted in an unreasonably increased load on the quality assurance lab.

It was decided instead to stay with the five samples and to conduct each test until burnout, or up to 20 seconds, whichever came first. The time to burnout for each sampled unit was then recorded and subsequent evaluations were based upon these observed times. The new measurements added only slightly to the cost and were much more informative in comparing the flammability of different materials than the previous binary data.

11.6.3 Strive to Avoid Systematically Unrecorded Observations

Unplanned data are often deficient due to systematically unrecorded observations. These arise, for example,

- When information is recorded on failed units (or on people with some disease) only and little is known about unfailed units (or about healthy people).

- Information is available on only those units of a product that are still under warranty.

- Observations felt to be outliers are excluded from the data, or changed, without recording that this has been done and the rationale for the exclusion or change.[13]

In some instances (e.g., information available only on units under warranty), what can be done to restore systematically unrecorded observations is limited by practical constraints. In many situations, early planning of the data to be gathered can avoid subsequent ambiguities.

11.6.4 Strive for Precise Measurements

An important part of planning to get good data is paying attention to the quality of the measurements that are obtained. Instruments used for measurement might have varying levels of precision, may be biased if not calibrated, and may drift over time. It is therefore important at the outset, to evaluate and, if needed, improve the precision and/or accuracy of the measurements and to establish procedures for periodic assessment and recalibration; see Hahn and Doganaksoy (2008, Section 3.4).

11.6.5 Strive for Consistent and Accurate Data Recording

Missing Values. In addition to unrecorded observations (Section 11.6.3), unplanned data may contain missing values (i.e., missing results on some variables for some observations). Some general guidelines for addressing these are

- Understand the reasons for a value (or group of values) being missing (and, if possible and appropriate, restore them to the database).
- Use this understanding to help minimize the future occurrence of missing values.
- Appropriately handle in the statistical analysis those missing values that still remain (Gelman and Hill, 2006, Chapter 25).

Missing values could be due to poor communication of what needs to be recorded or indifferent attitudes by those responsible for taking or recording the measurements or many other reasons. Such situations can likely be avoided with good planning and management support.

In other situations, the fact that values on some observations are missing might relate directly to their values. For example, a reading might be incorrectly omitted because it is outside the measurement range. Such situations call for clear recording of this fact and for treating the observation as censored in the subsequent analysis (See O-ring example in Section 4.5.2).

Missing values may also be due to some completely extraneous reason, such as a malfunction of the measuring equipment. This reason also needs to be recorded.

---

[13] If the outliers are due to recording errors, they most likely warrant exclusion or correction. If they are real, they might be providing important information that needs to be factored into the analysis of the data.

Inconsistent Data Recording. When good recording practices are not specified, data may be recorded haphazardly, or even arbitrarily. For example,

- Some people are more conscientious than others in taking measurements and recording information.

- Different individuals might interpret a checklist differently. For example, classification of the condition of a part on a scale from 1 to 5 might be interpreted differently by different operators, especially if there are no detailed instructions on how to classify a part.

- Different individuals might without further instructions use different recording conventions. A recorded date of 2/8 generally designates February 8 in the United States, but means August 2 in many other countries. Confusion between conventional U.S. and metric measurement systems has led to serious blunders.

Inconsistencies can be minimized by developing and documenting data recording protocols and by appropriate training. Instructions should be as specific and as user-friendly as possible. If, for example, the condition of a part is to be classified on a numerical scale (say from 1 to 5), it might be helpful to have photographs that illustrate specific parts that meet each of the values on the scale.

Avoiding Errors in Data Recording Some errors in data recording are clearly obvious (e.g., a date of 15/32 or a male pregnancy) and can often be corrected, especially if they are scrutinized in a timely manner. Other observations may be suspect (e.g., a temperature of 60°F in January in Schenectady, NY), but not obviously incorrect. In other situations, errors in recording the observation (e.g., a moved decimal point, a missing minus sign, and failure to denote that an observation is censored) might not be readily discernible.

Careful planning and documentation of the data recording process and periodic audits can help to minimize data recording errors.

Data Collection Instrument Considerations. There are various mechanisms for collecting data. These range from manual entry on paper forms to automated instrumentation, and include such vehicles as data-entry screens and touch-tone telephones. Modern data collection instruments can refuse to accept invalid data entries (e.g., ineligible dates) and might provide user prompts to promote entry of valid data.

The sooner that data-entry errors or inconsistencies are detected, the greater is the likelihood that they can be corrected. One of many arguments against manual recording on paper forms is that there is often significant elapsed time between when the data are recorded and when their consistency and credibility are evaluated (if evaluated at all).

Data Cleaning. Careful up-front planning should help to reduce the onerous and time-consuming task of data cleaning, but does not generally eliminate its need altogether. Data cleaning efforts are typically twofold: the detection of (possibly) erroneous recorded data and their (to the best of your ability) correction.

Some hints for effective data cleaning are provided in Sidebar 11.12.

SIDEBAR *11.12*

SOME HINTS FOR EFFECTIVE DATA CLEANING

- Check numbers for consistency and credibility; for example,
  - Are dates (say of various steps in the production of a product) in a logical sequence?
  - Do the individual recorded values appear reasonable?
  - Do the data, when taken together, make sense (e.g., do all the ingredients of a mix add up to 100%)?
- Check for missing information or codes such as 0 or 999 that might be used to designate missing information (and ensure that such codes are not taken literally in the analysis).
- Plot the data (e.g., in histograms and scatter plots) to identify outliers that might suggest recording errors (such as incorrect decimal places).
- Query the data recorders to obtain explanations for, possibly, incorrect values; correct identified errors.
- Perform the preceding checks as soon as possible after the data have been recorded to minimize the impact of memory loss (or data dumping).
- Automate as much of the preceding as possible.

In summary,

- Data cleaning is a reactive process. The best strategy is to avoid or minimize the need for it by establishing, documenting, and providing training early on of complete, consistent, and easy-to-follow data gathering procedures.
- A disciplined, thorough, and timely data cleaning effort can help to detect and correct remaining errors.
- Dubious or missing data may still remain—despite your best efforts. Some of these may be addressed in the statistical analysis (e.g., methods for handling missing data). We also suggest that you conduct analyses under various, possibly extreme, assumptions with regard to such dubious or missing data—and assess the sensitivity of your conclusions to such assumptions.

11.6.6 Resist Purging of Potentially Useful Data

Data are sometimes purged when no longer felt to be useful for their original purpose. Information on credit payment history may be removed when applicants close their accounts. Some data are routinely removed from the database after a specified elapsed time. Such data purging might be a holdover from earlier days when data storage capabilities were more limited than they are now.

Care needs to be exercised to retain data that, though no longer of interest with regard to their original purpose, might still be useful in future analyses.

11.6.7 Strive for Compatibility and Integration of Databases

In many studies, the relevant data may be spread over different data sets or databases with different formats that do not readily "talk" to each other. In developing a credit scoring model payment history, demographic data, and information on past credit performance may all reside in different databases. Traceability may be impeded by people who have common names; who have changed names or sometimes use nicknames; who on some occasions use their middle names, at other times use only a middle initial, and at still other times exclude their middle name altogether; or who have changed addresses.

Government statisticians dealing with official statistics need to be especially facile in understanding and integrating various types of data—including primary data (such as from several surveys and geospatial data) and secondary data (such as administrative records and synthetic or model-based data) in order to provide policy makers sound and real-time knowledge to make informed and/or evidence-based decisions.

Traceability is also a key concern in medical studies that involve tracking patients whose records are likely to be stored in various facilities (and where integration of such records may raise confidentiality issues).

All of this makes it important to ensure early on that the data sets that are to be used in a study are as compatible as possible. Whenever feasible, the best solution is generally to have a single common database. When this is not possible or practical, provisions need to be made to help ensure compatibility of different databases.

11.6.8 The Challenge

Throughout this chapter, we urge early and detailed planning of the data gathering process. This is often easier said than done for a variety of reasons:

- **Added Cost and Delayed Payoff**: Obtaining good data almost always costs money. Moreover, the benefits for getting the right data may be realized only in the distant future (and may be difficult to quantify), while the costs are incurred now (Section 9.3.4).

- **Added Bureaucracy**: The data gathering process adds bureaucracy and can slow down operations. If repair staff have to conduct inspections beyond those needed for an immediate repair, to document their work, and to send back failed parts for engineering assessment, they have less time to do repairs.

- **Diversity of Ownership**: In many situations, the data analyses are sponsored and undertaken by parties different from those in charge of the data gathering process. The needed communication between all involved—that is required to ensure that data on the correct variables are collected using proper data collection methods—does not always take place.

You need to be keenly aware of such obstacles and work to overcome them. This requires you to be sensitive to the perspectives of plant managers or managers of service departments whose priorities may be different from yours and to understand their possible misgivings. You must carefully think through what you request, consider how it may impact and possibly interfere with operations, ensure that it is practical and doable, and be able and ready to justify it.

Keep in mind that the quality of data is generally more important than the quantity. A relatively small but carefully selected and well-documented sample may often meet your needs—recalling the comment in Section 7.7.6 that the actual, rather than the relative (to the population), size of a random sample is what counts in determining statistical precision.[14]

11.7   A PROCESS FOR DATA GATHERING

We now propose a five-step process for getting the right data. We then illustrate these steps by applying them to the assessment of the reliability of a newly designed washing machine, an example of a systems development study. Details will vary for different types of applications, such as observational studies. We feel, however, that the procedure that we propose, although it will require some modification to fit any given situation, has much general applicability.

Steiner and MacKay (2005) present a somewhat similar process, which they characterize as QPDAC (question, plan, data, analysis, and conclusion).

11.7.1   The Proposed Five-Step Process

Step 1: Define the Problem.   The first step calls for seeking a clear definition of the purpose of the study. This includes your determining

- The specific questions to be addressed by the study, by when, and with what precision.
- The type of actions that will be undertaken as a consequence of the findings.
- The specific population(s) or process(es) about which it is desired to draw conclusions.
- The data that you would gather if you could readily obtain whatever data you wanted (and how you would use such data)?[15]

Step 2: Evaluate the Existing Data.   The second step involves your getting a good understanding of what is already known and what is not known. It includes

- Gaining an appreciation of the process or operation under study and its intricacies.

---

[14] There are some exceptions. One example arises in assessing a highly unusual, but very important, event, such as the occurrence of a rare disease or an infrequent, but costly, product failure. In this case, you might want to strive to use the entire population to estimate the occurrence probability, get detailed information on each of the diseased or failures, but obtain such information on only a random sample of the healthy or unfailed members of the population.

[15] We thank our former colleague, Wayne Nelson, for suggesting this question.

- Understanding and possibly analyzing the existing data and reporting the findings.
- Identifying gaps and inadequacies in the existing data in addressing the needs of the study.

Sometimes this step reveals that the existing data allow one to respond satisfactorily to the question(s) at hand. Often the available data are found to be insufficient. Gaining an understanding of the inadequacies of the existing data may also provide an initial appreciation of the opportunities and limitations for future data gathering.

Step 3: Understand the Data Gathering Opportunities and Limitations. The next step is to obtain a detailed understanding of the opportunities and methods for acquiring needed added data and the associated costs and limitations. This includes your determining

- The nature of the data that can be gathered and how such data would be obtained.
- The errors in measurement associated with obtaining future data. If the measurement error appears to be unacceptable, work is required to reduce it.
- The practical considerations and limitations in gathering new data, such as budget and timing restrictions and facility limitations.
- The "representativeness" of the data that can be gathered and the limitations on the more general inferences that might be drawn from such data. Identification of these limitations can help in their mitigation and in gaining a realistic understanding of what can and cannot be determined from further data gathering.

Step 4: Plan the Data Gathering and Analysis. You are now ready to prepare a specific plan for data gathering. This involves working with the project team to

- Develop the details of the proposed data gathering process. The nature of these varies depending upon the type of study. In some situations, such as the washing machine example, it includes specifying the test conditions and operational environments for the study as well as the sample size and sample selection method. In other cases, it might call for the design of an experiment or of a random sampling study or the specifics of gathering data from an existing database. For some studies, it also includes specifying the measurements to be taken, as well as the data collection instruments and the data recording process to be used.
- Outline at least the general nature of the analyses to be performed on the resulting data. In some cases, such as many clinical trials, the details of such analyses need to be provided.
- Assess the statistical properties of the proposed plan to determine whether the inferences that are to be drawn from the results will have a sufficient degree

of statistical precision to meet the goals of the study. Revise the plan if necessary.

- Establish any further required specifics of the testing protocol and the operational details.

- Develop a pilot study (i.e., a miniature version of the proposed study) to identify and remove possible bugs in the plan.

The proposed data gathering process is presented to and evaluated by the project leadership and by management and may be amended based upon their assessments.

Step 5: Monitor, Clean Data, Analyze, and Validate. Depending upon the situation, you may be directly engaged in the study implementation and data gathering. As a minimum, you need to be sufficiently on top of things to ensure that the data gathering plan is meeting its goals and that unforeseen issues are addressed speedily. This will require you to

- Monitor the implementation of the study to ensure that the specified data gathering process is understood and being followed, to identify any unanticipated problems, and to speedily amend the plan if this appears necessary.

- Ensure needed data cleaning. Most of the effort up to this point has been to get the best possible data so as to minimize the required data cleaning. The degree to which this is feasible depends upon the nature of the study. Significant data cleaning is still likely to be required, especially for observational studies.

- Conduct a preliminary analysis and possibly subsequent interim analyses of the data and speedily feed back the results to all concerned. Amend the data gathering process if this is suggested by the initial analyses; this is especially important in clinical studies in which the health of humans may be at stake.

- Perform the "final" data analyses and report the findings. We will not elaborate further here—this is what most books and courses in statistics are about.

- Propose added data gathering for further validation or extension of the findings and/or to monitor future performance (or to launch the next phase of the study).

11.7.2 Application to Washing Machine Example

A manufacturer has designed a new washing machine. It provides customers with useful new features, such as devices for sensing and adjusting for different types of fabrics and for varying degrees of soiling to give a better wash. Various design changes were made to improve reliability by addressing some previously discovered failure modes. A particular concern is to ensure that these improvements and other changes, or interactions between them and existing features, do not introduce new failure modes.

High reliability (i.e., failure-free operation over the life of the product) is an important customer concern and a key criterion in deciding which brand to buy. Many potential buyers consult consumer magazines[16] to obtain information on quality, reliability, and other features of a product versus those of competitors.

The manufacturer wishes to validate that the new design has an acceptable level of reliability (e.g., at least 80%) over a 10-year lifetime. To meet the production schedule, the desired validation needs to be completed within an elapsed time of 6 months (plus a short time for pilot testing).

Appropriate data need to be gathered to assess the reliability of the new design and to ensure that it meets requirements (and if not, to act accordingly). In addition, it is desired to establish an ongoing process to obtain timely and relevant updated information about the reliability of the machine immediately after manufacture and during field use.

Step 1: Define the Problem. The manufacturer wishes to show with 95% confidence and within an elapsed testing time of 6 months (preceded possibly by a 1-month pilot study) that the new machine meets or exceeds the following reliability goals:

- 95% reliability during the first year of operation.
- 90% reliability during the first 5 years of operation.
- 80% reliability during the first 10 years of operation.

Reliability is defined as the washing machine requiring no repair or unusual servicing over the specified time period.

If the reliability goals appear to be met, full-scale production will proceed. If the test reveals any problems, the results should provide information about such problems to help design engineers correct these in a speedy manner.

The population of interest is the approximately 6 million washing machines to be built over the next 5 years. Thus, the ideal (and obviously unattainable at this time) information would be knowledge of the 10-year reliability performance for each of the 6 million future units—and information on the nature and root causes of any failures that might take place.

Step 2: Evaluate the Existing Data. To assess the reliability of the new design, an understanding of the changes from the previous design is needed. Engineering analyses and design reviews were conducted to assess the degree to which the design changes addressed past reliability problems and their potential for introducing new failure modes. The statistician participated in the reviews.

Three potentially useful sources of existing data were identified:

- Results of in-house testing on the previous design. The component and subassembly testing on the previous design was similar to that to be conducted on the new design, except that improved physical evaluation methods and

---

[16] Such magazines typically base their evaluations on information obtained from a sample of past purchasers of the product and/or testing a typically small number of units of the product. In either case, the results are likely to come from units from early production, providing the manufacturer further motivation to ensure high quality and reliability from the start.

instrumentation are now available. Systems testing on assembled previous design washing machines had also been performed.

- Field data on previous designs. The data identified some failure modes that had not been caught by the earlier in-house testing and allowed quantification of the field reliability of the previous design, subject to the limitations of the data. It also permitted an assessment of how the field reliability of the previous design differed from that estimated from in-house test data at the time of product launch.
- Results from some initial component and subassembly tests for the new design.

Analysis of the available data led to the following conclusions:

- The earlier design failed to meet current reliability goals. The need to correct previous reliability problems was one of the reasons for the redesign.
- The component and subassembly tests, taken together with the failure modes and effects analysis (FMEA), indicated that the new design should eliminate or mitigate the reliability problems encountered in the previous design. One newly introduced failure mode was discovered during early testing and was corrected by some further redesign.

The major missing needed information is on the performance over time of the fully assembled (new design) washing machine in the use environments.

Step 3: Understand the Data Gathering Opportunities and Limitations. Information on system reliability can be obtained through testing assembled washing machines using in-house testing facilities. Such testing involves use-rate acceleration, that is, using the product (much) more frequently than in normal operation; see Doganaksoy et al. (2007). By running washing machines continuously (with only short breaks between usages to allow the machines to cool down), it is possible to simulate 3.5 years of product life in each month of testing, and 21 years of field operation in 6 months, assuming, conservatively, four washes per week in normal use.[17]

It was also determined that, with good instructions and a precise definition of what constitutes a failure, time to failure can be measured precisely. In addition, a measure of the degree of product degradation (and therefore an estimate of how much longer an unfailed unit would be expected to run before failing) could be obtained from disassembling various critical parts of unfailed washing machines and measuring the wear on them.

Thirty-six stands are available to conduct such testing. Three prototype production lots, involving about 30 machines each, had been manufactured and were available for testing. A fourth prototype lot was to be built in about 3 months. Budgetary provisions had been made for conducting such tests over a 6-month period (plus a short time for pilot testing).

The statistical analyses likely to be conducted on the system test results assume that

---

[17] A more sophisticated evaluation would take the probability distribution of product usage into consideration in assessing actual product usage and evaluating product reliability.

- Units from the four prototype lots are from the same "population" (at least with regard to their reliability) as those from high-volume production over the forthcoming 5-year period.
- Time to failure is determined by operating cycles of product usage, and not by elapsed time.[18]
- A "harsh" field operating environment can be realistically simulated by in-house testing.

All concerned were apprised of these assumptions and the importance of making the study as broad as possible. (This resulted in increasing the number of prototype lots to four, instead of the one lot that had originally been suggested.)

Step 4: Plan the Data Gathering and Analysis. Based upon the previously gathered background information and discussions, the following data gathering plan and analysis was proposed to the project leader and (after some amendments) to management:

- A test program, calling for running washing machines with a full load of soiled towels, mixed with sand, wrapped in a bag to accentuate moisture, is to be performed. This is the "harsh" environment referred to previously. In particular,
  - Select 12 units at random from each of the three initial prototype lots and place on test.
  - After 3 months, remove from the test four randomly selected unfailed units from each of the three lots, and disassemble these to obtain degradation measurements. Place 12 randomly selected units from the fourth prototype lot on test to replace the removed units.
  - Continue testing for 6 months from the start of the test program.
  - Replace each failed unit with a randomly selected unit from the prototype lot from which the failed unit came.
  - Obtain parts degradation measurements on all units removed from test.
- The following major analyses are to be performed on the results of the study:
  - **Statistical Evaluation**: Estimate 10-year reliability and associated lower 95% confidence bound using standard statistical procedures for such data; see Meeker and Escobar (1998a). Conduct various more detailed analyses, as appropriate.
  - **Physical Evaluation**: Conduct physical evaluations of all failed units to understand the root causes of failure and act thereon.[19]
- Assessment of statistical properties of the proposed plan. This involves an evaluation (using simulation or advanced statistical methods) of the statistical

---

[18] This would be an incorrect assumption if, for example, corrosion were a significant failure mode.

[19] It is also frequently useful to conduct such evaluations on a sample of unfailed units. Such evaluations might show one or more impending failure modes. Also, they might provide a useful comparison with failed units, giving insights into what has been done right in the design and manufacture of the product.

uncertainty that is anticipated from an analysis of the resulting data and an assessment of whether the proposed plan provides the desired level of statistical precision.

- Development of test protocol and operational details. These include
  - A definition of product failure.
  - A plan for the weekly evaluations of the state (i.e., failed or unfailed) of all test units.
  - A data recording process geared to entering data rapidly in a computer file established for this purpose.
  - A process for operators to obtain and record degradation measurements.
  - A procedure for immediately notifying the responsible engineer of all failures and the details thereof.
- Pilot study. An initial run of three washing machines (one from each of the three initial production lots) is to be conducted over a period of 1 month to identify possible bugs in the proposed study and to permit these to be addressed.

Step 5: Monitor, Clean Data, Analyze, and Validate. The statistician is to review the implementation of the test initially and monitor it periodically thereafter, paying particular attention to the quality of the data being gathered. Special consideration is to be given to any missing or questionable data. Particular attention is to be given to ensuring that the failure times on failed units and the running times on still functioning units are recorded accurately.

Preliminary analyses are to be conducted 1 week, 1 month, and 3 months after the start of the testing and the findings reported to all involved. A final analysis is to be conducted and reported after 6 months.

In light of the previously identified limitations of this study, special attention is to be given to validating the results. Specifically,

- A random sample of 6 of the 36 test units is to be continued on test for up to an additional year to identify long-life failure modes and to gain improved reliability estimates (especially for longer lifetimes).
- An additional 100 washing machines randomly selected from prototype production are to be given to selected company employees who were expected to be heavy users, under the condition that they record usage, that periodic inspections be allowed for up to 5 years, and that all malfunctions and observations on performance be reported immediately. Ten of these machines are to be randomly selected to be returned for teardown and further evaluation after 3 months. Another 60 randomly selected washing machines are to be installed in laundromats and similarly monitored.
- Six more washing machines are to be randomly selected each week from production. Five of these are to be life tested for 1 week under the previous testing regimen; the remaining unit is to be tested for 3 months. The results are to be compared with those from earlier tested units both to assess whether the production units maintain the previously established reliability performance and to provide early warning when this is not the case.

- A process was developed and put in place for rapidly capturing, analyzing, and feeding back to responsible engineers and management information on all failures.

11.8   MAJOR TAKEAWAYS

- Any statistical analysis can be only as good as the data upon which it is based. Careful advance planning of the data gathering is often a statistician's most important (and often insufficiently recognized) contribution. The earlier the statistician becomes involved in the study and in planning the data gathering, the better.

- A designed experiment is a planned study in which one employs randomization methods to systematically assign various treatments or conditions to a group of objects or participants so as to evaluate the impact of potential explanatory variables on one or more measured response variables. Randomized controlled trials are the gold standard in evaluating a new drug or medical device.

- Random sampling (and census) studies are typically conducted to characterize a defined population with respect to one or more variables of interest. Public opinion polls are well-known sampling studies that generally use some form of random sampling.

- Systems development (or action-oriented) studies, such as in reliability testing, typically lead to the development or validation of a system and generally result in a specific action or decision that relies heavily upon analysis of the gathered data. Such studies frequently involve at least some degree of control over the data acquisition process.

- It is often impossible, impractical, or even unethical to conduct a randomized study and one is limited to observational studies on existing or forthcoming data. Such studies cannot provide definitive cause and effect conclusions due to the existence of often unmeasured, and possibly unidentified, lurking variables, which, in fact, might be the true explanatory variables. When such lurking variables are believed to be known and measured, they can and generally should be included in the analysis.

- Observational studies, when analyzed appropriately—and their results reported responsibly—can lead to useful findings and open the door to important further investigation. Case–control and cohort studies are two approaches, used extensively in medical and social science applications, for leveraging observational studies to provide the best possible analyses.

- Observational studies can also be effective in predicting future outcomes (under comparable conditions).

- We provide the following hints for getting the right data:
  - Strive to obtain information on key variables and events.
  - Strive to obtain continuous data.

- Strive to avoid systematically unrecorded observations.
- Strive for precise measurements.
- Strive for consistent and accurate data recording.
- Resist purging of potentially useful data.
- Strive for compatibility and integration of databases.
- We propose the following five-step process for data gathering:

  - Define the problem.
  - Evaluate the existing data.
  - Understand the data gathering opportunities and limitations.
  - Plan the data gathering and analysis.
  - Monitor, clean data, analyze, and validate.

DISCUSSION QUESTIONS

(* indicates question does *not* require any past statistical training)

1. *In discussing questionnaire construction, we asserted that the phrasing of questions can impact the results. Provide an example of how alternative phrasings of a question (e.g., concerning the imposition of a new tax) can impact the findings. Then suggest (seemingly) neutral phrasing.

2. Presidential elections in the United States are based upon an electoral college of 538 delegates selected by a statewide winner-take-all system in most of the 50 states and the District of Columbia. As a consequence, the (winning) candidate with the highest number of electoral college votes will not necessarily be the one with the most total actual votes (as was the case in the 2000 election). Most public opinion polls typically report the estimated percentage of the total actual vote, rather than the estimated electoral college vote, for each candidate. Propose a process for using the results of a national public opinion poll to project the electoral college winner (including an estimate of statistical uncertainty). Why do you think pollsters often do not provide this kind of estimate?

3. *Provide one or more examples of situations in which the media descriptions of the results of an observational study have suggested cause and effect relationships. What are some other possible explanations for the observed relationships in these studies?

4. We state that "models constructed from observational data can be effective in providing useful predictions as long as the structure of past relationships holds into the future, irrespective of whether or not the true causal variables have been included." Explain and justify this statement.

5. Comment on some of the things that you might suggest in both the data gathering and the data analysis in the hormone replacement therapy study or

some other specific example to enhance the ability to uncover cause and effect relationships.

6. In the flammability evaluation described in Sidebar 11.11, it was decided to stay with five samples but conduct each test until burnout or 20 seconds, whichever came first. How would you estimate the average and standard deviation of the distribution for time to burnout from the resulting data?

7. The manufacturer of a razor blade wishes to conduct a study to determine the effective life (in number of shaves) of a disposable razor. The findings are to be used to make an advertising claim that the razor is "good for x shaves" so as to withstand possible attack by a regulatory agency or by a competitor. Develop a detailed plan for conducting such a study, perhaps using the five-step process described in Section 11.7.

8. *Assume that you are working for a telephone call-in center that takes customer orders. You have been asked to evaluate whether or not the center should install a system that advises callers who cannot be serviced immediately as to how long they will have to wait for service and provides them the option of an automatic callback. The major criterion is maximizing revenue. Propose an investigation to provide data to make this assessment, perhaps using the five-step process described in Section 11.7.

9. Identify a problem of personal interest to you that will require data gathering. (For some typical examples, see Hunter, 1977). Develop a plan to obtain data that will be responsive to your problem. If feasible, conduct the study, analyze the results, perhaps graphically, and present these for discussion and critique.

10. In Step 3 in the washing machine example in Section 11.7.2, we state in a footnote "A more sophisticated evaluation would take the probability distribution for product usage into consideration in assessing actual product usage and evaluating product reliability." How would you do this?

11. In Step 4 in the washing machine example in Section. 11.7.2, we state that the assessment of the statistical properties of the proposed plan "involves an evaluation (using simulation or advanced statistical methods) of the statistical uncertainty that is anticipated from an analysis of the resulting data and an assessment of whether the proposed plan provides the desired level of statistical precision." How would you do this?

CHAPTER *12*

CAREER PATHS

12.1 ABOUT THIS CHAPTER

Are we there yet?

—Sally Morton (2009b)

In this chapter, we elaborate on career paths for applied statisticians. These paths are generally less formally defined than those for statisticians in academia that, in many cases, are heavily dependent upon the tenure system (see Chapter 13).

In the first two sections, we describe roles for those who choose to remain applied statisticians (outside of academia) and for those who decide to move into academia, an application area, or management. We then describe typical career paths for applied statisticians and comment further on statistical leadership and statisticians' contributions beyond the workplace. We illustrate with the career paths of three statisticians.

12.2 SOME ROLES FOR APPLIED STATISTICIANS

We differentiate among the following roles or steps in the career ladder that statisticians may take during the course of their careers within the field: statistical analyst, applied statistician, senior statistician, statistical manager, and private statistical consultant. This breakdown, however, is somewhat arbitrary, and there is often no clear delineation as individuals move from one role to the other.

12.2.1 Statistical Analyst

Many trained in statistics, and especially those with a bachelor's, or possibly master's, degree, begin their careers as statistical analysts. This principally entry-level (or foot-in-the-door) position typically involves conducting data analyses, often under the supervision of an applied or senior statistician—or a knowledgeable practitioner—who might also serve as a mentor and role model. Over time, many statistical analysts strive to move from principally "backroom" positions to taking on increasingly more responsibility, conducting more advanced technical tasks, and working more independently.

A Career in Statistics: Beyond the Numbers, Gerald J. Hahn and Necip Doganaksoy.
© 2011 John Wiley & Sons, Inc. Published 2011 by John Wiley & Sons, Inc.

12.2.2 Applied Statistician

Applied statisticians bear prime responsibility for everything from helping ensure that the right data are collected to analyzing the data (or directing such analyses) and reporting the findings, interacting closely with other technical staff. Ideally, the applied statistician is an integral member of the project team (Section 5.4.2).

12.2.3 Senior Statistician

Senior statisticians, in addition to taking on all or most of the roles of the applied statistician, also assume broader responsibilities. They look at problems holistically and strive to relate them to the general goals of the organization. Senior statisticians also play a proactive role in proposing new projects and ventures that will benefit their organizations or clients in the future.

Senior statisticians are especially involved in the early and final stages of a project. Initially, they help to define a problem quantitatively and present the path forward to senior management. At the conclusion of the work, they participate in putting together and presenting the findings. Senior statisticians, thus, provide statistical leadership—a topic that we will return to in Section 12.5. They are sometimes considered by their colleagues and clients as the ultimate source of knowledge and wisdom in all statistical matters.

12.2.4 Statistical Manager

In organizations that have statistical groups, a possible career path for some is to aspire to leadership of such a group. Managers of statistical groups are typically involved in securing work projects—especially for the more junior members of the group who have, as yet, not built up their own clientele—and helping define such projects. They then frequently select one or more members of the group to do the work, provide guidance, as needed, and take on overall responsibility for successful implementation.

The manager of the statistics group also has various administrative responsibilities such as recruitment, development, and evaluation of staff, and ensuring that the group has state-of-the-art computational and other needed resources. Group managers also periodically apprise management of the group's technical accomplishments, help foster the interests and aspirations of the group, and develop a vision for the future.

The number of such positions is limited.

12.2.5 Private Statistical Consultant

Many successful applied statisticians, possibly nearing the end of their formal careers, retire or just decide to leave their current employer and go into business on their own as private statistical consultants. This carries with it the obvious advantages of not needing to satisfy the possible whims and timetables of an official manager (your customers now become your sole managers) and provides more opportunity to work on those projects that interest you most.

Being a private statistical consultant also carries with it added uncertainty—an important factor if you are relying on such work as a major source of income. You may be the first to be cut in bad times. But you might also be the first to be back to work in times of economic recovery since often the internal statistical skills of organizations have been depleted as a consequence of a downturn.

Your employment might be on a retainer, calling for you to overview an organization's work periodically. Or it might be on a project-by-project basis. As an outsider, you might be asked to provide a broad assessment of a problem and have appreciable liberty to think proactively, ask probing questions, and voice opinions. Also, as an external expert, your assessment may be given greater credence.

At other times, you might be asked to respond to narrow technical questions. As a consultant, you are unlikely to be a member of the team and might have limited opportunity to fully understand the business environment. And, although your advice might be given more weight initially, it is also more likely to be forgotten, due to lack of reinforcement, once you leave the premises.

As a consultant, you may also be called upon to review the work of practitioners or of other statisticians. A frequent use of statistical consultants is in lawsuits (or potential lawsuits), possibly as expert witnesses[1] (Section 4.6.4).

See Cabrera and McDougall (2002) and Goldstein (2008) for some further comments on the life of a statistical consultant.

12.2.6 Overview of Roles

We show in Table 12.1 some typical tasks and interface opportunities for statisticians in various roles.

12.3 OTHER MAJOR ROLES TO WHICH APPLIED STATISTICIANS MIGHT ASPIRE

In this section, we briefly describe three further roles that, during the course of your career, you might pursue and that may take you beyond being an applied statistician.

12.3.1 Moving into Academia

Some statisticians, after a stint in an application area, take a position in academia. They might do so early in their careers after getting a taste of life in business, industry, and/or government. Others make this move later, possibly while also working as statistical consultants. Academia might recognize extensive experience, and especially research accomplishments, by offering a higher entry-level appointment and in some cases even tenure.

Going into academia may be appealing in light of its somewhat less stressful pace (see Chapter 13). Learning about "real-life" experience—directly from

---

[1] Organizations like to hire external consultants as expert witnesses because of the unpredictable timing in which lawsuits arise and the supposed greater objectivity of an "outside voice," even though this voice is being paid by an interested party.

TABLE 12.1 Typical Tasks and Interface Opportunities for Statisticians in Business, Industry, and Government

Role	Typical tasks	Interface opportunities
Statistical analyst	Combines data from multiple databases (e.g., raw material properties, processing conditions, and final product measurements) and cleans data; conducts exploratory data analyses, or analyses of results of designed experiment, sampling study, or observational study, typically with guidance from applied or senior statistician or practitioner	Often has limited contact, initially, with others than task assigners. Gradually interacts directly with technical teams, participates in presentations, and becomes more broadly involved
Applied statistician	Develops statistical plan for life test of a new product, follows program during its implementation, conducts (or directs) analysis of resulting data, and reports findings. Develops new or revised questions on surveys or works on redesign of questionnaires typically as member of a multidisciplinary team	Works closely with other team members and builds ever-increasing credibility with colleagues and management
Senior statistician	Helps define a new measurement system (e.g., credit scoring system for a new market segment), develops and oversees data gathering process, specifies and supervises algorithm development, helps to achieve successful implementation, and provides periodic status reports to management. In government, designs major data products (e.g., characterization of high schools in the United States) and oversees evaluations of key aspects of official statistical systems (e.g., the quality of the decennial census)	Has close contact with technical staff and management. Responsible for specific deliverables. Looked to as a generator of ideas to move the organization forward and as a guide in project implementation
Statistical manager	Negotiates group's role in design and analysis of projects (e.g., test program for product reliability assessment) and designates who will do the work. In government, oversees all aspects of major official statistical products (e.g., the American Community Survey or a toxic waste monitoring system study)	Relates to management at all levels. Helps develop and sell technical vision, and understands both business and technical aspects of problems
Private statistical consultant	Develops test plan or sample study or reviews plan developed by others, and/or guides analysis of the resulting data	May have access to top management

somebody who has lived it—may also be attractive to the institution and its students, making it a "win–win" situation for all.

A possible first step might be taking on an adjunct professorship (Section 13.3.2). This may involve teaching a course, often in the evening, at a local college or university. It can provide you a taste of academia while you continue your day job and an opportunity to determine whether you would eventually like to move into academia.

12.3.2   Moving into an Application Area

Some statisticians, and especially those with past training in an application area, may decide to move into that area. Likely positions include jobs as quality assurance or reliability engineers, epidemiologists, economists, market research analysts, and risk analysts. The switch might be a gradual one or it might involve a formal job change. It is often an outgrowth of your extensive involvement in one or more projects in a particular area, as well as your past training. For example, following up your work in developing and implementing a credit scoring system and leveraging your past training in business administration, you might be asked to oversee the system's continued growth and further evolution. Your major tasks may become more and more applications oriented and less focused on the statistical aspects. This evolution may eventually be reflected by a change in your formal job title. You might continue to be regarded as a statistical expert and on occasion called upon in that capacity.

12.3.3   Moving into Management

Still others use their positions as statisticians as a springboard for a management position. Their new roles may not call for them to do much statistics, although their ability to deal with numbers and to explain empirical facts in context are highly useful skills for an administrator and leader.

For higher level management and executive jobs, the role of statistics is mostly about applying statistical thinking at a conceptual level rather than in the application of statistical methods per se. This includes an appreciation of the importance of effective data gathering and of the ever presence of variability. One consequence of statistical training is the ability not to overreact to the most recent numbers, such as last week's sales, and instead to look at the bigger picture.

In Section 1.10.3, we briefly described the career of Jim Goodnight, a statistician who became a successful entrepreneur. In Sidebar 12.1, we summarize the career of another statistician (Lawrence Garfinkel) who distinguished himself by first applying his statistical skills in an application area (epidemiology), and then moving into that area and ending his career as a research administrator. A further example, this one from academia (Albert Bowker), is provided in the next chapter.

SIDEBAR 12.1

THE CAREER OF LAWRENCE GARFINKEL[2]

Lawrence Garfinkel received a bachelor's degree in statistics from the City College of New York and a master's from Columbia. He started his career as a "statistical clerk" with the

---

[2] Much of this description is taken directly from an obituary, published in the *New York Times* on January 27, 2010.

American Cancer Society on a "temporary job" that lasted for 43 years. He advanced to the role of biostatistician and, together with colleagues, was instrumental in designing a massive study that demonstrated the correlation between smoking and lung cancer.

Garfinkel subsequently collaborated in an examination of tissues from deceased smokers that showed that the degree of precancerous change was related to how much and how long an individual had smoked. This supplemented the earlier statistical evidence by providing physical evidence of what smoke could do to the lungs and gained Garfinkel recognition as an epidemiologist.

His work eventually led Garfinkel to roles in management and he was promoted to become the society's research director, a position he held for the last 10 years of his career.

12.4 SOME CAREER PATHS

Figure 12.1 shows some likely entry-level career opportunities in statistics for graduates with degrees in statistics at various levels.

Figure 12.2 shows some typical paths of progression among the roles described in the previous two sections. The complex nature of this chart demonstrates the wide variety of paths statistically trained individuals can travel.

As suggested in Section 8.2, the path that any particular individual takes may be the consequence of a deliberate plan, based on carefully laid out aspirations. At other times, it may be principally the result of circumstances or even happenstance. Frequently, it is a combination of all of these. Making a transition may call for additional training, for example, to move from statistical analyst to application area analyst.

Often the transition from one role to the next is gradual rather than sudden. It may reflect growth within a particular job, rather than a change of jobs. When a change of jobs is involved, this may be within an organization (especially for those in large

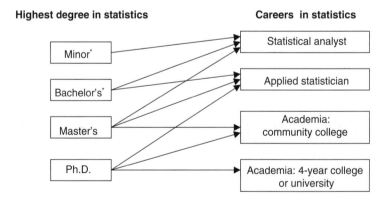

* Might also take position in an application area
  that is likely to require some statistical training.

Figure 12.1 Entry-level career opportunities for statistics graduates.

Careers in applied statistics   **Careers beyond applied statistics**

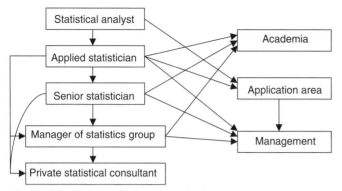

Figure 12.2   Some typical career paths for applied statisticians.

organizations), or in a new organization; see Sidebar 12.2 for comments on job mobility. You may, moreover, wear two or more hats at any given time. Or you might zigzag in your path, at least for a while, in probing what you feel is the most suitable eventual role for you.

SIDEBAR *12.2*

COMMENTS ON JOB MOBILITY

At one time, people joined an organization upon graduation and—especially those who entered a large organization—with the expectation that if things went well they would be spending the rest of their careers with that organization. Those with an organization for over 5–10 years were likely to remain there until retirement. This level of loyalty, both on the part of the organization to the employee and on the part of the employee to the organization, tends to be a thing of the past in our highly mobile society. Your relationship with your employer today is a contract that is subject to constant review and renewal by both parties.

Employers, especially in business and industry, continuously examine what their employees have done for them recently, while employees ask themselves whether their current position is still in their best interest or whether they could do better in another capacity or somewhere else.

Employees today, thus, constantly "review the field" to determine whether there are other opportunities that might be better for them. Those in government might seek out opportunities either within their own agency or in other agencies. As a result, some résumés show a relatively large number of different positions, often over a relatively brief time period. This might be especially the case for statisticians because of the large choice and diversity of attractive alternatives open to them.

Job changes might also be triggered by changes in the environment, as reflected by changes in business conditions, changes in governmental administrations and funding policies, corporate takeovers, or even innovation. Some statisticians in the telecom industry have, over the years, moved to Internet-based businesses.

An advantage of frequent job changes is that it is likely to result in a wider diversity of experiences. Also, promotions and salary increases often accompany such changes. Thus, job change can accelerate your growth.

Frequent job changes, on the other hand, might be taken as a negative by a potential employer. They might suggest that you are difficult to please and cannot be relied on to stay on the job for very long. It might also raise the question as to whether the frequent changes are an indication of poor performance on the job and to what degree they were voluntary.

The general message conveyed by Figures 12.1 and 12.2 is that statistically trained individuals have the opportunity to leverage their training and experience in a wide variety of ways. Some aspire to take on increasingly greater technically focused roles and responsibilities, perhaps culminating their careers as their organization's recognized statistical guru. Others may wish to become more deeply involved in an application area and, perhaps even, in a management role.

12.5   MORE ON STATISTICAL LEADERSHIP

The preceding discussion and our comments in earlier chapters have made clear our opinion (and that of many others) that a career goal for today's statisticians should be to exert leadership and be proactive.

Deming wrote and spoke extensively about the concept of a statistical leader. He wanted statisticians to be advisors to top management, reporting directly to the company CEO, and given a great deal of independence; see Sidebar 12.3.

SIDEBAR *12.3*

DEMING'S STATISTICAL LEADER

W. Edwards Deming had strong ideas about statistical leadership. In Deming (1986), he asserts (with our apologies for his lack of gender sensitivity)

There will be a leader of statistical methodology, responsible to top management. He must be a man of unquestioned ability. He will assume leadership in statistical methodology throughout the company. He will have authority from top management to be a participant in any activity that in his judgment is worth his pursuit. He will be a regular participant in any major meeting of the president and staff. He has the right and obligation to ask questions about any activity, and he is entitled to responsible answers. The choice of application for him to pursue must be left to his judgment, not to the judgment of others, though he will, of course, try to be helpful to anyone that asks for advice. The non-statistician can not always recognize a statistical problem when he sees one.

What would be the minimum qualifications for this job? (1) equivalent of a master's degree in statistical theory; (2) experience in industry or in government; (3) authorship of published papers in theory and in practice of statistical methodology; (4) demonstrated ability to teach and to lead top management toward constant improvement of quality and productivity. He himself will constantly improve his education.

In addition, Deming shows an organizational chart under which the position of "Leadership in statistical methodology" reports directly to the company president. Moreover, in a subsequent section entitled "Where may you find the right man," Deming states

The combination of knowledge and leadership is exceedingly rare, and will require patience and earnest prayer to find . . . You may have to interview many applicants to find the right one.

An incumbent as head of statistical methodology will command a high salary. (A recommendation that we heartily endorse). The problem will be to find someone competent, not what his pay will be.

Deming's ideas concerning the role in organizations for statisticians, although suitable for a person of his stature, are, in our opinion, too ambitious for most others and are not well suited to our times (or quoting Lynne Hare, "If the rest of us acted like Deming, we would have been fired long ago."). But Deming was right on target when he urged statisticians to exert leadership and suggested that they be proactive. Statisticians today are more engaged in the projects in which they are involved and in taking a leadership role than they were in the past—even though this role may be from a more modest base in the organization than that envisaged by Deming.

12.6 CONTRIBUTIONS BEYOND THE WORKPLACE

Statisticians as they move forward in their careers have the opportunity to contribute beyond the workplace. This can involve technical contributions or "giving back" to their profession or to society as a whole.

12.6.1 Technical Contributions

Statisticians advance their profession technically by publications in the statistical and application area literatures and by presentations at professional society meetings. Publications, especially in scholarly journals, are a key requirement for advancement in universities and 4-year colleges, as we will see in the next chapter. They may also be important for those in some research organizations, such as national laboratories, outside academia.

In most other organizations, you are not expected to publish. Should you choose to do so, you are likely to contribute your own time for this activity—although your employer might permit and possibly even encourage you to present your results at professional meetings and pay your expenses.

Publications can help you gain increased stature and recognition on the job and in your dealings with colleagues and customers. They are usually regarded positively, but may not carry much weight, in your job performance evaluation. Publications, however, are likely to be highly beneficial if you are planning a move into academia.

Your technical contributions may be of a theoretical or methodological nature. Many recent Ph.D.'s publish work based on their dissertations after graduation (sometimes due to the time lag between submittal of a paper and its publication). Later in your career, your publications might be work that you conduct independently from the job or, more likely, are based on your ongoing work. They might deal directly with an interesting application or case study and may be principally expository.

Your publications need to be reviewed and approved by your employer and may not include any information deemed confidential. This presents a barrier to some professionals, such as many physical scientists. Statisticians can often readily disguise the specific application, and thereby its confidential aspects, without seriously compromising the usefulness or "publishability" of their work. This gives them an advantage over colleagues in many other areas.

12.6.2 Professional Participation

> Professional societies provide an excellent venue in which to contribute and to become known in your profession.
> —Robert Starbuck (2009)

Statisticians can through the course of their careers contribute to the profession in a variety of ways, in addition to their direct technical contributions. Some examples are

- Membership on an editorial board of a professional journal or willingness to referee papers submitted to such journals (a moral obligation for those who publish).
- Participation in sections of a national statistical society such as the Section on Government Statistics or on Quality and Productivity of the American Statistical Association (ASA), or of an application area-oriented society.
- Contribution to the local activities of national associations such as your local chapter of the ASA.

Over time, you are likely to be asked to take on leadership roles in such organizations, such as editorship of a journal or as an officer in a national society or its local chapter.

Participation in such activities is again likely to be mostly on your own time, with your employer possibly supporting associated travel expenses, and giving you modest credit in your performance evaluation.

12.6.3 Contributions to Community and Society

> Statisticians hope to understand enough about nature and the underlying laws to improve the lot of mankind.
> —William Golomski

As we have tried to make clear throughout this volume, the applications of statistics are almost endless. You may be asked, on different occasions, to volunteer your time and expertise to a wide variety of worthy community activities. Some examples are helping design a membership satisfaction study for a social or religious organization and analyzing the results, helping measure teacher effectiveness for a local school district, or guiding a candidate seeking political office in conducting a public opinion survey.

Your contributions might extend beyond your own community and could have national or even international implications. Scheuren (2007) describes five applications in which his volunteer work helped enhance human welfare in Cambodia, Switzerland, (the former) Yugoslavia, Afghanistan, and the United States. The first of these dealt with compiling and evaluating evidence of civilian casualties resulting from the hostilities in Cambodia. Statisticians have also been involved since the late 1970s, through the auspices of the ASA, in projects such as helping obtain accurate counts of human rights violations (see Asher, 2009).

Statistics Without Borders, an ASA affiliated group of volunteers, was formed in 2008 to provide "pro-bono statistical consulting and assistance to organizations and government agencies in support of these organizations' not-for-profit efforts to deal with international health issues (broadly defined)"; see *Amstat News* (2010b). One of the initial projects was to help develop reliable estimates of the extent of damage to homes and displacement of people (and the nature of that displacement) resulting from an immense earthquake in January 2010 in Haiti; see Cochran (2010).

A further international study co-led by a statistician and dealing with estimating the number of people living with AIDS is described in Sidebar 12.4.

SIDEBAR *12.4*

A STATISTICIAN STUDIES THE GLOBAL AIDS PANDEMIC

Roger Hoerl, Manager of the Applied Statistics Laboratory of GE Global Research, in collaboration with research partner Presha Neidermeyer, used a 6-month sabbatical (fully supported and paid by his company) to study the global AIDS epidemic. Some of their findings are reported in Hoerl and Neidermeyer (2009).

One of the key goals of the study was to estimate the number of people living with the HIV virus in different parts of the world. Getting good estimates is of critical importance in helping decide how to allocate AIDS relief funding between and within countries, and in addressing the broader question of how to balance funding for AIDS relief with funding to fight other preventable diseases; see Hoerl (2008b). In doing this work, Hoerl found his statistical background to be invaluable, enabling him to take an objective view of a very emotional topic, to gather relevant data firsthand, and to weight conflicting and sometimes contradictory evidence to arrive at actionable conclusions.

12.7 SOME MORE CAREER EXAMPLES

In Chapter 1, we provided some examples of how different individuals might enter a career in statistics. In this section, we describe the different career paths (to date) of three statisticians, principally after they entered the workforce.

Statistician A. Statistician A received an undergraduate degree in mechanical engineering. Upon graduation, he took a job with a semiconductor company. This position got him to appreciate the importance of statistics. He was an excellent mathematician and returned to school to get a Ph.D. from the Statistics Department of a large university. In his graduate work, he became involved in the statistical aspects of oil exploration and filed a patent application in that area. He accepted a job as an applied statistician in the then small statistical group of a large company and became involved in a wide variety of applications. His engineering background, his love for real problems, his extraordinary statistical knowledge, his ability to apply this knowledge in actual situations, and his interest in working with people made him an outstanding technical contributor. Over time, he took on more and more challenging projects and his services were in constantly increasing demand, as he built a reputation as a technical expert.

As the group to which he belonged grew in size, Statistician A also took on the role of technical mentor to less experienced statisticians. In this manner, he evolved into a statistician's statistician, as colleagues came to him for guidance on tough technical problems.

Statistician A published papers in statistical journals, based upon his work, and actively participated in the yearly Joint Statistical Meetings (Section 14.5.1). He was elected a Fellow of the ASA (in recognition of his substantial contributions to spreading the use of statistical methodology successfully in a wide variety of application areas) and as a member of the International Statistical Institute. In addition, his contributions to his company were recognized by a number of prestigious internal awards.

Statistician A stayed in essentially the same job, and grew steadily in it over a period of over 25 years until his retirement. He gained satisfaction and recognition by continuingly increasing his skills as a senior statistician, applying these skills to help address more important and challenging problems, and doing so in a steadily more proactive manner.

Statistician B. A career choice came easily for statistician B. His father, who had started his career as a mechanical engineer, took up statistics and became world renowned for his work. Statistician B also went into statistics and got a Ph.D. in the field. He was always strongly interested in practical applications and interned with a large chemical company while still in school.

Statistician B was always keen at looking at problems conceptually and holistically. He also had outstanding personal and communication skills. He focused on statistical thinking and regarded statistical tools as an important means to an end, rather than as an end in itself. After graduation, his first job was with a chemical company where he divided his time between research and development and the corporate quality group. After 4 years in that position, he accepted a job with a paper goods company and worked on a variety of quality improvement applications, rising to become the manager of a small statistical group.

When his company was bought out, Statistician B took a job with the same statistical group as Statistician A. In this position, he got to learn the company culture and became acquainted with many of its businesses. His capabilities were rapidly recognized and, as the Six Sigma initiative emerged, Statistician B became a master black belt.

He then joined the company's corporate audit staff—an organization responsible for improving operations throughout the company's businesses worldwide. Although he was formally a manager in this group, Statistician B's major responsibilities involved hands-on support to businesses that were learning how to apply Six Sigma successfully. He specialized in, but did not limit himself to, statistical applications.

Statistician B subsequently returned to the statistical group that he had joined originally—but this time as its manager. In this role, he maintained his statistical roots, while guiding a group of other statisticians in a broad range of applications. Among other things, he advanced the stature of the group by ensuring that it becomes involved in key company problems to which it can contribute significantly and by drawing its accomplishments and future potential to the attention of higher management. He frequently is engaged directly in the start-up phase of important new projects. This, together with his broad business knowledge, makes him an important resource to those in his group who subsequently work on such projects. His contributions have also been recognized by a prestigious company award.

Statistician B has also achieved high professional visibility. He has written extensively about his work. He is also a highly sought keynote speaker at statistical and quality assurance meetings, at which he talks, among other matters, about what statisticians should aspire to be. He, too, was elected a Fellow of the ASA (for his substantial contributions and insights in furthering the role of statistics in industry), a Fellow of the American Society for Quality, a member of the International Statistical Institute, and an Academician in the International Academy of Quality, and is the recipient of numerous important professional and company awards. He has also actively involved himself in projects, beyond his work, to help society.

Statistician B is still professionally (very) active. We look forward to updating his progress in forthcoming editions of this book.

Statistician C. Statistician C also holds a Ph.D. in applied statistics (others with similar career paths might be at the master's degree level). She, too, has outstanding personal and communication skills and upon graduation joined the statistical group of the same company as Statisticians A and B. Her ability to address business problems and to respond to them effectively became apparent immediately. She also became a Six Sigma master black belt—and was besieged by job offers from the company businesses with which she worked.

Like Statistician B, she also went to work for the company's corporate audit group and assumed Statistician B's managerial role when he left. The job provided her valuable experience and business insights that were rare for a person only a few years out of school.

Statistician C, however, aspired to take a different, and less statistically oriented, career path than Statistician B. Her next position was as Six Sigma leader for one of the company's businesses, reporting directly to the CEO of that business.

In her next advancement, Statistician C left the company to assume a position in another large organization where she led company-wide Six Sigma implementation and quality efforts.

Unlike Statisticians A and B, Statistician C no longer works as a statistician or statistics manager. She is rarely involved in professional statistical activities. However, she has leveraged statistical thinking and her personal skills into a highly successful early career in management—which is still evolving. Her statistical background and knowledge of the role of uncertainty continue to serve her well.

Statistician C, in some ways, illustrates Deming's concept of a statistical leader. She, however, also bears significant day-to-day responsibilities as a leader in her own right (rather than as an advisor) and no longer is regarded principally as a statistician by her colleagues.

12.8 PUTTING IT ALL TOGETHER

Ask yourself what you like and don't like about your job and act on those that you can do something about.

—Leslie Fowler

So what should be your career path as a statistician? That, of course, is a question that only you can answer. It depends heavily on your personal aspirations and skills, and on how you prioritize different criteria for success. Ideally, your career will lead not only to fame and fortune but also to a high degree of job satisfaction.

Fortune is the easiest of these criteria to quantify since it can be measured by your paycheck and other job benefits.

Your fame as a statistician is reflected by your reputation as an effective on-the-job contributor and, somewhat more quantitatively, by your publication (and speaking) record, the professional awards and honors bestowed upon you (e.g., election to fellowship of the ASA), and the offices in professional societies to which you have been elected. Your reputation could be national or even international, as evidenced by such statisticians as Bowker (Section 13.14.2), Deming, Garfinkel, and Goodnight (Section 1.10).

Job satisfaction is more personal and the hardest to quantify. You might seek a high level of predictability and stability in your job (which might, perhaps, eventually lead you to seek a tenured position in academia), or you may thrive on change. Many statisticians gain high satisfaction in successfully addressing real-life challenges and in the variety of their work and contributions. Others get their proverbial kicks from solving tough technical problems. Yet others want their work—perhaps in addition to all of the preceding—to be beneficial to their communities and to society.

The beauty of our field is that it can accommodate a great variety of aspirations and provide career paths that have at least the potential of coming close to satisfying all of these criteria.

In any case, as we indicated in Section 8.2, it is helpful for you to define initially your personal career goals as a statistician—even though you will likely modify these as time goes by. With these in mind, you can then seek tasks and strive to drive them to successful completion that will help you in advancing your goals. And, on a more immediate basis, it is always a good idea for you to keep your résumé up to date, just in case some new attractive, and perhaps unexpected, opportunities should arise.

12.9 MAJOR TAKEAWAYS

- We have somewhat arbitrarily categorized statisticians (outside of academia) during the course of their careers as statistical analysts, applied statisticians, senior statisticians, statistical managers, and private statistical consultants.
- Some applied statisticians might choose to move into academia, into an application area, or into management.
- The specific path followed by any particular individual is typically gradual and may be a consequence or combination of a deliberate plan, circumstances, or even happenstance.
- We urge statisticians to take increasingly proactive leadership roles as they move up the career ladder.
- In addition to their direct on-the-job roles, many statisticians also make important technical contributions through publications and presentations, work to advance their profession, and apply their knowledge to help their communities and society.

DISCUSSION QUESTIONS

(* indicates that question does *not* require any past statistical training)

1. *We suggest that the specific path followed by any particular individual may be the result of circumstances or even happenstance. What are some of the special circumstances (or happenstances) that might significantly impact an individual's career path?

2. *Suggest some ways in which some of the career paths suggested in Figure 12.2 might come about in practice.

3. We state "Deming's ideas concerning the organizational role for statisticians, although suitable for a person of his stature, are ... too ambitious for most others and may not be well suited to our times." Discuss.

4. *What are you doing now to lay the foundation for achieving your career goals and aspirations? What else might you do now and in the future to improve your chances of success?

5. Comment on how statisticians can be more proactive in each of the roles (i.e., statistical analyst, applied statistician, etc.) that we have described.

6. What are some ways that statisticians are contributing—or should be—in addressing problems that face your community and/or society today?

7. *Contrast the different ways in which Statisticians A, B, and C exert leadership. Which of these paths seems most appealing to you?

PART IV

FURTHER PERSPECTIVES: CAREERS IN ACADEMIA AND PROFESSIONAL DEVELOPMENT

IN THIS final part of the book, we consider two topics:

- Chapter 13 puts the spotlight on careers in academia for statisticians.
- Chapter 14 describes strategies for lifelong learning (and other activities) that allow statisticians, through the course of their careers, to maintain and build upon the momentum that they acquired through their education and earlier experiences.

CAREERS IN ACADEMIA[1]

13.1 ABOUT THIS CHAPTER

Throughout this volume, we have made occasional reference to a career in statistics in academia. We feel that this subject is sufficiently important—and certainly sufficiently different from most careers of nonacademic statisticians—to warrant an entire (and lengthy) chapter of its own. A career in academia may be selected by a statistician upon graduation. Or, as we have noted in Section 12.3.1, it may be chosen after a stint in a nonacademic setting.

In this chapter, we consider how subjects similar to those discussed in earlier chapters apply to academia:

- What statisticians in academia do.
- Types of positions.
- Location within the institution.
- More on tenure.
- Life beyond tenure.
- Teaching challenges.
- Research challenges.
- Consulting challenges.
- Administrative service challenges.
- Professional service challenges.
- More on the academic environment.
- Training to become an academic statistician.
- Career paths.
- Downsides of a career as an academic statistician.
- Bright sides of a career as an academic statistician.
- A career as a statistician in academia: a summary comparison.

Our discussion is heavily centered on practices in the United States; these may be somewhat different in other countries.

---

[1] This chapter was written largely by contributing authors Josef Schmee and William Meeker.

We provide more detail in this chapter than in most others. We hope this will be useful to those who have decided on a career in academia. Readers who wish a comprehensive summary of how life in academia compares with that in business and industry and government might want to look at Table 13.1 and at the major takeaways section at the end of this chapter and skip some of the other detail on a first reading.

13.2   WHAT STATISTICIANS IN ACADEMIA DO: AN OVERVIEW

The term university, derived from a Latin phrase for "the community of teachers and scholars," suggests the two most important responsibilities of an academician: to teach and conduct research. In addition, the collegial nature of modern academic institutions suggests that faculty participate, mostly through service on advisory committees and student advisement, in the academic administration of their institution. To these three basic responsibilities should be added the derivative ones of consulting (including community service) and service to the profession. These five areas often compete for an academician's time. The specific mix depends on the type of institution and, to some degree, where within the institution you are "housed." The mix will also vary among similar types of institutions, and can vary between, and even, within departments in the same institution. We provide a brief overview below and discuss in further detail later in this chapter.

Research Universities. A large research university may, for example, have three openings for statisticians. One is in the statistics department with a very strong and highly regarded Ph.D. program. The second is in a separate biostatistics department (which might be part of the university's medical school), and the third is in a school of business.

The position in the statistics department will likely have a high statistical research orientation. With increasing seniority, the supervision of master's theses and Ph.D. dissertations will be added. The position will also require teaching graduate, and possibly undergraduate, courses.

The teaching focus of the position in the biostatistics department is on master's and Ph.D. students. Research is also expected, but should be oriented toward medical or biological applications of statistics and is often done in collaboration with medical professionals.

The position in the school of business is strongly focused on teaching a large number of undergraduate business majors and MBAs, even though the school may offer a Ph.D. Some research is also expected, but this is likely to be directed at applications in business and economics.

Four-Year and 2-Year Colleges. In contrast to research universities, 4-year colleges generally place more emphasis on teaching and may expand their definition of research to include a wider variety of activities. For example, papers and talks about teaching innovations and literature reviews may be acceptable and treated as equal to more traditional research in statistical theory or methodology.

TABLE 13.1 Summary Comparison of a Career as a Statistician in Academia Versus Business and Industry and Government

Topic	Academia	Business and industry	Government
Individual characteristics			
Need for Ph.D.	Essential for tenure-track positions and research universities	Important in some areas, not in others	Helpful in some areas, not in others
Need for communication ability (oral)	Essential for teaching; important in other contexts	Essential	Essential
Need for communication ability (written)	Important; emphasis on scholarly writing	Important; emphasis on clarity, conciseness, and impact	Important; emphasis on clarity, conciseness, and impact
Entrepreneurship	Frequent need to get research grants	Frequent need to sell services and programs	Varies with organization and level. May need to sell services and programs
Nature of work			
Freedom to select work projects	Generally extensive	Restricted to business needs	Restricted to governmental needs (especially the specific agency)
Marketing needs	Often requires extensive proposal writing	Often requires selling of technical discipline and ideas	Often requires selling of technical discipline and ideas
Independence	Generally appreciable	Need to be responsive to direct manager, higher level managers, and project managers	Need to be responsive to direct manager, higher level managers, project managers, and various public officials
Depth of work	Can focus deeply on narrow area; often limited only by (often broad) grant and publication criteria	Can be broad and multidisciplinary; limited mainly by practical situation and by 80/20 rule: quit upon getting 80% of answer for 20% of cost	Can be broad and multidisciplinary; limited mainly by practical situation and by 80/20 rule: quit upon getting 80% of answer for 20% of cost
Pressures/deadlines	Often self-disciplined and long range in research work	Frequent; dictated by business needs	Dependent on political needs, laws, and program requirements
Predictability of work	Generally predictable	Often quite unpredictable	Sometimes predictable, sometimes not
Publications and professional society involvement	Essential, especially in research universities	Usually nice to do—sometimes gains "extra credit" (but often done on own time)	Usually nice to do—sometimes gains "extra credit" (but often done on own time)

(continued)

271

TABLE 13.1 (*Continued*)

Topic	Academia	Business and industry	Government
Consequences			
Major criteria for success	For research: peer recognized research, evidenced by peer-reviewed scholarly publications; for teaching: teaching ability, evidenced by positive course evaluations and other assessments.	Success in solving company problems in timely manner, with demonstrable "payoffs" and ability to communicate these to management	Success in solving agency problems, production of statistical analyses on time, ability to communicate these to management and the public
Compensation for similar background (12 months)	Median salaries fairly similar at start, but more opportunity to do very well for those in business and industry, especially for persons willing to go into management		
Job security	Excellent after achieving tenure for tenure-track positions; generally limited otherwise; often defined by contract	Dependent on performance, as well as fortunes of business, acquisitions, and the economy	Excellent after probationary period, but dependent on type of appointment, size of organization, and the economy
Some possible frustrations	Transitioning research results to real-world applications	Inability to pursue problems to their ultimate completion	Inability to pursue problems to completion; some assignments can become routine
Other positive elements	Opportunity to work on intellectually challenging problems; interact with students and faculty; high level of prestige; great environment	Opportunity to work on real problems and business issues; dynamic and challenging nature of work; broad interactions; team and leadership opportunities; sometimes to do research on intellectually challenging problems.	Opportunity to improve public policy and decisions and impact lives; sometimes to do research on intellectually challenging problems

Note: These are "on the average" comparisons; there is appreciable "within-group" variability.

Two-year community colleges are typically not research-oriented and generally require professors[2] to teach more hours per week than 4-year colleges and universities. They may require little or no research.

Teaching Loads. There are usually no hard and fast rules for specifying a particular professor's course load during the academic year (running from late August to early May for institutions that operate on a semester basis). But some institutions, and especially community colleges, have strict rules (often as part of a union contract) on the number of credit hours taught.

Teaching loads at a university or 4-year college using a semester system typically range from three to eight courses or 9 to 24 credit hours, per academic year, depending on the research orientation of the institution. At colleges with no research expectations, and especially community colleges, a professor may sometimes teach up to 32 credit hours per year.

Professors whose work is heavily supported by grants or external consulting agreements may be allowed course buyouts. Some institutions offer course reductions to professors with a highly distinguished ongoing record of research or who are guiding a large number of graduate students. Course releases may also be offered for performing certain academic duties, such as temporarily stepping into an administrative assignment within the institution or serving as an editor of a prestigious research journal. Other professors may teach more than their normal course load, possibly for extra pay.

The *effective* workload on professors also depends upon the level of the courses they are teaching and their frequency; effective load tends to be reduced when professors are teaching multiple sections of the same course or teaching a course repeatedly.

During the semester, professors typically divide their time between classroom preparation and teaching, grading papers, helping students who need added attention, guiding graduate students in their research, conducting their own research and consulting, and attending faculty and committee meetings. At other times, such as between semesters or during the summer, when professors do not have to teach or attend formal meetings, their day may be devoted mostly to their research, future class preparation, and consulting.

Because time is limited—and different undertakings invariably conflict with one another—it is critical that you understand specifically what is required in any academic position that you are considering, and how this mix factors into your performance evaluation.

13.3 TYPES OF POSITIONS

13.3.1 Tenure-Track Positions

Tenure-track appointments come with the expectation of future consideration for tenure. Faculty members who are granted tenure can be removed from their positions

---

[2] Following standard practice, we will refer to all faculty involved in academic endeavors as professors, irrespective of their official rank.

only under very special circumstances that are generally specified in the institution's faculty manual. Most tenure-track appointments are made to individuals who are near the beginning of their career, usually at the rank of Assistant Professor. Typically, an initial contract for a period of, for example, 3 years is granted and, upon review, extended to 7 years. Granting of tenure is considered at approximately the end of the sixth year (extensions are occasionally given under extenuating circumstances). Most universities promote Assistant Professors to Associate Professor when tenure is awarded.

In some institutions, individuals with previous experience may be offered a tenure-track position at the Associate Professor or even the (full) Professor level. This might occur, for example, in hiring a renowned scholar from outside academia (such as from a national laboratory) or somebody who had been previously tenured at another institution.

Some, but not all, institutions limit the number of tenure-track positions, either by department or by college.

13.3.2 Nontenure-Track Positions

Nontenure-track positions may be long-term or temporary. Many long-term nontenured appointments are at the rank of lecturer or instructor. Others are designated as "clinical" or by some other term to distinguish them from tenure-track appointments. Temporary appointments are appropriately called visiting (or fixed-term or temporary) or, when applicable, postdoc. In addition, there are adjunct appointments that usually involve teaching half-time or less.

Some institutions make nontenure-track appointments when it is not possible to add tenure-track positions or, perhaps, to save money. Others use nontenure-track positions as a trial period and then encourage promising appointees to apply for tenure-track positions. Many community colleges, instead of tenure, offer long-term contracts to their full-time faculty members.

We provide a brief description of the various types of appointments, noting that their definition and prominence vary from school to school.

Lecturer/Instructor Appointments. In U.S. universities,[3] lecturers are usually nontenure-track appointees, chosen for their teaching skills and engage principally in teaching and, perhaps, in providing some departmental services. Lecturers are particularly common in statistics departments that are responsible for teaching introductory statistics courses, often with a large number of students, to nonstatistics majors. Lecturers may also coordinate such courses and supervise graduate students who act as teaching assistants and may be responsible for lab sections or even teaching courses on their own.

Clinical Appointments. The term clinical professor, as the name suggests, originated in medical schools, but has found its way into other institutions. Other names, such as

---

[3] The term lecturer has a different meaning in some countries, such as the United Kingdom, from what it does in the United States.

research professor, are also used. Such positions may or may not involve teaching responsibilities. The job may be quite similar to that of a tenure-track position, without providing the same potential job security.

Clinical professor appointments are, sometimes, linked to specific externally funded projects. When the grants for such projects are not renewed, the positions may be terminated. Holders of such positions may be invited to apply for tenure-track positions as these become available.

Visiting Appointments. Visiting appointments are mainly for a specified short time period, typically one or two semesters, but might extend up to 2 years. They may be offered to replace professors on leave or to hire professors with specialties that are needed only for a short time. Visiting appointments provide a mechanism for bringing in experts, perhaps in new areas, or individuals with special skills and new ideas. Some institutions use visiting appointments in the same manner as clinical appointments or lecturer/instructor appointments.

Visiting appointments may be attractive to experienced statisticians, possibly on sabbatical from other institutions. They allow them to engage in specific research, to interact with other professors who are experts in a field in which they want to conduct research, or simply to experience life in a different institution. Also, visiting appointments occasionally provide a refreshing short-term change and broadening of perspective (and, possibly, a first step in a potential career change) to applied statisticians in government or business and industry.

Postdoc Appointments. Postdoc positions are similar to visiting positions, but tend to be aimed at recent Ph.D.s and are research-focused. The major purpose of a postdoc position is to allow individuals to demonstrate high potential for research, independent of their dissertation supervisors. Some postdoc positions also provide opportunities to obtain teaching experience.

In many scientific disciplines (other than statistics), a postdoc appointment is the most common path to a tenure-track position at a research university. Some academic departments will not even consider individuals for tenure-track positions unless they have successfully finished a stint as a postdoc.

Postdocs seem to be less common in statistics—perhaps because of the nature of research funding and because top students from top programs are often able to secure tenure-track positions without such experience. On the other hand, a recent statistics Ph.D. with a successful 2-year record as a postdoc will often make the short list of tenure-track candidates at many universities.

Adjunct Appointments. Adjunct positions are generally part-time and entail some well-defined—but limited—duties, usually the teaching of a specific—and sometimes specialized—class or classes. Such positions may be offered at ranks ranging from adjunct instructor to adjunct professor. They are attractive to many institutions, especially community colleges, because of their relatively low costs.

Adjunct professors often teach courses in the evening and hold full-time jobs. (Like visiting professorships, such positions provide outsiders a glimpse into academia.)

Some adjuncts are retired. Others take multiple appointments until an appropriate full-time opportunity at a single institution presents itself.

Adjunct professors are generally not asked to perform duties beyond teaching[4] and do not share in the financial and other benefits accorded to most other professors. Their research credentials may not be equivalent to those of full-time faculty, and they may not be required to have Ph.D. degrees. However, their practical experience may compensate for this and make them good role models for students.

13.3.3 Nine-Month Versus 12-Month Positions

Because academic institutions tend to have much smaller educational programs during the summer months, most appointments are for the academic year (9 months) only, although benefits are on a 12-month basis. This is not to imply that professors get an extended summer vacation. Especially at research universities, professors generally find the summer months, when they often do not teach, the most productive time of the year for research and consulting. One important motivation for seeking external funding is, in fact, to help support summer research.

13.3.4 The Formalities

Job offers are made by letters of appointment; see Sidebar 13.1. Their terms can protect professors against an institution reneging on its promises, especially when changes of administration occur.

SIDEBAR 13.1

LETTER OF APPOINTMENT

Academic institutions make job offers to candidates in a letter of appointment. While many of the features of the appointment letter (e.g., specification of salary) are similar to those in industry and government, an academic appointment letter typically also specifies the academic rank associated with a position and whether or not it is tenure track. The appointment letter also provides a general description of the job and the department and school in which it is located. The letter should also indicate the teaching load expected (or refer to the faculty manual for this information). If the appointment is time-limited, the letter specifies its length.

The letter also needs to spell out any special employment arrangements negotiated with the administration, such as extra resources or course reductions during the initial break-in years. As a legally binding contract, the appointment letter also describes or refers to general employment conditions, such as those specified in a faculty manual. Finally, the appointment letter indicates whether the appointment is for the academic year or the calendar year.

---

[4] Occasionally (and especially in Canada), adjunct appointments are granted, somewhat similar to visiting professorships, to eminent researchers who visit an institution periodically or are engaged in supervising graduate students.

13.4   LOCATION WITHIN THE INSTITUTION

Typically, economists have appointments in economics departments, psychologists in psychology departments, engineers in engineering schools, and so on. Similarly, many statisticians in research-oriented institutions have appointments in statistics departments. Such departments offer courses and majors in statistics at the graduate, and sometimes the undergraduate, level and engage in statistical research. In addition, department members often teach introductory courses aimed at students in other departments—for example, students in the school of liberal arts, the business school, or the engineering school—or for the institution as a whole.

However, not all universities and few 4-year and 2-year colleges have statistics departments. In such cases, statisticians may be faculty members in one of a number of departments, notably the mathematics department, a combined mathematics and statistics department, a department of information science, or an application area school or department. At some institutions, statisticians are grouped into academic units together with operations research and other decision sciences faculty or with computer science under, possibly, a general umbrella name such as informatics. Such groupings also sometimes offer degree programs in statistics or, more often, in an area such as data mining, with important statistical overtones.

In institutions that do not have statistics departments or quantitatively-oriented groupings, as well as some that do, statisticians have appointments in a variety of different schools or departments. For example, business schools may have statisticians in a department specializing in quantitative methods or a psychology department may hire statisticians into their department to teach courses in statistics and research methods. Also, some institutions that do not have formal statistics departments might offer majors or minors in statistics through one of a variety of other academic departments or, possibly, through interdisciplinary programs that are linked across two or more departments.

Most medical schools have statisticians on their staff, often organized in a department of biostatistics within a school of public health, working on research, assisting with grants, and teaching courses. Some business or engineering schools hire statisticians to teach introductory statistics courses and to complement their offerings with such topics as design of experiments, times series analysis, or forecasting.

In some institutions, statisticians may be formally listed as members of more than one department or of an interdisciplinary program. In such cases, however, there is often one department with which the faculty member is primarily associated.

Sometimes, especially in smaller or 2-year community colleges, statisticians may be the only members of their profession within the department, or even in the entire institution.

13.5   MORE ON TENURE

Tenure was introduced to help guarantee academic freedom. It was felt that, for the institution to thrive, academicians need to have the right to openly disagree with predominant opinions and theories and be able to pursue their research without interference.

Gaining tenure is a key early goal of many statisticians who seek a career in academia.

13.5.1 The Stakes and the Consequences

Tenure provides a high level of job security and independence. If a professor is denied tenure, employment is typically terminated no later than at the end of the following academic year. Denial of tenure can be a substantial career setback and even the end of a planned academic career. There are, nevertheless, many examples of prominent statisticians who though denied tenure at one institution went on to thrive at another.

The period leading up to the tenure decision can be especially harrowing, even for those considered shoo-ins. Tenure-track candidates need to pursue the goal of achieving tenure with acuity and a single-minded determination from the very beginning of their appointments. Their teaching and research efforts, especially, need to match the institution's expectations and be of top quality, as judged by their peers, both inside and outside their departments.

Also, just as outside of academia (Section 5.8.2), expectations may change over time with, for example, a change in administration. You need to be prepared for such changes by constantly seeking answers to such questions as "has the recent appointment of an influential new dean altered the ground rules or focus" or "has the institution changed its research emphasis in, perhaps, seeking a new form of accreditation or new contract opportunities?" Also, there is much subjectivity in the assessment of a candidate's teaching and research performance. The best strategy is to anticipate unexpected developments, strive to exceed expectations, and, like outside academia, try to follow a strategy that is robust to change.

13.5.2 Criteria

The three main criteria for tenure are performance in research, teaching, and service to the institution. At many universities, the ability to generate external funding for your (and others') research may also be an important criterion for promotion and tenure. Different institutions weight the importance of each criterion differently, although, with the possible exception of 2-year community colleges, service is typically the least important of the three criteria. Some prestigious research universities value research so highly that tenure is awarded only to candidates who have achieved some degree of national stature in their fields. Other well-regarded universities and some 4-year colleges have more balanced requirements and may assign approximately equal weight to research and teaching in awarding tenure. Yet other 4-year colleges and community colleges tend to be focused on teaching and may require only limited demonstrated research or other scholarly achievements.

There is much variability in how different institutions, and schools and departments within an institution, make the determination of what constitutes a good research record and a good teaching record. A school with focus on teaching might consider a total of two to six publications over a period of 6 years as sufficient. But to be considered a good teacher, a candidate is required to have attained very good teaching evaluations from students and positive assessments from peers who are assigned to evaluate teaching performance.

In contrast, research universities typically require more published papers and/or papers in highly ranked journals for demonstration of superior research. At the same time, evidence of good teaching at such institutions might require only a preponderance of reasonably good teaching evaluations.

Expectations for tenure are typically stated in general terms in the institution's faculty manual. However, important specifics, such as the number of publications required and the minimum average teaching performance score on student questionnaires, will not be stated at most institutions. This lack of specificity allows the institution to judge the entire person applying for tenure and leads to some leeway in applying the criteria.

13.5.3 Process

It is clearly important for an aspiring tenure-track assistant professor to understand the mechanics involved in the tenure-granting process. We summarize this process on the book's ftp site.

13.6 LIFE BEYOND TENURE

The path forward in academia does not end with the granting of tenure. Quite to the contrary, tenure offers a more unconstrained pursuit of professional goals offered by few other career paths. Newly tenured professors have many reasons to keep up the good work—if for no other reason than professional pride and personal satisfaction. They are also now in a position to assume greater risks, such as taking on more challenging and longer term research projects in which success may be less likely, but the payoff, if successful, is higher.

There are also important tangible incentives for maintaining high performance after tenure. Annual merit reviews by department chairs, deans, and, in some cases, committees are designed to determine whether a professor maintains a level of performance that meets the expectation of the institution. These reviews typically focus on the professor's teaching and research record, but may also consider collegiality and service and are used in determining merit increases in a professor's salary. They may also initiate serious conversations with professors who fail to meet expectations.

Further in-depth post-tenure reviews are conducted periodically, such as every 5 years, as a basis for recommendations of promotions in rank. Promotion to full professor, and especially the award of an endowed or distinguished chair, requires a record of substantial achievement—but one to which most academics aspire. Evaluations for promotion are not unlike tenure evaluations in that outside reviews and support are frequently sought and accomplishments are carefully scrutinized.

The respect accorded to you by your colleagues and the chances of your work being judged exemplary are enhanced by the external acclaim accorded to your work through continued and increased funding, professional invitations to give presentations, awards, and various other recognitions.

13.7 TEACHING CHALLENGES

Statistics professors tend to teach a mix of introductory (so-called "service") courses to nonstatistics majors and more advanced courses to students with a greater interest in a particular aspect of statistics or—in those institutions where they exist—to statistics majors (or minors).

13.7.1 Teaching Introductory Courses

> The first courses in statistics, regardless whether at the high school or college level, are absolutely critical, and we need substantially more passionate and skilled educators in order to maximize their positive impact.
>
> —X. L. Meng (2009)

The Challenge. Many academic departments require, or strongly encourage, a course in introductory statistics, and for good reason. As discussed in Section 1.7, statistics helps people better understand numerous matters that impact their daily lives. It is equally important in a large number of professions and, especially, for disciplines in which data and empirical models play an important role.

Unfortunately, many students find their introductory course in statistics less than exciting. Poor experience in such courses drives students away from taking additional statistics courses and, more significantly, might close their minds to statistical thinking. Some applied statisticians, in fact, claim that the bad memories associated with the introductory statistics course to which their clients were exposed represent the first obstacle that they must overcome in their dealings with them. One of the causes for such low esteem may be that statistics, being methodological in nature, can be somewhat dry when taught abstractly.

It is helpful to differentiate among three situations: an undergraduate general "liberal arts" course, a more professionally or applications-oriented course directed at students in a particular subject matter area, and an omnibus course that tries to combine the two. The latter two courses might be at either the undergraduate or the graduate level.

The General Audience Introductory Course. Many institutions and departments require their students to take one or more courses in quantitative thinking as undergraduates. A general statistics course that emphasizes statistical literacy and statistical thinking is often one of the courses that will satisfy this requirement. Such courses have been characterized as being aimed mainly at consumers of statistics. Their focus should be on illustrating statistical issues in the information provided to the general public by the media and on using statistical concepts and statistical thinking in daily decision making. The resulting course should emphasize statistical concepts, assumptions, interpretations, and common misuses. When appropriate, actual case studies with real data should be included.

The Application Area-Oriented Introductory Course. As we have tried to make clear in the first part of this book, statistics and statistical concepts have relevance to most application areas and call for, at least, a minimal level of understanding by the numerous professionals that work in these areas. Such understanding has, moreover, become increasingly important with the democratization of statistics (Section 5.2).

Many institutions offer courses in statistics targeted at specific application areas as part of students' professional educations. Sometimes, these courses are developed and taught by nonstatisticians from within the department. But frequently the department calls upon statisticians, often from outside the department, to do so.

The resulting courses need to be closely attuned to students' backgrounds and application areas. There is typically a big difference in past coursework and skills, especially in mathematics, between, say, engineering students and social science students. Moreover, the type of problems encountered most often and the statistical approaches used for addressing them tend to differ among application areas. Thus, courses for economics and business majors need to emphasize regression and time series analysis, those for psychology majors would stress association among qualitative variables and experimental design, and those for engineers might have significant focus on methods for quality and reliability improvement. Also, such courses need to be made more appealing to students by using examples from their particular application area.

The Omnibus Course. Specialized courses directed at a specific application area are typically practical only in institutions with large enrollments in such areas. The hardest situation that you are likely to face in teaching an introductory statistics course is that in which those taking the course are a mix of students from the liberal arts and various application areas and where there is appreciable variability in analytical skills among students.[5] In such cases especially, you need to gain a good appreciation of the backgrounds and interests of students, and then to develop a course that is best suited and most beneficial to the class as a whole.

Large Class Challenges. Many introductory courses, and especially general and omnibus courses in large institutions, are often taught in a large lecture hall environment, and then broken down into laboratory or recitation sections, often led by graduate student teaching assistants. Thus, one of the important responsibilities of the professor is to help ensure that their teaching assistants are knowledgeable and

[5] Some 4-year colleges, and even some research-oriented universities, now offer two introductory omnibus courses—one that has elementary algebra as its main prerequisite and the other requiring a course in introductory calculus.

effective communicators. This requires meeting with them periodically to make sure they understand what they need to communicate, foster a common approach, provide pointers for explaining difficult technical concepts, coordinate instructional tasks, ensure that grading is fair and consistent, and resolve issues.

The electronic classroom (Section 13.7.3) has helped make teaching large classes more manageable. "Clicker technology" allows a lecturer to obtain on-the-spot reactions from students, including level of comprehension, and to act accordingly. This provides some compensation for the inability to communicate with most students individually.

Some large lecture classes employ machine grading of multiple-choice examinations. This reduces the amount of time required for grading, but limits the scope of questions that can be asked.

Meeting the Challenge. To improve the quality of statistics courses, teachers must communicate the relevance and excitement of the subject by using enlightening and relevant examples that demonstrate its importance—without being bogged down in details that are not essential to understanding.

Fortunately, the statistical community, in general, and various statistical organizations, in particular, have become keenly aware of the need for improving the teaching of the introductory statistics course. Recently published texts have become more user-friendly and provide a rich collection of important and relevant examples of the practice of statistics. Also, numerous papers have been written on how to design and teach more effective statistics courses (e.g., Brown and Kass, 2009; Easterling, 2010; Meng, 2009) in *The American Statistician* and elsewhere.

We highly recommend that new teachers of statistics regularly skim issues of the four major journals devoted to the teaching of statistics:

- *The Journal of Statistics Education.*
- *Teaching Statistics.*
- *Statistics Education Research Journal.*
- *Technology Innovations in Statistics Education.*

All of these journals (except *Teaching Statistics*) are free and easily accessible on the Internet. Moreover, the *Teaching Statistics* web site has a wealth of free information related to introductory statistics.

In addition, the American Statistical Association (ASA) has two sections that focus on teaching: the Section on Statistical Education[6] and the Section on Teaching Statistics in Health Sciences.[7] Internationally, the major organization concerned with teaching statistics is the International Association for Statistics Education.[8] Finally, two magazines (*Chance* and *Significance*) and various Internet sites (such as Chance News) provide numerous examples of the use and misuse of statistics in everyday life; see Sections 14.4 and 14.2, respectively.

[6] http://www.amstat.org/sections/educ/.

[7] http://www.bio.ri.ccf.org/ASA_TSHS/html/index.html.

[8] http://www.stat.auckland.ac.nz/~iase/.

The availability of student versions and institution-wide site licenses of sophisticated statistical software has meant that most introductory statistics courses have become more problem and data oriented, instead of formula based. Such software also allows striking graphical and animated demonstrations of data and statistical concepts. In addition, instructors need to develop relevant case studies and data sets that require limited explanation and that readily lend themselves to out-of-the-classroom analysis (see Shutes, 2009). The various resources mentioned in the previous paragraphs are sources for such case studies and data sets.

Teaching introductory statistics might seem easy from a technical point of view. Yet it usually presents formidable pedagogical challenges in explaining relatively complex statistical concepts, such as sampling error and confidence intervals, to beginners. Making the introductory course interesting and relevant to students often with a diversity of backgrounds requires considerable work. It should not be taken lightly or regarded as secondary to, say, teaching advanced seminars and doing research.

13.7.2 Teaching Beyond the Introductory Course

A further responsibility for some statisticians in academia is teaching a second course in statistics, often directed at a specific methodology area. Some schools or departments within schools might encourage, and occasionally require, students to take such a course. A business school may, for example, offer a second statistics course devoted principally to regression analysis, quantitative finance, or quality management. There may, in fact, be a variety of such courses offered within a school or department, aimed mainly at nonstatistics majors with different specialties.

In addition, statisticians in universities with a major or specialization in statistics teach more advanced courses at the graduate and sometimes undergraduate level. In fact, this is, in some cases, their major teaching responsibility. The wide, and ever-increasing, spectrum of statistics courses was discussed in Section 7.7.

Teaching more advanced courses in statistics might, in some ways, be easier than teaching introductory courses because you can generally expect more motivated students and less variability in their abilities. Instead, a major challenge is to provide students the knowledge and understanding to be effective applied statisticians and/or teachers of statistics and researchers. This requires not only imparting technical knowledge but also, and most importantly, helping students learn how to adapt that knowledge to address new problems, differing from standard classroom examples, that they may encounter in the course of their careers. In addition, you need to prepare students for the nontechnical challenges that they are likely to face and that we have strived to describe throughout this volume, such as building written and oral communication skills.

13.7.3 Some Further Tips for Effective Teaching

> If all students do not get a grade of A, the professor flunks.
>
> —W. Edwards Deming

While the preceding statement, attributed to Deming, is extreme, it expresses the need for a good teacher to be driven to engage all students at their highest individual levels.

In addition to being a good communicator, a successful teacher must have a sense of what students can absorb and how rapidly they can absorb it. Students have much variability with regard to both their innate ability and their interest in the subject—particularly, if they are required to take the course.

Good teachers, especially of the introductory course in statistics, are able to simplify complex subjects, reduce the material to essential concepts, and possess much patience. They need to be willing to help students succeed and to strive to motivate those who are less interested to realize the importance of statistics and statistical thinking in their daily lives and careers.

Planning and Organization.   Effective teaching requires more than just lecturing in class and interaction with students. Before a course can be offered, it needs to be planned. The teaching approach and the selection of a textbook and coverage of topics must match the course prerequisites and requirements, and anticipated student abilities and needs. Relevant examples and/or case studies must be assembled and problem sets that support learning developed. Such advance planning culminates in a syllabus containing, at a minimum, weekly topics and assignments, dates of tests, and grading policy. For most courses, students will require affordable and easy access to statistical software.

Assignments, tests, and papers need to be graded in a timely manner. For courses with large enrollments especially, some or most grading may be done by a graduate assistant, who, in turn needs to provide appropriate feedback to the professor.

The Electronic Classroom.   The emergence of the electronic classroom has aided communication between students and teachers, especially in large classroom situations.

Internet-based software makes it easier to communicate and update course information. It allows posting and downloading of supplementary course notes, homework assignments, and associated data sets and clarification of issues that arise both in class and outside of class. Completed assignments may be submitted electronically and returned graded. Online quizzes are used in some classes. Discussion groups can communicate with each other on group projects, allowing all group members to work on a single up-to-date version of the project document.

Soliciting Timely Feedback.   One proactive move for improvement (in addition to the feedback provided by clicker technology) is for a professor to administer periodically a short questionnaire, perhaps electronically. This might ask students to comment on how well they understand the material, topic by topic, and solicit suggestions for improvement in the way the class is being taught. Questions should be formulated so as to lead to constructive suggestions. The resulting information allows the instruction to be recalibrated in a manner that lets students benefit immediately from the feedback they provide. Also, reviewing the results with students can be used as an example of statistics in action.

All of this does not mean that professors need to act on every suggestion made; students' short-term goals are, after all, not always purely to improve the eventual learning experience—but all should be considered.

Summary. No simple formula defines a successful teacher of statistics. You need to communicate your enthusiasm about the subject to students in both introductory and more advanced courses. In return, you should derive much satisfaction from motivating students and from observing their growth. You need to focus on the essential concepts that students must grasp—whatever their level—to sense what they are able to do within a reasonable time, and to challenge them to do their best.

13.7.4 Teaching Evaluations

How is a professor's teaching effectiveness evaluated in making tenure and other promotion decisions? A short answer may be "with some difficulty."

Questionnaire Evaluations. Student course evaluations are a common tool for assessing teaching performance. Some institutions, in fact, require students to complete such questionnaires before releasing course grades. Questionnaires are typically completed anonymously at the end of a semester by students rating the professor on, say, a score from 1 to 5, and providing further comments in response to such statements as

- The professor was effective in teaching the class.
- The professor presented the material in a manner that raised my interest in the subject.
- The professor explained concepts clearly.
- The professor graded my work fairly.
- The weekly workload for the course was appropriate.
- Overall, this course was worthwhile.

    The response scores are then typically summarized in the form of histograms and averages, and communicated, together with comments, to the professor and the department chair.

    There is much controversy as to what student questionnaires actually measure and how they reflect and impact teaching quality. But they are easy to administer and consequently are widely used. See the book's ftp site for further discussion.

Using Questionnaires as a Tool for Improvement. Teaching evaluations can serve as an important tool for *improving* teaching effectiveness. This may be helped by adding questions whose answers can provide specific suggestions on how improvements might be achieved. Two questions that often lead to useful insights are

- What three things did you like most about this course?
- What three things did you like least about this course?

Other Evaluation Tools. A further, but somewhat time-absorbing, teacher assessment tool, sometimes used in tenure evaluations, is to have an independent skilled person (e.g., someone from an institutional research office) conduct personal interviews with former students. More commonly, junior professors' in-class performance may be directly observed by one or more senior colleagues, say, once a year or before tenure or other performance reviews. In addition, there may be an accumulation of anecdotal evidence, such as letters from students or parents. How an institution handles such information varies greatly, but has become increasingly transparent in most institutions by being specified in faculty manuals and similar documents.

13.8   RESEARCH CHALLENGES

> Publish or perish.
>
> —Familiar saying

Research-oriented statistics departments, especially in research-oriented universities, require their faculties to produce research of both high quality and quantity. A college dedicated to teaching, although usually requiring evidence of some scholarship, might regard research accomplishments as secondary to inspired and committed teaching. A 2-year community college may acknowledge research by its faculty, but often does not require it for faculty promotion.

13.8.1   Types of Statistical Research

Just as the relative emphasis given to research versus teaching and other responsibilities varies from one institution to the other, and even between departments within the same institution, so does the type of research undertaken.

Professors in a research-oriented statistics department often conduct work that is highly theoretical and methodological. Examples of some topics that appeared during 2010 in the theoretically motivated *Annals of Statistic*s are

- High-dimensional Ising model selection using ℓ_1-regularized logistic regression.
- Adjusted empirical likelihood with high-order precision.
- Exact properties of Efron's biased coin randomization procedure.
- Quantile calculus and censored regression.

At the same time, many statistics departments place a high premium on having their faculty engaged in applications, as well as theory and methods development. This often makes for better teachers and opens a wider range of grant funding opportunities. In addition, statisticians outside of statistics departments in, say, a business or medical school or in an engineering department might be engaged in research targeted at their department interests or interdisciplinary research. The results may be published in one of the journals of the ASA or in a highly regarded journal associated with some other professional organization. Some examples, also published in 2010, are

- How Many People Do You Know? Efficiently Estimating Personal Network Size, *Journal of the American Statistical Association.*
- Statistical Methods for Fighting Financial Crimes, *Technometrics.*
- Optimal Binary Prediction for Group Decision Making, *Journal of Business and Economic Statistics.*
- Joint Inference on HIV Viral Dynamics and Immune Suppression in Presence of Measurement Errors, *Biometrics.*

There are again no hard and fast rule governing the type of research conducted in a particular department and institution. It is not unusual for a theoretically oriented professor in a statistics department to be engaged on an important practical problem in collaboration with academic colleagues from another discipline. Also, many professors actively seek and work on problems that are motivated, and sometimes supported, by government, nonprofit organizations, educational institutions, or business and industry.

13.8.2 Identifying Research Projects

An important part of professors' jobs in a research-oriented environment is often to identify potential areas of research for themselves and graduate students. In making a selection from among alternative areas of research, you need to ask yourself questions such as

- Does the proposed research address an important (applied or theoretical) problem?
- Are you confident that others have not already solved the problem (or are about to do so)? Have you conducted an adequate search of the relevant literature?
- What are the chances that you (or one of your students) will succeed in making an important contribution in a reasonable amount of time? Is the problem too hard to be tractable or too simple to be regarded important?
- How "publishable" will this work be, when completed, and in which journals?
- How much time and effort will doing the work take?
- How much help will you need from others to complete the work successfully?
- Who are some candidates who might help you and how likely is it that they are willing and able to do so? Can you enlist their help in preparing a proposal for the work?
- Do you need funding to perform the research? If so, which agencies might you approach (or respond to) and how likely is it that you will be able to secure such funding or other needed support?

- Is successful completion of the work likely to generate more (funded) research?

The answers to the preceding questions should help you target in on the specific research topics that you will decide to pursue. For example, the need for funding can significantly narrow the list and focus of potential research topics.

Hamada and Sitter (2004) provide further guidance on how to get started in research.

13.8.3 Securing Research Grants

Importance. The need for obtaining external funding for the research conducted by professors and their students plays an increasingly important role in many research-oriented universities; in many cases, it is a prerequisite for doing substantial work on a problem. Many universities rely heavily on funding to support graduate programs, especially in science and engineering. Sometimes, such funding also supports summer salaries and attendance at conferences by faculty members[9] and their students.

The ability to attract external funding, moreover, provides tangible evidence of the importance of the work and how highly it is regarded. Thus, all research-oriented universities consider success in procuring funding, along with publication of the research, in tenure and merit evaluations. The award of a large research grant can be a major boost to a professor's career.

As a consequence, the development and writing of proposals for funding future research is a significant, and often highly time-consuming, activity for many university professors.

Funding Agencies. Statisticians in academia typically seek grants from government agencies that have ongoing research support programs, such as the National Science Foundation and the National Institutes of Health in the United States. Some U.S. organizations have Young Investigator Programs, described on their web sites, specifically designed for researchers with less experience who need to establish themselves.

Other possible sources of support might be a business or other agency with a particular interest in the solution of a particular problem. Typical examples might be a proposal to an appropriate agency for a new method for the analysis of data to assess the quality of health care or for handling nonrespondents in census or sample studies. In some other cases, statistical research or other work may be conducted in support of funded projects within the university.

Preparing the Proposal. Our comments about the preparation of proposals for project funding in Section 9.2.5 were mainly addressed to business and industry. But

---

[9] Many universities or departments pay expenses for attendance at conferences only for faculty who present papers or perform some important professional function. Others may have some faculty development funds available for this purpose. Since attendance at conferences is often extremely important to advance careers, some professors decide, if need be, to shoulder personally part, or even all, of the expenses.

many of these comments also hold for academic research. Proposals developed in academia and addressed to government agencies, however, are typically more formal, and often longer, than those for, say, securing internal funding in business and industry.

Due to the highly competitive nature of many research grant programs, awards often can be made to only a small fraction of those who apply. However, the success rate varies across agencies and across programs within agencies. In some situations, you might consider submitting a revised proposal for a particular segment or extension of the work, if your original proposal is not funded—and if the feedback that you receive from the funding agency appears to provide you encouragement to do so.

In light of the potentially extensive time commitment that many proposals require, it is imperative that you be judicious and realistic in the selection of the research projects that you pursue. Rather than proposing a new line of research, your chances might be better if you respond to an agency's request to conduct statistical research in support of some project known to be of interest to the agency. For example, in developing a proposal directed at an agency that is strongly concerned with reliability assessment, you might focus on work dealing with improving the information that can be gained from nondestructive testing.

Many universities and colleges have offices that can help you identify funding sources (and sometimes also project opportunities), write proposals, prepare budgets, and help ensure that your proposal is maximally effective and properly submitted. Do not be discouraged if you do not succeed initially—and always have an alternative Plan B in mind or, better still, underway.

13.8.4 Assessment of the Quality of Research

You need to know how the quality and quantity of your research will be evaluated by your institution in tenure, promotion, and other performance evaluations.

Success in Securing Research Funding. Success in securing funding is measured by the prestige of the funding agency,[10] the scholarly nature of the work, and the size of support that you were able to obtain.

Success in Publishing Research. Published research provides a tangible way of measuring your academic research accomplishments. What counts as published research, however, and how much weight it is given, based upon its perceived level of recognized and distinguished scholarship, varies from one institution and from one department to the other.

Publication or acceptance of papers in high-quality peer-reviewed refereed journals is used as prima facie evidence of successful research. Publication of a scholarly book, or, possibly, of a chapter in a book, is also generally regarded highly, but is a route that is likely to be premature, too risky, and too time-consuming for most seeking tenure.

---

[10] Grants from the National Science Foundation tend, for example, to be rated especially highly.

Other work that is generally *not* taken as prima facie evidence of research or scholarly contributions by a professor at a research-oriented institution, but might be given varying degrees of weight elsewhere include.

- Invited keynote address at a national conference.
- Invited paper (or participation in a panel) at a national research conference.
- Incorporation of a method (developed by the professor) in a well-known software package.
- Contributed paper presented at a national research conference.
- Report of research findings to a funding agency.
- Publication in a nonrefereed journal or conference proceedings.[11]
- Invited presentation to a local chapter of a professional society, such as the ASA or the American Society for Quality.

The preceding activities are listed in approximate sequence of importance, but this also varies between and within institutions. Some universities may not count any of them as evidence of successful research.

Moreover, even refereed journals are often tiered based upon their perceived intellectual level with heavier weight given to publications in journals that are regarded as prestigious (see Sidebar 13.2).

SIDEBAR *13.2*

TIERING STATISTICAL JOURNALS

The medical profession tends to recognize publications in a few journals, such as the *Journal of the American Medical Association* and the *New England Journal of Medicine*, as being more noteworthy than those in some others. Similarly, statistics departments tend to regard highly publications in, for example, the *Annals of Statistics*, the *Journal of the American Statistical Association*, *Biometrika*, or the *Journal of the Royal Statistical Society Series B*. University administrators may seek a list of the most prestigious journals, via a web site ranking, and use these in evaluating the worthiness of an applicant's research in tenure and promotion assessments.

Another alternative might be to rank journals by "impact factor" metrics, such as how frequently papers in the journal are referenced. However, articles may be cited for a variety of reasons—such as ringing endorsements, minor refinements, and critical assessments—some of which are not indicators of high quality or importance. There may, in addition, be differences among disciplines and biases introduced by some journals that tend to favor publishing papers that reference articles published in that journal.

In addition, many applied journals are held in high regard by, at least, the researchers who work in these areas. Examples are *Technometrics* for engineering statistics, *Biometrics* for biological statistics, and the *Journal of Computational and Graphical Statistics* for computing statistics.

---

[11] Some conference proceedings are refereed, but the rigor of the refereeing processes varies and tends to be less stringent than that used by highly regarded journals.

When publishing in lesser known journals, it may be the author's responsibility to demonstrate the reputation of the journal. The same may hold, even for highly regarded statistical journals, for statisticians located in an application area department. Some business schools use the list of the *Financial Times* Top Academic Journals in the Business to establish top tier status. Of the 40 journals currently on this list, the *Journal of the American Statistical Association* is the only statistics journal. The scholarly credentials of other journals might need to be demonstrated. This may involve providing evidence of the strictness of the refereeing process or pointing to publications in the journal by distinguished colleagues.

Actual Impact of Research. One way of assessing the perceived contribution of your research is how frequently it has been cited by others in the published literature. Such information is provided by several citation indexes.[12] This, however, carries with it some of the same problems as cited in Sidebar 13.2 and it usually takes several years before an article becomes widely cited.

Another measure of the potential importance of research in some scientific areas is the application and/or issuance of patents resulting from the work. Although not unheard of for statistics, this has greater applicability in some other disciplines, such as engineering. It is also more conducive to research in an industrial or business environment in which a major goal is for a company to benefit monetarily from the research, as opposed to making the work readily accessible to as many others as possible.

A further measure of the impact of a (applications oriented) professor's research is the degree to which it has actually been used by the sponsoring agencies and others. Such information is, however, often difficult to obtain and even more difficult to measure. In addition, especially for highly theoretical work, there may be a significant time lag between when the research was conducted and its successful use in applications or in the development of new theory.

Summary. Some specific issues concerning how your research is to be evaluated that you need to have a good feel for, from the start of your appointment, include

- What is the importance of publications in prestigious journals versus success in securing research grants?
- What is the weight given to different journals in which you might wish to publish (both in statistical and in application area journals)?
- To what degree can lack of publication in perceived top-tier journals be compensated for by a higher number of lower tier publications?
- How much does writing chapters in other authors' books, edited collections, and encyclopedias count? What are the weights, if any, assigned to writing textbooks and research monographs that are not subject to the usual peer-review process?

---

[12] For example, see http://www.harzing.com/pop.htm.

- How is (possibly multiple) coauthored work assessed versus solely authored work?

- What weight, if any, is given to citations, evidence of applications of your work, and possible other criteria?

- Are other scholarly activities, such as serving as a reviewer of manuscripts submitted to journals, given any consideration, and if so, how much weight do they receive in rating your research/scholarship? If these activities do not count as research/scholarship, do they count as service to the profession?

13.9   CONSULTING CHALLENGES

We have already commented on some of the opportunities and challenges of within-the-institution (or internal) consulting from students' perspectives in Section 7.10 and of private statistical consulting in Section 12.2.5. In this section, we briefly discuss consulting—or, more broadly, project collaboration—from a faculty member's perspective.

If you are to be engaged in consulting, you should strive to settle up front (preferably in writing) the terms of your involvement, such as authorship and compensation, and understand how such work impacts tenure and promotion evaluations.

13.9.1   Internal Consulting

Statisticians in academia are frequently asked to guide students working on a thesis or help colleagues from throughout the institution.

Such activity can arise in various ways. For example, a faculty member in another department may request help on a paper that referees have criticized for inadequate statistical analysis. You may be asked to rectify the problem expeditiously without being given needed background information on the purpose of the study and how the data were collected. As is the case outside academia (Section 5.2.4 and Chapter 11), a more productive arrangement arises when you are invited, at the outset, to participate as a collaborator with responsibilities that include helping to plan all aspects of the data gathering.

Some universities, especially those with departments of statistics, operate statistical consulting centers to help graduate students or faculty from throughout the institution in the design of statistical studies and the analysis of the resulting data (Section 7.10). As a faculty member, you may be part of that center or even be leading its activities. Frequently, rather than addressing problems directly yourself, you may funnel them to students, whom you would guide and supervise.

13.9.2   External Consulting

As a statistician in academia, you may be called upon to help a business, government agency, or other external organization as a consultant, or possibly collaborator. You might also be asked to be an expert witness in a lawsuit (Section 4.6.4) or to provide instruction (e.g., through a short course) to your clients' staff in a specialized area, such as product life data analysis, analysis of spatial data, or sampling.

Most universities have provisions that allow professors to spend a limited amount of time (e.g., 2 days per month) on such external activities. Many professors find it useful to involve students in such work—although this is not always possible in light of confidentiality considerations.

13.9.3 Some Pros and Cons of Academic Consulting

Consulting can be beneficial to you as an academician because of the resulting contacts across campus and externally, the contributions that you might make by such work, and the goodwill and enhancement of your reputation that it can create. Interesting research challenges often grow out of such consulting. These can help broaden your horizon, provide practical examples for class discussion, and result in publications. External consulting may also lead to added funding that can supplement, or serve as a substitute for, research funding or, in some cases, provide additional personal income. Many such projects involve a healthy mix of consulting and applied research.

It is also possible for consulting, and especially external consulting, to become a distraction from the traditional and immediate requirements in academia of teaching and research; see Sidebar 13.3. Before getting too heavily engaged in external consulting, you need to understand fully what you will gain from such work, how it might add to and detract from your other responsibilities, and how it can be best scheduled so as to minimize the interference with other commitments.

SIDEBAR 13.3

THE CONFLICT BETWEEN CONSULTING AND OTHER ACADEMIC RESPONSIBILITIES: A CASE STUDY

A lawyer, defending a client who is potentially liable for a large sum of money for alleged improper conduct, engages a professor of statistics to review a sampling procedure that had been performed by the prosecution on the client's records. She finds important defects in how the survey had been conducted and is asked to present these in court. Some strict time deadlines are associated with the work and the court's calendar is beyond the professor's control. At the last minute, the hearings are delayed and the professor needs to sit around waiting for her turn to testify.

As a consequence, the professor is forced to miss classes and cancel office appointments. Colleagues have to pitch in to answer student questions. Time originally planned for putting together a conference presentation and finalizing a paper for publication is lost. The professor begins to wonder, but somewhat too late, whether the benefits of her participation are sufficient to compensate for the associated disruption.

13.10 ADMINISTRATIVE SERVICE CHALLENGES

Academic institutions generally have a high degree of self-governance. Overall policy is typically set by a governing board, often called the Board of Trustees. The details and implementation are largely in the hands of the administration—as represented by the president, academic vice presidents, deans, department chairs, and so on. In addition, in many institutions, the faculty has considerable influence, especially in some areas, through committees or governance bodies such as the faculty senate.

At most institutions, committee service is required of all permanent faculty, including junior faculty. The quantity of such service is measurable; its quality is more difficult to assess. One common mistake made by junior faculty is to take on too much committee (and similar) service. We suggest that you talk to senior colleagues before agreeing to committee service, especially outside of your department.

13.10.1 Responsibilities of Faculty Committees

Academic policies frequently begin with an initiative from the administration. The appropriate committee then studies and discusses the proposal and makes recommendations to the faculty of the involved department, school, or, even, the entire college or university. Faculty committees are generally responsible for making recommendations about the academic curriculum and related matters (see the next subsection). Institution administrators often are ex-officio committee members to ensure that the committee's recommendations conform to practical constraints and to institutional policy and in light of their subsequent responsibilities in implementation. Thus, such administrators often assert much influence on committee recommendations.

13.10.2 Types of Committees

In many institutions, two of the most important standing committees are those on setting curricular policy and on granting tenure and promotion. The committee on the curriculum (other names such as academic affairs are quite common) examines and approves (or disapproves) requests for new courses. Such a committee also generally reviews proposals for a new major or minor and makes recommendations to higher administrative units in (and in some cases beyond) the institution.

The committee on promotion and tenure assesses teaching, research, and other performance. It may, for example, develop guidelines on what type of work counts as research and how various types of research are to be weighted in tenure and other decisions. Based upon these guidelines, such committees then make the all-important recommendations on granting tenure or promotion of individual faculty members. In some institutions, such committees also conduct periodic merit reviews of faculty members. In most institutions, administrators, such as department chairs and deans, then use such information to make the final decisions or recommendations concerning, say, salary increases and promotions.

Other committees deal with such diverse matters as deciding whether under-performing or troublesome students should remain in the institution, how the intellectual life of students can be enhanced, athletics, and a host of other topics. Much of the work of standing committees is conducted in subcommittees. For example, the curriculum committee may have a subcommittee for the approval of new courses. Faculty may also serve on the faculty senate or a similar body. In addition, faculty may be asked to be on search committees for open faculty positions or for important positions in the administration, such as Dean and Provost.

13.10.3 Appointment to Committees

The composition of standing committees is generally specified in the institution's governance agreement. If you wish to be on a particular committee, you need to determine what you have to do or whom you need to contact to make this happen. Most important faculty committee positions are filled through elections. Members of some committees in some institutions are appointed by the administrators such as a provost, dean, or department chair. In other cases, appointments are made by elected faculty leaders. Some committee positions may be reserved for junior faculty.

13.10.4 The Pros and Cons of Administrative Service

Administrative service can provide you useful insights into—and a (small) say about—the workings of the institution that you serve. It also provides you the chance to get to know senior faculty outside your department who might be helpful to you as you seek tenure and promotion.

The biggest disadvantage of committee service is that it can be time-consuming and could, possibly, detract you from your research and teaching. As a consequence, some junior professors limit their service on committees, at least initially, only to that which is expected of them.

13.11 PROFESSIONAL SERVICE CHALLENGES

Service to the profession (Section 12.6), through organizations such as the ASA, editorial work, and organizing conferences or sessions at conferences, can enhance your professional life and visibility. In addition to providing satisfaction in "paying back" the profession that has nurtured you, such involvement can help you identify potential areas of research and new teaching approaches, as well as potential future research collaborators. Such exposure might also lead to invitations to present or publish your research. Involvement in the publications process can, moreover, provide you insights that may be helpful in getting your own research published.

Professional service, however, typically counts only modestly, at best, in most tenure evaluations. It may be more important when you are seeking promotion to full professor. Significant leadership roles such as election to be an officer of a professional society and, especially, editorship of a prestigious journal may count more.

Such leadership opportunities are, however, not readily available to junior professors and typically require prerequisite work in the trenches.

13.12 MORE ON THE ACADEMIC ENVIRONMENT

13.12.1 Relationships with Colleagues

The collegial nature of academic institutions brings faculty members together within and across departments, especially in their service on administrative or thesis committees. At the same time, in conducting their research, professors often work on selected projects alone or in small teams, often with a small group of graduate students.

Professors often need to coordinate their teaching with other instructional activities inside and outside the department. This is especially important when different professors are teaching the same course. In teaching more advanced courses, professors need to ensure that important prerequisites are covered in earlier courses. Also, coordination is required to eliminate redundancy and to maintain consistency in program offerings, and to adhere to departmental academic policy.

Collegiality works best when all show mutual respect. As a flourishing academician, you can and should have your own ideas, but you should try to ensure that they have merit in the eyes of others and that, while voicing them strongly, you also do so constructively. However, you should be willing to compromise. As a young faculty member, you should especially strive to avoid conflict with colleagues and administrators, keeping in mind, if nothing else, that some of your fellow committee members (or their close colleagues) may sit in judgment of you during the tenure, promotion, or contract renewal decision process.

13.12.2 Relationship with Students

Students come to an "institution of higher learning" to be educated and to graduate. Contact with students is an essential part of your job as a professor and is not limited to classroom teaching. It requires making yourself available to students during office hours and beyond, responding to e-mails concerning class work, and, in general, helping students become motivated, successful, and ethical statisticians or discriminating and savvy users of statistics.

Your relationship differs among different student groups. It is impossible for you to get to know each student personally in a large lecture class. However, if your department offers a statistics major, you do get to know well a, usually relatively small, cadre of undergraduate and/or graduate students, and especially those who are conducting research with you or writing a thesis under your direction. You may, in fact, be an important influence, and even a character building force, for such students.

You might also get to know students personally through department and institution-sponsored social functions and clubs. Many students enjoy participating in the social life that universities and colleges offer and like to mingle with faculty outside the classroom. Thus, many departments and colleges set up events to enhance the sense of community between students and faculty, such as dinners, parties, and other gatherings. In such more relaxed settings, you need to continue to serve as a role model to students—be it in leading discussions or in showing proper conduct. Moreover, you need to ensure that your relationship remains professional and does not create a potential conflict of interest.

After students graduate, the situation may well change. Former graduate students might eventually become collaborators—or even lifelong friends—with their ex-professors.

13.12.3 Isolated Statisticians

The Challenge. Some major universities have large and thriving statistics departments. But at other universities, many 4-year colleges, and most 2-year community colleges, statisticians might find themselves professionally isolated—at least within their departments—among, say, business faculty, medical researchers, mathematicians, or engineers.

The result is a very different environment from that in, say, a large statistics department. Not having statistical colleagues may make you feel lonely, especially if you have recently graduated from a school with a significant statistical presence. You may miss having technical discussions with colleagues on whom you can bounce off ideas, and participating in statistical seminars. The library may be reluctant to buy advanced statistics books or subscribe to journals on statistics. Other faculty may not readily understand the importance and quality of your research or appreciate the significance of your accomplishments; see Sidebar 13.4 for an example.

SIDEBAR *13.4*

RECOGNITION FROM NONSTATISTICAL COLLEAGUES: AN EXAMPLE

Election to fellowship in the ASA is a unique honor and may be the pinnacle of your career. Your feeling of euphoria upon receiving this award may be dampened by the "so what" response from colleagues outside your field. This is because in some other professions becoming a fellow might amount to little more than being certified as professionally qualified and payment of dues,[13] and does not require the rigorous process of evaluation that precedes such bestowment on a small fraction of statisticians.

---

[13] This is also the case for some statistical societies outside the United States, such as the Royal Statistical Society.

Addressing the Challenge. Statisticians strive to overcome isolation by expanding their horizons. In some universities, there may be other statisticians, or member of a closely related discipline, such as operations research, possibly elsewhere on campus. Or there may be statisticians in nearby institutions or professional activities outside academia. Many metropolitan areas in the United States, for example, have local chapters of the ASA.

Attending conferences is another good way to meet other statisticians and learn what they are doing. Sessions sponsored by the ASA Section on Statistical Education are a good place to start. Also, an informal group of Isolated Statisticians[14] organizes get-togethers at the Joint Statistical Meetings and engages in subsequent communication.

Finally, we note that there are also advantages to being embedded in an environment that is not dominated by statistics. Contact and collaboration with nonstatistical colleagues may expedite identifying important practical research opportunities, enhance communication skills, and, in general, help get you involved in addressing important real problems.

13.13 TRAINING TO BECOME AN ACADEMIC STATISTICIAN

What background and training should an academic statistician have? The short answer is one that supports all aspects of an academic career; that is, whatever is necessary to conduct research, teach, and, on occasion, consult.

Much of what we say in Chapter 7 also applies for academic statisticians. As just one example, judicious selection of your dissertation topic is highly important since it might define, at least, your early career research and publications.

13.13.1 Degree Requirements

As previously indicated, universities and most 4-year colleges require full-time faculty to hold doctoral degrees for tenure-track positions.[15] Two-year community colleges generally do not require a Ph.D., although some faculty members have such degrees.

Most statistics professors hold their final degrees in statistics. Degrees in other fields are also sometimes acceptable or even desirable. For example, individuals with majors in operations research, applied mathematics, computer science, decision sciences, or a similar discipline and strong training and/or experience in statistics may be hired as statistics professors.

---

[14] http://www.lawrence.edu/fast/jordanj/isostat.html.

[15] Offers are often made to candidates close to finishing their degrees, subject to successful completion.

13.13.2 Courses to Take

It is essential for academic statisticians, and especially those who expect to be heavily engaged in research and in extending the frontiers of statistics, to have a solid background in statistical theory. While a 1-year sequence in mathematical statistics may be sufficient to teach an introductory statistics course, it is not enough to read and fully understand some published statistical research or satisfactorily conduct your own research. Consequently, many Ph.D. programs in statistics require added courses in statistical theory.

Like their colleagues outside academia, prospective statistical academicians also need to study the core subjects suggested in Table 7.1, such as Monte Carlo simulation, (advanced) regression analysis, time series, and sampling. Furthermore, early exposure to specialized areas, such as multivariate analysis and nonparametric theory, can prove useful throughout your career. Finally, recent trends in statistics indicate an increasing importance of certain aspects of statistical computing and numerical analysis that many will want to explore in formal coursework.

13.13.3 Application Area Knowledge

A new Ph.D. with well-honed knowledge in a very narrow field may do well in a research environment, usually in the statistics department of a large university that seeks that individual's highly specific skills. But for many going into academia, narrow training in statistics is not sufficient. A professor in a business school or a school of public health will be much more effective with some substantive background in the field of application (e.g., economics and marketing for a business professor and biology and genetics for a professor in a school of public health).

A good understanding of application areas also allows researchers to identify important unsolved problems, thus presenting potential research topics. Application area knowledge will also make you a more effective collaborator and consultant, as well as help improve your classroom examples and your ability to motivate students.

To further such understanding, we urge you, like your colleagues who are planning on nonacademic positions, to seek internships in application areas while still a student. It may, after all, be the only opportunity you will have in your career to get close to the source of real problems.

13.13.4 Teaching Experience

Teaching experience is, understandably, also highly important, especially for positions with a strong teaching emphasis and may be weighted heavily in hiring new professors.

Many students gain useful experience by serving as teaching assistants and helping their professors with many of the chores that come with teaching. In some universities, you may also be asked to teach a course or help conduct sessions or lead discussion sections, using your professor as a coach and role model.

Some Ph.D. programs also offer or recommend useful courses on effective teaching.

13.13.5 Achieving a Compromise

We prescribe a proverbially large order in suggesting that, even though bound for academia, you should strive to have not only a solid theoretical and methodological background in statistics but also application area knowledge and teaching experience. Gaining all of these, and also graduating on time, may not be possible. You will, instead, need to tailor your priorities to best accommodate your career plans. You may, for example, have built significant application area knowledge as an undergraduate. In that case, you might, as a graduate student, choose to just nourish and update this knowledge by informal reading and participation in relevant seminars throughout campus.

You also need to recognize that your education, by no means, ends upon graduation (see Chapter 14). As a young professor, you may still audit courses in important subjects with which you are not familiar. Also, we urge you to consider a summer internship in industry or government, particularly if you did not do this earlier. At the same time, especially if you are in a tenure-track position, you need to ensure that such activities add to, and do not seriously detract from, the success of your research and teaching.

13.14   CAREER PATHS

Much of our discussion in this chapter has focused on achieving early success in an academic position. We now briefly consider some alternative career paths for statisticians who have established themselves in academia, extending our earlier discussion in Section 13.6.

13.14.1 Senior Academician

Many statisticians who have chosen to go into academia continue their research, teaching, administrative, and consulting activities throughout their careers. They strive to do bigger and better work, address greater intellectual challenges, and gain continued and higher recognition as authorities in their field. Such recognition may be within their own institution, such as being named a distinguished professor or gaining an endowed chair (which typically also comes with added funding). It may also be external to the institution, such as being elected to an important position or as a Fellow in the ASA, or being appointed to a prestigious government advisory panel.

13.14.2 Academic Leader

> Managing academics is like herding cats.
> —Popular saying

Another alternative is that of taking on an administrative leadership position in (or outside of) academia.

The First Step: Department Chairperson. Senior professors are sometimes asked to take on the role of department chair, often for a specified period of time. This position calls for leading, and often being deeply involved in, a wide variety of administrative functions, such as hiring new faculty, preparing evaluations of faculty performance, and interfacing with higher levels of the administration. It might also include making teaching assignments, recommending new courses for consideration by the curriculum committee, negotiating sometimes contentious issues with deans and chairs of other departments, representing the department at various internal and external functions, promoting the department and publicizing its activities, fund-raising, and addressing a wide variety of, often unpredictable, issues. To find the needed time, department chairs are almost always relieved of some, or perhaps most, of their normal teaching and research responsibilities.

Many academic statisticians regard being a department chairperson as an intrusion on what they really like doing most and do best: research and teaching. But they still take on the position for a limited time because they regard it their responsibility and also might like the change of pace and insights it provides. Some chairs still manage to undertake constructive research, teaching, and student guidance and, in general, ensure that, while doing their share for the department, they do not fall behind in pursuing their ultimate career goals. After completing their tour of duty, many happily return to their previous positions.

Some departments rotate the chairmanship among qualified senior faculty for, say, 3-year stints. In other departments, successful chairs may be reappointed for multiple terms.

Although department chair is the most common administrative position in which a faculty member might serve, other opportunities, such as associate dean or associate provost, may also present themselves.

Higher Leadership Positions. Other statisticians in academia enjoy their first academic leadership experiences as, say, department chairs. They may have made some widely recognized contributions, enjoy the challenges of academic administration, and feel ready to take on further leadership responsibilities. Such individuals may respond positively to invitations to consider, or may even apply for, jobs in, say, the dean's or the provost's office, either within their current institution or at a different one. In such positions, they will be dealing with challenges that cut across the campus. For example, associate deans may need to resolve a sensitive issue in a department whose faculty view them with suspicion in light of their earlier association with a different, and perhaps competing, department. A successful associate dean will also have an inside track to apply for a full dean position.

Some may eventually rise to become the institution's president or beyond, and will then have to assume such further responsibilities as setting long-range direction and raising funds. The career path of Albert Bowker, a highly regarded statistician who took on important leadership roles in academia, is described in Sidebar 13.5.

SIDEBAR *13.5*

ALBERT H. BOWKER: FROM RESEARCH STATISTICIAN TO UNIVERSITY CHANCELLOR

Albert H. Bowker earned a B.Sc. in mathematics at MIT and a Ph.D. in statistics from Columbia University. He joined Stanford University as an assistant professor in 1947 and wrote numerous scholarly articles and various books, including coauthoring an early, popular text on engineering statistics (Bowker and Lieberman, 1972).

Bowker helped establish Stanford's Statistics Department in 1948, and served as its chair until 1959. He developed and directed the Applied Mathematical and Statistics Laboratory from 1951 to 1963. His leadership was recognized in 1958 by his appointment as Stanford's Dean of Graduate Studies. In 1963, Bowker took on the position of Chancellor of the City University of New York. In 1971, he returned to California to become Chancellor of the University of California at Berkeley.

During his administrative career, Bowker found time to serve as President of the Institute of Mathematical Statistics (1961–1962) and of the American Statistical Association (1964).

At the 50th anniversary of the founding of the Stanford Statistics Department, Bowker said: "I am prouder of the formation of this department than anything else I have done."

Some Consequences. The greater and the more extensive your administrative responsibilities, the less time you will have to teach or perform research. In fact, if you try, for an extended period, to be an administrator and a statistics teacher/researcher, you may end up failing at both. Therefore, at some point, you will need to decide what you want your top priority to be. This is not much different from the decision that statisticians in industry and government make when they are offered management positions.

Permanent administrators generally do not, per se, have tenure and serve at the pleasure of, say, the institution's president or Board of Trustees. Those who progress to an administrative position after receiving tenure may be able to retain their tenure—and possibly use it as fallback should they lose interest or favor in their administrative roles. Often a tenured faculty member or administrator at one institution is accorded a tenured position when moving to an administrative position at a new institution.

Administrative positions in academia can be both satisfying and challenging. They place you at the crossroads of many diverging institutional interests and provide a fulfilling experience for those with the right temperament and leadership qualities.

13.14.3 Moving Beyond Academia

We noted in Section 12.3.1 that some statisticians, after spending some or much of their careers in business or government, leverage their knowledge and insights to move to a second career in academia. The reverse is also possible. For example, some academicians move to a research-oriented organization or advisory position outside academia, such as an industrial or government research lab or a private research foundation. There have also been statisticians at all levels who have taken positions in industry or government in a less research-oriented environment. Still others, and especially those who had been heavily involved in consulting, might, like colleagues in business and industry, become private consultants and/or teachers of short courses.

13.15 DOWNSIDES OF A CAREER AS AN ACADEMIC STATISTICIAN

A career in academia is not appealing to all statisticians. We summarize here various downsides, some of which have already been suggested.

13.15.1 Research

As we have seen, many academic statisticians have to actively seek, or sometimes postulate, research problems and then return to their offices to work out the solutions. This may not be as satisfying as, for example, being part of a hands-on team that helps build a more efficient and reliable jet engine or is responsible for approval of a new drug that can potentially treat a deadly disease.

The requirements to secure funding for research and to support graduate students' research—together with the high level of competition for grants and frequent low success rate—are also unattractive to some, especially when they find themselves devoting nearly as much time to this activity as to doing the research itself.

Finally, given a peer-review system that invites critical reviews, academics, like many other professionals, need a thick skin and high endurance and energy levels to be successful researchers. Despite their somewhat flexible official schedule, they may—like their colleagues in business and industry or government—have to work long hours to be successful.

13.15.2 Teaching

The challenges in teaching an introductory course to nonstatisticians and getting students interested in statistics are inspirational to some. To others, teaching students with limited motivation who just want to get through a required course with a passing grade can be a less than satisfying experience. This is especially so if you have to teach large classes without much opportunity to get to know students individually, and have to do so term after term. Some also feel that, despite its importance, teaching introductory statistics courses interferes with their prime interest of conducting research and teaching advanced courses to more motivated students.

13.15.3 Service

Some professors enjoy committee work, regard it as a change of pace, and appreciate its importance and the interesting insights into the operation of the institution that it can provide. Others, however, find such activities to be somewhat of a drudgery and overly political, and resent the time it takes away from their research and teaching.

13.16   BRIGHT SIDES OF A CAREER AS AN ACADEMIC STATISTICIAN

It should be evident from our earlier discussion that there is also much to be said in favor of a career as an academic statistician. We reiterate some of the reasons in this section.

13.16.1 Freedom of Research

Academia can be a world in which one can engage in theoretical work, selected from one's own broad list of potential topics without necessarily having to be concerned about immediate—or, in some cases, even eventual—applicability. The major criterion for success of your work is having your results published in an esteemed journal. Beyond that, you can enjoy the thrill of discovery. This is attractive to many scholars. Your freedom is, however, on occasion, constrained by the requirement to secure funding to conduct the research, and, subsequently, to follow the guidelines and expectations of your funding agencies.

13.16.2 The Joys of Teaching

> One of the invigorating things about academia is that even though I keep on getting older, the students remain the same age.
>
> —Stephen Berk

For many professors, teaching and interacting with learners at all levels—and especially young and fresh minds—is a major attraction of academic life. Teaching statistics challenges you to explain statistical concepts in a way that makes sense to students, while still being true to your subject. This often leads you to new insights and increased understanding. Most importantly, seeing students succeed as a result, at least partially, of your teaching and guidance can be a great source of satisfaction.

13.16.3 Service Opportunities

For many professors, the free administrative structure of academia is a major attraction. Each professor, irrespective of rank, can have a say in determining the direction of the institution. In your committee work, you can contribute your share in shaping your institution.

13.16.4 Limited Supervision

Department chairs, deans, and even sometimes the institution president have some say in what you do and especially the courses that you teach (and how many). But, unlike the nonacademic world, there are usually no strong bosses or managers in academia that dictate your workday or, in general, tell you what to do—and especially so, after you have gained tenure. Moreover, being your own boss to a great degree also offers you much flexibility in arranging your work schedule.

13.16.5 Sabbaticals

In many institutions, there is a long-standing practice that faculty may apply period-ically (e.g., every 7 years) for a sabbatical, ranging in length from a few months to a year.[16] Faculty members usually receive full salary (often for a one-semester leave) or a substantial fraction of their regular salary (typically for a full-year leave).

A sabbatical provides you a respite from normal duties and allows you to focus on a particular project, such as writing a book or completing a major research task. It also presents an opportunity to pursue new interests, visit other academic institu-tions, or work in industry or government to gain insight into application areas (see Iglewicz, 2009)—an option that, in our opinion, not enough academicians consider.

Award of sabbaticals is often competitive and might be restricted by resource limitations. Eligible faculty typically submit a proposal describing, in some detail, the purpose of the sabbatical, the proposed plan of execution, and the expected results. In many institutions, there is a further requirement to provide a summary report of accomplishments at the end of the leave.

13.16.6 Prestige

Academics are, in light of their positions, typically accorded a high level of public recognition and respect—at least, outside of their own university.

13.16.7 Great Environment

Last but not least, in academia you can enjoy the advantages and inspiration of an academic environment and the company of colleagues, many with great minds, who, like you, have chosen a scholarly career.

13.17 A CAREER AS A STATISTICIAN IN ACADEMIA: A SUMMARY COMPARISON

Table 13.1 (Section 13.1) provided a detailed summary comparison of a career as a statistician in academia (as described in this chapter) with that in other areas.

13.18 MAJOR TAKEAWAYS

- Statisticians in academia may be located in a university statistics department, a closely related department—such as mathematics or information science—or in an application area, such as the business school or the school of public health. Some large schools have many statisticians. Smaller institutions, and especially

---

[16] Some institutions give different names to such programs to convey their purpose, for example, Faculty Improvement Leave or Faculty Professional Development Assignment.

community colleges, may have only a single statistician in a department or on campus.

- Teaching and research are the two major responsibilities of statisticians in academia, followed by participation in administration through committee activities. Statisticians are also engaged, to varying degrees, in consulting and in service to the profession. The specific mix and emphasis varies greatly between institutions and schools within an institution.

- Tenure-track positions are typically sought by those who aspire to a career as statisticians in academia. It is essential that you understand the process and criteria for granting tenure, such as the relative emphasis on research versus teaching, at your institution and how your performance on each will be assessed.

- Granting of tenure opens new horizons and allows you to engage in riskier and longer term activities.

- Major nontenure-track positions include lecturer/instructor, clinical, visiting, postdoc, and adjunct appointments.

- Statisticians teach introductory statistics courses (presenting some important challenges) to the general student body and advanced courses (especially in schools that offer degrees or minors in statistics). Course evaluations are an important tool for assessing teaching performance.

- The conduct of successful research is a key requirement in many universities. Publication in scholarly journals is a major measure of success. Obtaining funding from a government or other agency to support research has become increasingly important.

- Statisticians in academia are engaged both in consulting within the institution (perhaps through a statistical consulting center) and with external customers in business, industry, or government.

- Much of the governance of an academic institution is channeled through committees of faculty members—such as those on setting curricular policy and on granting tenure and promotion.

- Universities and most 4-year colleges require full-time faculty to hold Ph.D. degrees, especially for tenure-track positions. All academic statisticians need a firm foundation in statistical methodology. Those who expect to become engaged in research, in addition, need a strong background in statistical theory. Knowledge of an application area and teaching experience are also important.

- Many statisticians in academia aspire to be recognized for their sustained and continually more significant research, teaching, and professional accomplishments. Some aspire to leadership positions.

- The disconnect from real-world problems, the frequent requirement to seek research funding, the peer-review process, the repeat teaching of introductory courses, and the occasional drudgery of committee work have been cited as downsides of a career as a statistician in academia.

- The freedom and research opportunities, the challenge of engaging students, the opportunity of having a say through committee activities, the limited supervision, the opportunity for sabbaticals, the associated prestige, and the inspirational nature of the campus environment are among the bright sides of a career as a statistician in academia.

DISCUSSION QUESTIONS

(* indicates that question does *not* require any past statistical training)

1. *What is the mix of academic appointments within the institution that you are attending or recently attended? Based on your perspective, what recommendations do you have for changing this mix?

2. *Discuss the pros and cons of a statistics professor being located within versus outside a statistics department. What would be your preference?

3. *What are the criteria for granting tenure, and how are they weighted, at an institution to which you might apply (or that you currently or recently attended)?

4. Critique your first course in statistics. In which of the three categories described in Section 13.7.1 did this course fall? How would you improve the course?

5. *Consider one or more course evaluations that you have been asked to complete. How might you change these to help improve the course and the professor's performance?

6. Identify one or more research topics that you might want to pursue in a tenure-track position and agencies that you might approach to support such research.

7. For the research topic(s) that you identified in response to the preceding question, how would you respond to the questions that "you need to ask yourself," raised in Section 13.8.2.

8. *How does consulting in academia tend to differ from that in business and industry and government?

9. *What is the faculty committee structure in the university or college that you are attending or recently attended? How are appointments to committees made?

10. How should the training for a statistician aiming to go into academia differ from that for one who is planning to go into business and industry or government?

11. *Should you decide to go into academia, what career path is most attractive to you and why?

12. *What aspects of academia do you personally find most and least appealing?

13. *Consider the comparison of the environment in academia versus that in business, industry and government in Table 13.1. How does this correlate with your experience to date? What changes and additions would you suggest in this comparison?

14. *In the comparison of the environment in academia versus that in business, industry, and government given in Table 13.1, for each topic rate your preference according to the following score:

2: Strong preference for business, industry, and government.

1: Modest preference for business, industry, and government.

0: Indifferent.

− 1: Modest preference for academia.

− 2: Strong preference for academia.

Then, for each item listed, assign a weight from 1 (unimportant) to 5 (highly important) that measures the importance that you would assign to that item, and obtain a weighted average score for your overall preference. What does that tell you?

CHAPTER 14

MAINTAINING THE MOMENTUM

14.1 ABOUT THIS CHAPTER

> Your formal schooling may be over, but your education never ends.
>
> —Popular saying

Statisticians, no matter how well trained, need to invest in continuing self-development. We are in a rapidly advancing field. Those who do not stay abreast will be left behind. In this way, statisticians do not differ from professionals in other rapidly advancing fields, such as medicine, biology, and computer science.

New methods and tools are being developed all the time. Those who received their formal coursework some time ago were, most likely, not trained in such areas as simulation-based inference, statistical learning theory, spatial analysis, and Markov chain Monte Carlo methods for Bayesian analysis, which are now part of the standard curriculum at many graduate statistics departments. In addition, there are numerous emerging applications—in such areas as biology and genetics, environment and ecology, atmospheric sciences, and Internet social networks—that were essentially unknown some years ago, and in which practicing statisticians are involved today. In the same way, you can expect numerous new technical developments that you will need to keep up with through the course of your career.

In addition, you will need to learn about specialized tools that pertain particularly to your selected application area and/or job that you did not study during your formal education, as discussed in Section 7.7.5. And if all of that is not enough, you also need to stay abreast of new developments in software and hardware as well as new technology.

Less tangible, but equally important, is the need to network and to get to know others engaged in similar types of work, as well as technical leaders in your fields of application. This can provide you the chance to learn of emerging new directions, play off ideas on others, and exchange experiences. It may even lead to new career opportunities. Such networking is especially important if you are an isolated statistician with limited peer contact in your work environment.

There are various ways that you can acquire new knowledge. We will discuss

- Some Internet resources.
- Formal education.

A Career in Statistics: Beyond the Numbers, Gerald J. Hahn and Necip Doganaksoy.
© 2011 John Wiley & Sons, Inc. Published 2011 by John Wiley & Sons, Inc.

- Technical and professional journals.
- Technical conferences.

In the final section, we briefly speculate on some exciting future opportunity areas for statisticians. We apologize if much of this chapter sounds like a catalog of offerings. But we hope it does convey the wealth of learning opportunities available to you and will serve you as a useful reference guide for the future.

14.2 SOME INTERNET RESOURCES

The Internet provides a useful and constantly expanding source of information about statistical concepts, tools, and applications to update your knowledge base. The sites briefly described here are just a few examples of the diverse sources of available technical information. The web site of the American Statistical Association (ASA), as well as those of other professional societies, provides current information of interest to the statistical community.

Chance News.[1] Chance News reviews current issues in the news that use probability or statistical concepts. A typical issue included

- Thought-provoking quotations, for example, "Like dreams, statistics are a form of wish fulfillment."—Jean Baudrillard.
- Forsooth (i.e., some questionable assertions), for example, "People who live longer have a greater chance of developing cancer in old age." Heard on the "Today" news program on BBC Radio 4.
- Recent news items (many with discussion questions), for example,
  - How Did the Polls Do? (A discussion of 2010 preelection polls), *Huffington Post*, November 3, 2010.
  - Mammogram Benefit Seen for Women in Their 40s, *New York Times*, September 29, 2010.
  - Using Facebook Updates to Chronicle Breakups, *New York Times*, Bits Blog, November 3, 2010.

The Engineering Statistics Handbook.[2] The stated goal of this handbook is to help scientists and engineers incorporate statistical methods in their work as efficiently as possible. It was developed jointly by SEMATECH and the U.S. National Institute of Standards and Technology and contains many case studies, especially from the semiconductor industry. It provides a useful reference guide for practitioners and statisticians. The discussion is in eight sections: Explore, Measure, Characterize, Model, Improve, Monitor, Compare, and Reliability.

[1] http://www.causeweb.org/wiki/chance/index.php/Main_Page.
[2] http://www.itl.nist.gov/div898/handbook/index.htm.

The Current Index to Statistics.[3] The *Current Index to Statistics* (CIS), a joint venture of ASA and the Institute of Mathematical Statistics (IMS), has since 1975 provided a bibliographic index to publications in statistics, probability, and related fields. ASA members have free access to CIS. The online CIS extended database examines the entire contents of 160 "core journals" and selected articles from about 1200 additional journals and, to date, over 11,000 books, published since 1975.

14.3   FORMAL EDUCATION

> Recently I attended a lecture on biology and genetics. It was eye-opening for me to see how much of the material that is currently taught in high school was totally new to me.
>
> —Michael Chernick

14.3.1   Courses and Degree Programs

You can continue your education while on the job by taking selected courses at a local college or university, subject to the school's accessibility and your time constraints. Many schools provide part-time evening or distance education programs to attract those with full-time jobs. Online courses and programs offer more flexibility and continue to gain popularity.

Your studies may be part of an advanced degree program either in statistics or in an application area. Some with master's, and even Ph.D., degrees in statistics have gone on to obtain MBA degrees. Your goals may be less ambitious—such as taking a relevant course or two in an area of special pertinence to your current work or your desired career path.

14.3.2   Short Courses

Short courses on specialized topics provide another alternative for enhancing your knowledge. The duration of short courses typically ranges from half a day to a week. These generally deal with specialized subjects (e.g., design of experiments, reliability, and life data analysis) or emerging areas (e.g., hierarchical Bayes methods) and new software packages. Such courses are often given in conjunction with professional meetings—such as the Continuing Education courses offered at the Joint Statistical Meetings (Section 14.5.1). Others are offered by universities or by consulting, software, or other companies.

Short courses are also offered increasingly online in single shots or installments. For example, various ASA sections offer webinars on a regular schedule.[4]

---

[3] http://www.statindex.org/.

[4] http://www.amstat.org/education/weblectures/index.cfm.

14.4 TECHNICAL AND PROFESSIONAL JOURNALS

Your first exposure to professional journals will likely occur as a student. A good way of keeping yourself up to date is by regularly reading current issues of journals. Many journals are published by professional societies; others are issued by publishing houses or other companies or institutions. Online (only) journals are also rapidly gaining popularity.

14.4.1 General Interest Statistical Journals and Newsletters

The ASA issues general interest publications at various levels.[5]

Amstat News is issued monthly both in an abridged hard copy version, sent to all ASA members, and in a longer online version,[6] which is accessible free of charge to the general public. It provides news items of interest to ASA members, including announcements and descriptions of forthcoming meetings and events, chapter and section news, job postings, and a summary of articles appearing in other ASA journals. Additional features include messages from the ASA President, results of salary surveys, stories of successful statisticians, and numerous articles about the practice of statistics (many of which have been referenced or quoted in this book). There are also occasional articles dealing with issues of general interest (such as statisticians' comments on the status of climate change science; see Smith et al., 2010).

ASA members currently receive *Significance*, issued quarterly under a joint agreement between ASA and the Royal Statistical Society (RSS). This magazine-like journal aims "to communicate and demonstrate in an entertaining and thought-provoking way the practical use of statistics in all walks of life and to show how statistics benefit society." Articles in a typical issue included ones on

- London murders: a predictable pattern?
- Too many males in China: the causes and the consequences.
- Safety in numbers: how human beings react to fire.
- The flavor of whiskey.
- Big Brother and state control—a new career for statisticians.

ASA members also have free electronic access to

- *The American Statistician*, published quarterly: This journal publishes articles of general interest to the statistical profession. It includes sections dealing with Statistical Practice, General (Topics), History Corner, Teacher's Corner, Interdisciplinary (subjects), Statistical Computing and Graphics, and Reviews of Books and Teaching Materials.

---

[5] ASA also published 51 issues of *STATS, The Magazine for Students of Statistics*, up to 2009. *STATS* strived to provide "a lively and entertaining look at how statistics can be used to study everyday questions." This magazine also featured articles of particular interest to students. These past issues of *STATS* are worth looking at to get a better feel for the breadth of applications of statistics.

[6] http://magazine.amstat.org/.

- The *Journal of the American Statistical Association* (JASA), also published quarterly: This journal describes itself to be "long ... considered the premier journal of statistical science. Established in 1888, JASA focuses on statistical applications, theory, and methods in economic, social, physical, engineering, and health sciences and on new methods of statistical education." JASA also claims to be the most frequently cited journal in the mathematical sciences. In general, articles in JASA are more theoretical than those in *The American Statistician*. It also includes an extensive book review section.

Also published by ASA, and available by subscription, *Chance* magazine "is intended for everyone interested in the analysis of data. Articles showcase statistical methods in the social, biological, physical, and medical sciences." Typical articles deal with such subjects as

- Using statistics to uncover cheating on an exam.
- Predicting whether a kicker in (American) football will make the extra point.
- How to improve your odds of winning the office pool.
- The role of statistics in the analysis of functional magnetic resonance images.
- Finding cancer signals in mass spectrometry data.
- Using statistics in war crimes trials.

The RSS also publishes three general interest journals (*Series A: Statistics in Society*, *Series B: Statistical Methodology*, and *Series C: Applied Statistics*), as do individual national statistical societies and organizations, for example, *Sankhya* (from the Indian Statistical Institute), the *Canadian Journal of Statistics*, and the *Scandinavian Journal of Statistics* (from the statistical societies of Denmark, Finland, Norway, and Sweden).

The Institute of Mathematical Statistics publishes several journals. *Statistical Science* is mainly a review journal, containing both methodological and applied expository papers. *Annals of Applied Statistics* publishes papers in "all areas of applied statistics." Some other IMS journals (e.g., *Annals of Statistics*) tend to be highly theoretical and aimed mainly at statisticians conducting research, principally in academia.

Further, the International Statistical Institute (ISI) publishes the *International Statistical Review* (ISR). The ISR is the flagship journal of ISI and publishes a wide variety of articles from very applied to very theoretical submitted by authors from around the world. An added feature of ISR is its extensive "Short Book Reviews" section, which gives one- to three-paragraph reviews of recently released books in all areas of statistics. While full access to the journal is restricted to members of ISI and its sections, important articles are often made available to the general public.

14.4.2 Specialized Journals

From the ASA. The ASA also publishes (or copublishes) numerous journals that pertain to specific application or technical areas. Online access is currently provided to ASA members to

- *Statistics in Biopharmaceutical Research.*
- *The Journal of Agricultural, Biological, and Environmental Statistics.*
- *The Journal of Business and Economic Statistics.*
- *The Journal of Quantitative Analysis in Sports.*
- *Statistical Analysis and Data Mining.*
- *Journal of Statistics Education*

In addition, the ASA is involved in the publication of various other specialized journals, such as the *Journal of Nonparametric Statistics*, the *Journal of Statistical Software*, and *Technometrics* (published jointly with the American Society for Quality (ASQ)); see the ASA web site for a current complete list of such publications. ASA also partners with the Society for Industrial and Applied Mathematics (SIAM) on joint publications of books on "hot topics" in applied statistics under the ASA–SIAM Series.

From Other Societies or Organizations. Societies in various application areas also publish journals that frequently have extensive statistical content. The ASQ publishes the *Journal of Quality Technology, Quality Engineering*, and *Six Sigma Forum Magazine* and includes a Statistics Roundtable series *in Quality Progress*, its flagship publication.

Publications from the Institute for Operations Research and the Management Sciences (INFORMS) include *Decision Analysis, Interfaces, Management Science, Operations Research*, and *OR/MS Today* (its general magazine for its approximately 10,000 members). Articles in these publications frequently contain much statistical content.

The *Statistical Journal of the IAOS* (International Association for Official Statistics) publishes "articles to promote the understanding and advancement of official statistics and to foster the development of effective and efficient official statistical services on a global basis." Statistics Sweden's *Journal of Official Statistics* contains "articles on statistical methodology and theory, with an emphasis on applications."

Other professional organizations in various application areas also publish journals that contain articles of interest to statisticians working in these areas. Journals dealing with pharmaceutical applications include *Applied Clinical Trials, Biostatistics, Statistics in Medicine*, the *Drug Information Journal*, the *Journal of Biopharmaceutical Statistics, Pharmaceutical Statistics, Statistics Issues in Drug Development*, and *Statistical Methods in Medical Research*. Also, the Institute of Electrical and Electronics Engineers (IEEE) publishes more than 100 journals, including the *IEEE Transactions on Reliability, IEEE Transactions on Knowledge and Data Engineering*, and *IEEE Transactions on Neural Networks*.

Various sections of ASA (Section 14.5.1) issue newsletters that describe recent or forthcoming events of interest to their readers and often also include brief technical articles, as do section or division newsletters of other societies. For example, the ASQ Divisions on Reliability and on Statistics publish newsletters of interest to statisticians working in quality and reliability improvement.

14.5   TECHNICAL CONFERENCES

14.5.1   National Society Meetings

The yearly Joint Statistical Meetings (JSM) are the largest gathering of statisticians in the world, regularly attracting over 5000 attendees. It currently takes place over a 5-day period during early August in different locations in the United States and Canada. It is the annual national meeting of ASA and is typically held jointly with the International Biometric Society (IBS), the Institute of Mathematical Statistics, the Statistical Society of Canada, the International Indian Statistical Association, and the International Chinese Statistical Association. Formal meeting activities include oral presentations, panel discussion sessions, poster presentations, and ASA section business meetings, as well as continuing education courses. The meetings also feature an extensive exhibit hall, a job placement service (Section 8.3.2), and various social activities.

Based upon the authors' combined attendance at over 70 such meetings (and still counting), we present, in Sidebar 14.1, our assessment of the pros and cons of attending the JSM—and propose some strategies should you decide to do so.[7]

SIDEBAR 14.1

THE JOINT STATISTICAL MEETINGS: A CRITIAL ASSESSMENT

The Pros. The JSM provides a great opportunity for intellectual lubrication—a one-shot yearly glimpse at some of the new developments in the field and an opportunity to see the leaders in your profession in action. It also allows you, at least, a fleeting glance, hallway conversation, or even a meal with professional colleagues that you may not have seen since the last time that you both attended the JSM. The exhibition area provides you an update of recently published books and software. The placement service allows job seekers and employers to identify and get to know each other and often provides candidates a gateway to on-site job interviews. Also, you may avail yourself of the opportunity to present a contributed paper on some of your recent work or might even find yourself invited to give a paper or to participate in a session as a panel member or chairperson. This provides you professional exposure and a chance to meet others with similar interests.

The Cons. Critics of the JSM describe them as a madhouse with so much going on that it is difficult to focus on anything. There may be as many as 10 continuing education events, 8 invited technical paper sessions, 15 technical contributed sections, and 5 business meetings all going on at the same time—in addition to 120 exhibitors, the job placement service, and various other activities.

Our Recommendations. Both of the preceding assessments are correct; how to weight them is a matter of personal choice. We encourage you to attend, at least, one JSM early in your career, and use this experience as a basis for your future decisions. To gain maximum

---

[7] Also see http://www.chrisbilder.com/what/jsm.htm for further suggestions.

benefit, we propose that you plan your tentative schedule in advance—deciding from the program and online published abstracts of papers, which sessions or parts of sessions you plan to attend in each time slot. Make sure that your schedule also includes an opportunity to visit the exhibit area, participate in the business and social meetings of the ASA sections that interest you, and "visit" with colleagues. When you arrive at the meeting, get acquainted with the layout and how to get from one place to the next between meetings (or even between talks) expeditiously.[8] Try to keep good notes, especially of ideas and contacts that you wish to follow up. And go with the recognition that, invariably, you will not be able to cover everything that you would like.

Other national associations, such as ASQ, INFORMS, and the Decision Sciences Institute, hold annual meetings similar to the JSM, as do the statistical societies of various other countries. In addition, the International Statistical Institute World Statistics Congress meets in different locations throughout the world every 2 years.

The 3-day Annual Conference of the European Network of Business and Industrial Statisticians (ENBIS), held since 2001, aims to bring together statistical practitioners, academic statisticians, consultants, Six Sigma black belts, and other professionals who are involved in business and industrial statistics in Europe. This meeting typically involves a mix of plenary and parallel sessions.

14.5.2 Other National Meetings

There are numerous smaller and more specialized national or regional meetings worldwide of interest to applied statisticians. Such meetings are typically more limited and focused than the JSM. Many of these are sponsored or cosponsored by one or more sections of ASA (in the United States) or by a section of some other professional society; see Sidebar 14.2.

SIDEBAR 14.2

AMERICAN STATISTICAL ASSOCIATION AND OTHER PROFESSIONAL SOCIETIES

Chapter 1 provided a brief description of professional societies in which statisticians become involved—such as the ASA, the statistical associations of other countries, the ASQ, and INFORMS.

These and various other professional societies also have divisions, sections, or groupings of special interest to statisticians. Some examples are the Statistics Division of ASQ and the IAOS, a section of the ISI that strives to "bring together producers and users of official statistics ... to promote the understanding and advancement of official statistics and to foster the development of effective and efficient official statistical services on a global basis."

[8] Your path might, however, be slowed down as you encounter a former schoolmate or colleague along the way.

A few examples of such meetings are

- The Fall Technical Conference: Sponsored by the Chemical and Process Industry Division and the Statistics Division of ASQ and the Sections on Physical and Engineering Sciences and on Quality and Productivity of ASA. This meeting, conducted yearly since 1947, runs for 2 days. Three technical sessions take place simultaneously.

- The Quality and Productivity Research Conference: Sponsored by the Quality and Productivity Section of ASA. This conference, held each spring since 1984, typically features two or three sessions conducted in parallel over a 2.5-day period.

- The Annual Deming Conference on Applied Statistics: Cosponsored by the NY/NJ Metropolitan Section and the Statistics Division of ASQ and the Biopharmaceutical Division of ASQ. This 3-day conference, held yearly since 1946 and meeting in early December in Atlantic City, NJ, emphasizes pharmaceutical applications. It features two parallel sessions each with single half-day expository presentations.

- The Food and Drug Administration (FDA) Industry Statistics Workshop: Cosponsored by the FDA and the ASA Biopharmaceutical Section. This 3-day meeting features presentations and short courses on statistical topics of interest to statisticians supporting the pharmaceutical industry. It is usually held in September.

- The Spring Meeting of the Eastern North American Region of the International Biometric Society: This annual meeting (held usually in March) focuses on biostatistics.

For statisticians in government and official statistics, major conferences include

- The Annual Conference of the Association of Public Data Users (Section 3.8).

- Conferences organized by the Council of Professional Associations on Federal Statistics and the Federal Committee on Statistical Methodology, as well as meetings of the National Academy of Sciences' Committee on National Statistics.

In addition, some organizations sponsor sessions or satellite conferences in conjunction with meetings of national societies. The International Conference on Statistics in Business and Industry, sponsored by the International Society on Industrial and Business Statistics (a section of ISI), has, since 1999, held both its own meetings and satellite meetings in conjunction with ISI World Statistics Congresses.

Amstat News and the ISI online calendar[9] provide an update on meetings worldwide of interest to statisticians. The following is a sample of some recently listed meeting announcements:

- Workshop on Statistical Challenges and Biomedical Applications of Deep Sequencing Data, Ascona, Switzerland.

---

[9] http://isi.cbs.nl/calendar.htm.

- University of Florida Annual Winter Workshop: Emerging Methods in Environmental Statistics.
- Annual Symposium on the Interface: Computing Science and Statistics.
- International Conference on Design of Experiments.
- Federal Forecasters Conference.
- Living to 100 Symposium (our favorite).

There are also a variety of other yearly national meetings sponsored by various organizations that although not directed at statisticians are of interest to those engaged in different application areas. For example,

- The Reliability, Availability, and Maintainability Symposium: This 4-day symposium is jointly sponsored by 10 professional organizations, including the Electronics and Reliability Divisions of ASQ and the Reliability Society of the IEEE.
- The Drug Information Association Annual Meeting: This 3-day conference deals with topics in drug development and with a variety of other related subjects.
- The Society for Clinical Trials Annual Meeting: This 4-day meeting brings together individuals from various disciplines involved in conducting clinical trials.

14.5.3 Local Meetings

The ASA has over 75 chapters spread throughout the United States. These typically hold monthly meetings on selected technical topics. Some chapters also conduct symposia or special courses. Involvement in chapter activities provides an easy vehicle for updating yourself and interacting with local colleagues. Other professional organizations, such as the ASQ, operate similarly. IBS and IMS jointly organize the WNAR (Western North American Region) Annual Meeting. In Europe, ENBIS also sponsors periodic local meetings through its local networks (e.g., bENBIS in Belgium).

14.6 FUTURE OPPORTUNITY AREAS

> The sexy job in the next ten years will be statistician—and I'm not kidding.
>
> —Hal Varian

We have described in various places, such as in Chapter 2, the evolution of the application of statistics as new areas have emerged. This evolution, triggered by advances in information science and technology, will, undoubtedly, continue; see Lindsay et al. (2003/2004), Raftery et al. (2001), and Straf (2003).

In Hahn and Doganaksoy (2008), we briefly identify a number of new statistical application areas. These include bioinformatics and statistical genetics, nanotechnology, and automated tracking. We should add environmental science in general and climate change in particular to this list (Section 4.5.3).

And the list continues to grow. To maintain the momentum, you need, through the course of your career, to be on your toes to recognize new opportunity areas, as they arise, and even consider moving into them. Also, new resources will become available to you. We hope that the many channels that we have identified in this chapter will be helpful in bringing these to your attention rapidly and allow you to take advantage of them.

14.7 MAJOR TAKEAWAYS

- Statistics is a dynamic field. Advances will continue to be driven by progress in information technology and in application areas, as well as by new theoretical developments. Statisticians need to stay abreast so as not to fall behind.

- The Internet is an important source of information. There are numerous sites of special interest to statisticians.

- Some statisticians continue their education by taking courses (and even degree programs) or short courses in specialized areas.

- A large number of general interest and specialized technical journals and magazines provide updates on recent developments.

- Professional society meetings are an important vehicle for learning about recent advances and provide important networking opportunities. Such meetings range from small highly focused and local get-togethers to large society gatherings such as the Joint Statistical Meetings.

- New areas, such as bioinformatics and statistical genetics, nanotechnology, automated tracking, and environmental science, will provide new opportunities for statisticians. Others will continue to emerge and new resources will become available. You need to position yourself to take advantage of these.

DISCUSSION QUESTIONS

(* indicates that question does *not* require any past statistical training)

1. What specific steps are you taking, or contemplating, to keep abreast (and stay ahead) in your career?

2. What are some web sites that you have found to be especially useful for your professional growth?

3. What technical journals and statistical meetings seem most beneficial for your professional development both now and potentially in the future?

4. Describe any professional meetings that you have attended and your experiences and recommendations to others who are considering attending similar meetings in the future.

5. *What are some emerging technical areas that you especially want to keep abreast of in future years. Are there any that you would add to our list?

REFERENCES[1]

Albert, J., J. Bennett, and J.J. Cochran (Editors) (2005). *Anthology of Statistics in Sports*, ASA–SIAM Series on Statistics and Applied Probability, Philadelphia, PA: SIAM.

Albert, J., and R.H. Koning (Editors) (2008). *Statistical Thinking in Sports*, Boca Raton, FL: Chapman & Hall/CRC.

Amstat News (2009a). "A Statistician's Life (featuring J.M. Gonzales, N. Kannan, J.L. Moreno, J. Tooze and G. Verbeke)," *Amstat News*, September Issue (#387), 56–62.

Amstat News (2009b). "American Statistical Association to Begin Program of Voluntary Individual Accreditation of Statisticians," *Amstat News*, September Issue (#387), 8.

Amstat News (2009c). "Professional Ethics: Committee Updates Goals and Guidelines," *Amstat News*, December Issue (#390), 61.

Amstat News (2010a). "Statistician's View Given at Congressional Briefing on Climate Science," *Amstat News*, July Issue (#397), 14.

Amstat News (2010b). "Statistics Without Borders Pushes Forward," *Amstat News*, January Issue (#391), 31.

Andrade, C. (2007). "Confounding," *Indian Journal of Psychiatry*, Volume 49, Number 2, 129–131.

Aschengrau, A., and G.R. Seage III (2008). *Essentials of Epidemiology in Public Health*, Second Edition, Sudbury, MA: Jones and Bartlett.

Asher, J. (2008). "A Statistician at Home," *Amstat News*, June Issue (#372), 7–8.

Asher, J. (2009). "Statisticians Seen, Heard at Human Rights Coalition Launch," *Amstat News*, April Issue (#382), 17–19.

Asher, J. (2010a). "The Role of Statistics in Reforming U.S. Foreign Assistance Policy," *Amstat News*, January Issue (#391), 11–12.

Asher, J. (2010b). "Collecting Data in Challenging Settings," *Chance*, Volume 12, Number 2, 6–13.

Bacon, D. (2000). "Integrity in Statistics," W.J. Youden Memorial Address at the 43rd Annual Fall Technical Conference, *ASQ Statistics Division Newsletter*, Winter Issue, 5–10.

Banks, D. (2008). "Snakes and Ladders: Building a Career in Statistics," *Amstat News*, September Issue (#375), 34–37.

Beale, C. (2004). "Reflections on 50 + Years as a Demographic Statistician," *Statistical Research Division Seminar*, U.S. Census Bureau, February 25. (Reprinted in the April 2004 issue of *Amstat News*.).

Berk, K.N., and P.M. Carey (2009). *Data Analysis with Microsoft Excel: Updated for Office 2007*, Third Edition, Boston, MA: Brooks/Cole.

Best, J. (2001). *Damned Lies and Statistics: Underlying Numbers from the Media, Politicians, and Activists*, Berkeley, CA: University of California Press.

Best, J. (2004). *More Damned Lies and Statistics: How Numbers Confuse Public Issues*, Berkeley, CA: University of California Press.

Bisgaard, S., and R.J.M.M. Does (2009). "Health Care Quality—Reducing the Length of Stay at a Hospital," *Quality Engineering*, Volume 21, Number 1, 117–131.

Blumberg, C.J. (2009). "Opportunities in the Department of Energy for Graduates with Bachelor's Degrees in Science and Mathematics," Invited presentation at the *Joint Statistical Meetings*, Washington, DC.

---

[1] The web sites referenced here were in working condition as of April 2011.

Boen, J.R., and D.A. Zahn (1982). *The Human Side of Statistical Consulting*, Belmont, CA: Lifetime Learning Publications.

Bose, J. (2010). "Why Statistics," *Amstat News*, September Issue (#399), 27–29.

Bot, B.M. (2009). "What Can I Do with an Undergraduate Degree in Statistics?" Discussion, *Joint Statistical Meetings,* Washington, DC.

Bowker, A.H., and G.J. Lieberman (1972). *Engineering Statistics*, Second Edition, Englewood Cliffs, NJ: Prentice-Hall.

Box, G.E.P., J.S. Hunter, and W.G. Hunter (2005). *Statistics for Experimenters: Design, Innovation, and Discovery*, Second Edition, Hoboken, NJ: Wiley.

Bracewell, P.J., and K. Ruggiero (2009). "A Parametric Control Chart for Monitoring Individual Performance in Cricket," *Journal of Quantitative Analysis in Sports*, Volume 5, Number 3, Article 5.

Britz, G.C., D.W. Emerling, L.B. Hare, R.W. Hoerl, S.J. Janis, and J.E. Shade (2000). *Improving Performance Through Statistical Thinking*, Milwaukee, WI: ASQ Quality Press.

Brown, E., and R. Kass (2009). "What is Statistics?" (with discussion), *The American Statistician*, Volume 63, Number 2, 105–123.

Bruce, P., and J. Bose (2010). "Taking a Chance on Statistics," *Amstat News*, September Issue (#399), 24–29.

Bryce, G.R. (2002). "Undergraduate Statistics Education: An Introduction and Review of Selected Literature," *Journal of Statistics Education*, Volume 10, Number 2. Retrieved from http://www.amstat.org/publications/jse/v10n2/bryce.html.

Bryce, G.R. (2005). "Developing Tomorrow's Statisticians," *Journal of Statistics Education*, Volume 13, Number 1. Retrieved from http://www.amstat.org/publications/jse/v13n1/bryce.html.

Bryce, G.R., R. Gould, W.I. Notz, and R.L. Peck (2001). "Curriculum Guidelines for Bachelor of Science Degrees in Statistical Science," *The American Statistician*, Volume 55, Number 1, 7–14.

Buchanan, M. (2009). "The Prosecutor's Fallacy," Presented at the *Joint Statistical Meetings,* Washington, DC.

Buncher, C.R., and J.-Y. Tsay (2005). *Statistics in the Pharmaceutical Industry*, Third Edition, Boca Raton, FL: CRC Press.

Cabrera, J., and A. McDougall (2002). *Statistical Consulting*, New York, NY: Springer.

Churchill, G.A., and D. Iacobucci (2010). *Marketing Research: Methodological Foundations*, Tenth Edition, Mason, OH: South-Western Cengage.

Citro, C.F. (2009). "The Federal Statistical System: R&D to Bolster a National Treasure," *Amstat News*, May Issue (#383), 28–30.

Clark, C. (2008). "Official Statistics: Diverse Opportunities and Challenges," *Amstat News*, October Issue (#376), 39–40.

Cleophas, T.J., A.H. Zwinwemn, and T.F. Cleophas (2006). *Statistics Applied to Clinical Trials*, Third Edition, New York, NY: Springer.

Cobb, G., and S. Gehlbach (2006). "Statistics in the Courtroom," Chapter 1 in *Statistics: A Guide to the Unknown*, Fourth Edition, edited by R. Peck, G. Casella, G.W. Cobb, R. Hoerl, and D. Nolan, Belmont, CA: Duxbury Press.

Cochran, J.J. (2010). "Statistics Without Borders Assists with Haitian Data Collection Project," *Amstat News*, May Issue (#395), 18–19.

Coleman, D.E., and D.C. Montgomery (1993). "A Systematic Approach Planning for a Designed Industrial Experiment," *Technometrics*, Volume 35, Number 1, 1–12.

Coleman, S., T. Greenfield, D. Stewardson, and D.C. Montgomery (2010). *Statistical Practice in Business and Industry*, Hoboken, NJ: Wiley.

Crank, K. (2009). "What Do Statisticians and Biostatisticians Do?" *Amstat News*, January Issue (#379), 19.

Crank, K. (2010a). "Academic Salary Survey," *Amstat News*, December Issue (#402), 4–5.

Crank, K. (2010b). "More on Biostatistics," *Amstat News*, January Issue (#391), 19.

Crank, K. (2010c). "Funding Opportunities: Who Wants to be a Biostatistician (or Environmental Statistician, or Social Science Statistician, or . . .)," *Amstat News*, June Issue (#396), 31–32.

Crank, K. (2011). "Salary Survey of Biostatistics and Other Biomedical Statistics Departments," *Amstat News*, January Issue (#403), 8–10.

Crawford, C.G. (2008). "My Career in Environmental Statistics: Luck, Timing and Other People," *Newsletter of the American Statistical Association Section on Statistics and the Environment*, Volume 10, Number 1, 2–3.

Czitrom, V. (2003). "Guidelines for Selecting Factors and Factor Levels for an Industrial Designed Experiment," Chapter 1 in *Handbook of Statistics, Volume 22: Statistics in Industry*, edited by R. Khattree and C.R. Rao, Amsterdam, The Netherlands: Elsevier Science B.V.

Czitrom, V., and P.D. Spagon (1997). *Statistical Case Studies for Industrial Process Improvement*, ASA–SIAM Series on Statistics and Applied Probability, Philadelphia, PA: SIAM.

Dalal, S.R., E.B. Fowlkes, and B. Hoadley (1989). "Risk Analysis of the Space Shuttle: Pre-Challenger Prediction of Failure," *Journal of the American Statistical Association*, Volume 84, Number 408, 945–957.

Davis, W.W. (2010). "Statisticians Thriving at the Social Security Administration," *Amstat News*, April Issue (#394), 21–22.

DeGroot, M.H., S.E. Fienberg, and J.B. Kadane (Editors) (1994). *Statistics and the Law*, New York, NY: Wiley.

Deming, W.E. (1967). "Walter A. Shewhart, 1891-1967," *The American Statistician*, Volume 21, Number 2, 39–40.

Deming, W.E. (1972). "Code of Professional Conduct," *International Statistical Review*, Volume 40, Number 2, 215–219.

Deming, W.E. (1975). "On Probability as a Basis for Action," *The American Statistician*, Volume 29, Number 4, 146–152.

Deming, W.E. (1986). *Out of the Crisis*, Cambridge, MA: MIT Center for Advanced Engineering Study.

Deming, W.E. (1993). *The New Economics for Industry, Government, Education*, Second Edition, Cambridge, MA: The MIT Press.

Denis, J., D. Dolson, J. Dufour, and P. Whitridge (2002). "Preparing Statisticians for a Career at Statistics Canada," *Working Paper No. HSMD-2001-006E/F*, Ottawa, CA: Statistics Canada.

Derr, J. (2000). *Statistical Consulting: A Guide to Effective Communication*, Pacific Grove, CA: Duxbury.

DeVeaux, R.D., and D.J. Hand (2005). "How to Lie with Bad Data," *Statistical Science*, Volume 20, Number 3, 231–238.

Dias, J., K. Miller, and V. George (2009). "Salary Survey of Business, Industry and Government Statisticians," *Amstat News*, October Issue (#388), 5–9.

Doganaksoy, N., G. Hahn, and W.Q. Meeker (2007). "Reliability Assessment by Use-Rate Acceleration," *Quality Progress*, June Issue, 74–76.

Easterling, R.G. (2010). "Passion-Driven Statistics," *The American Statistician*, Volume 64, Number 1, 1–5.

Easton, G.S. (2006). "Expert Witness Testimony: A Discussion," *Conference on Making Statistics More Effective in Schools and Businesses*, Chicago, IL.

Ellenberg, J.H., M.H. Gail, and N.L. Geller (1997). "Conversations with NIH Statisticians: Interviews with the Pioneers of Biostatistics at the United States National Institute of Health," *Statistical Science*, Volume 12, Number 2, 77–81.

Emerson, J.W., M. Seltzer, and D. Lin (2009). "Assessing Judging Bias: An Example from the 2000 Olympic Games," *The American Statistician*, Volume 43, Number 2, 124–131.

Emerson, M.O., and D. Sikkink (2006). *Panel Study of American Religion and Ethnicity, 1st Wave*. Retrieved from http://www.ps-are.org/about/overview.asp.

Fecso, R.S. (2008). "A 'Random Walk': My Statistics Career in Federal Government," *Amstat News*, July Issue (#373), 27–29.

Fecso, R.S., and T. Olson (2006). "Statisticians in the Federal Government," Presented at the *Joint Statistical Meetings*, Seattle, WA. Retrieved from http://www.amstat.org/careers/statisticiansingovernment.ppt.

Fellegi, I.P. (1996). "Characteristics of an Effective Statistical System" (with discussion), *International Statistical Review*, Volume 64, Number 2, 165–197.

Fellegi, I.P. (2004). "Official Statistics: Pressures and Challenges," *International Statistical Review*, Volume 72, Number 1, 139–155.

Fenn-Buderer, N. (2000). "Written Communication Skills for Consulting Statisticians: Creating a Collaborative Environment with Clients," *The Statistical Consultant*, Volume 17, Number 1, 5–10.

Feuerverger, A., P. Hall, G. Tilahun, and M. Gervers (2008). "Using Statistical Smoothing to Date Medieval Manuscripts," in *Beyond Interdisciplinary Research: Festschrift in Honor of Professor Pranab K. Sen*, IMS Collections, Volume 1, Beachwood, OH: Institute of Mathematical Statistics.

Fienberg, S.E. (1997). "Ethics and the Expert Witness: Statistics on Trial," *Journal of the Royal Statistical Society, Series A*, Volume 160, Number 2, 321–331.

Finkelstein, M.O. (2009). *Basic Concepts of Probability and Statistics in the Law*, New York, NY: Springer.

Flanagan-Hyde, P. (2006). "Observational Studies: The Neglected Stepchild in the Family of Data Gathering," *Stats*, Issue 46 (Fall), 19–21.

Freedman, D.A. (1999). "From Association to Causation: Some Remarks on the History of Statistics," *Statistical Science*, Volume 14, Number 3, 243–258.

Gastwirth, J.L. (Editor) (2000). *Statistical Science in the Courtroom*, New York, NY: Springer.

Gastwirth, J.L., and Q. Pen (2010). "Careful Statistical Reasoning Can Provide Support for Supreme Court Decisions," *Amstat News*, July Issue (#397), 23–24.

Gauvin, J.L.S. (2007). "One Statistician's Perspective on Getting a Job," *Amstat News*, November Issue (#365), 31–32.

Gelman, A, and J. Cortina (Editors) (2009). *A Quantitative Tour of the Social Sciences*, New York, NY: Cambridge University Press.

Gelman, A., and J. Hill (2006). *Data Analysis Using Regression and Multilevel/Hierarchical Models*, Cambridge, UK: Cambridge University Press.

Gelman, A., N. Silver, and D. Lee (2009). "The Senate's Health Care Calculations," Op-Ed article, *New York Times*, November 19, A35.

Gibbons, J.D. (2009). "Membership Spotlight," *Amstat News*, November Issue (#389), 16–17.

Gibbs, A., and H. Reid (2009). Comment on "What is Statistics?" by E.N. Brown and R.E. Kass, *The American Statistician*, Volume 63, Number 2, 112–113.

Goldstein, R. (2008). "Hanging Out the Shingle: The Life of a Statistical Consultant," President's Invited Column, *Amstat News*, June Issue (#372), 2–4.

Good, P.I. (2001). *Applying Statistics in the Courtroom: A New Approach for Attorneys and Expert Witnesses*, Boca Raton, FL: Chapman & Hall/CRC.

Grodstein, F., M.J. Stampfer, G.A. Colditz, W.C. Willett, J.E. Manson, M. Joffe, B. Rosner, C. Fuchs, S.E. Hankinson, D.J. Hunter, C.H. Hennekens, and F.E. Speizer (1997). "Postmenopausal Hormone Therapy and Mortality," *New England Journal of Medicine*, Volume 336, Number 25, 1769–1775.

Groves, R.M., F.J. Fowler, M.P. Couper, J.M. Lepkowski, E. Singer, and R. Tourangeau (2009). *Survey Methodology*, Second Edition, Hoboken, NJ: Wiley.

Habiger, J., K. Lopiano, and C. Tot (2009). "Three Interns Share Insights from NISS-NASS Projects," *Amstat News*, December Issue (#402), 49–51.

Hahn, G.J. (1977). "Some Things Engineers Should Know About Experimental Design," *Journal of Quality Technology*, Volume 9, Number 1, 13–20.

Hahn, G.J., and N. Doganaksoy (2008). *The Role of Statistics in Business and Industry*, Hoboken, NJ: Wiley.

Hahn, G.J., N. Doganaksoy, R. Lewis, J.E. Oppenlander, and J. Schmee (2009). "Numbers in Everyday Life: A Short Course for Adults," *Amstat News*, February Issue (#380), 16–19.

Hahn, G.J., and W.Q. Meeker (1991). *Statistical Intervals: A Guide for Practitioners*, New York, NY: Wiley.

Hamada, M., and R. Sitter (2004). "Statistical Research: Some Advice for Beginners," *The American Statistician*, Volume 58, Number 2, 93–101.

Hamasaki, T., S. Evans, and G. Molenberghs (2009). "Biostatistics on the Rise in Japan," *Amstat News*, November Issue (#389), 25–26.

Hand, D.J. (2001). "Reject Inference in Credit Operations," in *Handbook of Credit Scoring*, edited by E. Mays, London, UK: Glenlake.

Hedin, H., and D. Vock (2010). "Advice, Professional Development Tips for Graduate Students, *Amstat News*, December Issue (#402), 19–20.

Hernan, M.A., A. Alonso, R. Logan, F. Grodstein, K.B. Michels, W.C. Willett, J.E. Manson, and J.M. Robins (2008). "Observational Studies Analyzed Like Randomized Experiments: An Application to Postmenopausal Hormone Therapy and Coronary Heart Disease," *Epidemiology*, Volume 19, Number 6, 766–779.

Hesketh, T. (2009). "Too Many Males in China: The Causes and the Consequences," *Significance*, Volume 6, Number 1, 9–13.

Hill, J. (2006). "Evaluating School Choice Programs," Chapter 4 in *Statistics: A Guide to the Unknown*, Fourth Edition, edited by R. Peck, G. Casella, G.W. Cobb, R. Hoerl, and D. Nolan, Belmont, CA: Duxbury Press.

Hoerl, R.W. (2008a). "Statistical Leadership: From Consultant to Effective Leader," Presented at the *European Network for Business and Industrial Statistics (ENBIS)*, Athens, Greece.

Hoerl, R.W. (2008b). "Speaking Out and Reaching Out on Global Health Policy—The Case of HIV/AIDS," *Amstat News*, July Issue (#373), 31–32.

Hoerl, R.W., J.H. Hooper, P.J. Jacobs, "Skills for Industrial Statisticians to Survive and Prosper in the Emerging Quality Environment," *The American Statistician*, Volume 47, Number 4, 280–291.

Hoerl, R.W., and P. Neidermeyer (2009). *Use What You Have: Resolving the HIV/AIDS Pandemic*, Bloomington, IN: Xlibris.

Hoerl, R.W., and R.D. Snee (2001). *Statistical Thinking: Improving Business Performance*, Pacific Grove, CA: Duxbury Press.

Hoerl, R.W., and R.D. Snee (2010a). "Moving the Statistics Profession Forward to the Next Level," *The American Statistician*, Volume 64, Number 1, 10–14.

Hoerl, R.W., and R.D. Snee (2010b). "Statistical Thinking and Methods in Quality Improvement: A Look into the Future" (with discussion), *Quality Engineering*, Volume 22, Number 3, 119–129.

Hooke, R. (1983). *How to Tell the Liars from the Statisticians*, New York, NY: Marcel Dekker.

Huff, D. (1954). *How to Lie with Statistics*, New York, NY: W.W. Norton & Company. Reprinted in 1993.

Huiras, S., K. Virkler, and L. Zahn (2009). "A Day in the Life of an Undergraduate Statistical Consultant," *Amstat News*, September Issue (399), 47–50.

Hulley, S.B., S.R. Cummings, W.S. Browner, G.D.G. Grady, and T.B. Newman (2006). *Designing Clinical Research: An Epidemiological Approach*, Third Edition, Philadelphia, PA: Lippincott Williams & Wilkins.

Hunter, W. (1977). "Some Ideas About Teaching Design of Experiments, with 2^5 Examples of Experiments Conducted by Students," *The American Statistician*, Volume 31, Number 1, 12–17.

Huovilainen, S. (2010). "Statistics Key in Financial Services Industry," *Amstat News*, March Issue (#393), 39.

Hutchinson, M. (2010). "A Finer Formula for Assessing Risk," *New York Times*, May 11, B2.

Iglewicz, B. (2009). "Faculty Sabbaticals at Government, Industrial Organizations: A Different Approach," *Amstat News*, October Issue (#388), 10–11.

International Statistical Institute (2010). *Declaration of Professional Ethics for Statisticians.* Retrieved from http://isi-web.org/images/about/DeclarationOfProfEthics.pdf.

Johnston, I. (2010). "Board Approves Accreditation Guidelines," *Amstat News*, June Issue (#396), 10–11.

Joiner, B.L. (1981). "Lurking Variables: Some Examples," *The American Statistician*, Volume 35, Number 4, 227–233.

Kadane, J.B. (2008). "Ethical Issue in Being an Expert Witness," in *Statistics in the Law: A Practitioner's Guide, Cases, and Materials*, edited by J.B. Kadane, New York, NY: Oxford University Press.

Keller, J.L., and D.J. Sargent (2010). "Getting a Job: What Distinguishes You?" *Amstat News*, September Issue (#399), 44–46.

Kenett, R.S., S. Coleman, and D. Stewardson (2003). "Statistical Efficiency: The Practical Perspective," *Quality and Reliability Engineering International*, Volume 19, Number 4, 265–272.

Kenett, R., and P. Thyregod (2006). "Aspects of Statistical Consulting not Taught by Academia," *Statistica Neerlandica*, Volume 60, Number 3, 396–411.

Kiernan, K. (2009). "Master's Notebook: Opportunities Abound for Statisticians in Tech Support," *Amstat News*, December Issue (#390), 35–36.

Kirk, R.E. (1995). *Experimental Design: Procedures for the Behavioral Sciences*, Third Edition, Pacific Grove, CA: Brooks/Cole.

Kish, L. (1995). *Survey Sampling*, Second Edition, New York, NY: Wiley.

Kotz, S., B. Campbell, N. Read, and B.V. Balakrishnan (2005). *Encyclopedia of Statistical Sciences*, Second Edition, New York, NY: Wiley.

Lakshminarayanan, M. (2008a). "Statistics in the Insurance Industry," *Amstat News*, August Issue (#374), 29.

Lakshminarayanan, M. (2008b). "Consulting in an Outsourcing Environment . . . Collaboration is the Key," *Amstat News*, November Issue (#377), 21.

Lampone, V. (2009). "A Quick Guide to Online Data Quality," *Stats*, Issue 51 (Fall), 5–9.

Landes, R.D. (2009). "Passing on the Passion for the Profession," *The American Statistician*, Volume 63, Number 2, 163–172.

Larsen, T., J. Price, and J. Wolfers (2008). Racial Bias in the NBA: Implications in Betting Markets, *Journal of Analysis in Sports*, Volume 4, Number 2, Article 7.

Ledolter, J., and A. Swersey (2007). *Testing 1-2-3: Experimental Design with Applications in Marketing and Service Operations*, Stanford, CA: Stanford University Press.

Lee, M.L. (2008). "Statistical Accreditation: The Royal Statistical Society Experience," *Amstat News*, October Issue (#376), 11.

Lesser, L.M. (2001). "Ethical Statistics and Statistical Ethics: The Experience of Creating an Interdisciplinary Module," *Proceedings of the Annual Meeting of the American Statistical Association*, Alexandria, VA: American Statistical Association

Levitt, S.S., and S.J. Dubner (2005). *Freakonomics: A Rogue Economist Explains the Hidden Side of Everything*, New York, NY: HarperCollins.

Levitt, S.S., and Dubner, S.J. (2009). *Super Freakonomics: Global Cooling, Patriotic Prostitutes, and Why Suicide Bombers Should Buy Life Insurance*, New York, NY: HarperCollins.

Lewis, M. (2004). *Moneyball: The Art of Winning an Unfair Game*, New York, NY: W.W. Norton & Company.

Lindsay, B., J. Kettenring, and D. Siegmund (2003). "A Report on the Future of Statistics," *Statistical Science*, Volume 19, Number 3, 387–413.

Lindsay, B., J. Kettenring, and D. Siegmund (Editors) (2004). *Statistics: Challenges and Opportunities for the Twenty-First Century*, National Science Foundation—American Statistical Association Report.

Lohr, S. (2010). *Sampling: Design and Analysis*, Second Edition, Boston, MA: Brooks/Cole Cengage.

Lott, J.T. (2008). "Statistical Opportunities Abound in Federal Government," *Amstat News*, January Issue (#367), 23.

Lu, J., S. Jeng, and K. Wang (2009). "A Review of Statistical Methods for Quality Improvement and Control in Nanotechnology," *Journal of Quality Technology*, Volume 41, Number 2, 148–164.

Mason, R.L., R.F. Gunst, and J.L. Hess (2003). *Statistical Design and Analysis of Experiments*, Second Edition, Hoboken, NJ: Wiley.

Maxwell, S.E., and H.D. Delaney (2004). *Designing Experiments and Analyzing Data: A Model Comparison Perspective*, Second Edition, Mahwah, NJ: Erlbaum.

McCullough, B.D., and D.A. Heiser (2008). "On the Accuracy of Statistical Procedures in Microsoft Excel 2007," *Computational Statistics and Data Analysis*, Volume 52, Number 10, 4570–4578.

McCullough, B.D., and B. Wilson (2002). "On the Accuracy of Statistical Procedures in Microsoft Excel 2000 and Excel XP," *Computational Statistics & Data Analysis*, Volume 40, Number 4, 713–721.

McDonald, G.C. (1999). "Shaping Statistics for Success in the 21st Century: The Needs of Industry," *The American Statistician*, Volume 53, Number 3, 203–207.

McKenzie, J.D. (2008). "The Use of Microsoft Excel for Statistical Analysis—An Update," *Proceedings of the 2008 Joint Statistical Meetings*, Alexandria, VA: American Statistical Association.

Meeker, W.Q. (2009). "The Effective Industrial Statistician: Necessary Knowledge and Skills," Presented at the *Quality and Productivity Research Conference*, Yorktown Heights, New York.

Meeker, W.Q., and L.A. Escobar (1998a). *Statistical Methods for Reliability Data*, New York, NY: Wiley.

Meeker, W.Q., and L.A. Escobar (1998b). "Pitfalls of Accelerated Testing," *IEEE Transactions on Reliability*, Volume 47, Number 2, 114–118.

Meng, X.-L. (2009). "Desired and Feared—What Do We Do Now and Over the Next 50 Years?" *The American Statistician*, Volume 63, Number 3, 202–210.

Millard, S.P., and A. Krause (2001). *Applied Statistics in the Pharmaceutical Industry*, New York, NY: Springer.

Montgomery, D.C. (2009). *Design and Analysis of Experiments*, Seventh Edition, Hoboken, NJ: Wiley.

Moore, D.S. (2001). "Undergraduate Programs and the Future of Academic Statistics," *The American Statistician*, Volume 55, Number 1, 1–6.

Morton, S. (2009a). "Statistics and Statisticians: Essential to Evidence-Based Decisionmaking," *Amstat News*, October Issue (#388), 3–4.

Morton, S. (2009b). "Presidential Address, American Statistical Association," *Joint Statistical Meetings,* Washington, DC.

Neagu, R., and R. Hoerl (2005). "A Six Sigma Approach to Predicting Corporate Defaults," *Quality and Reliability Engineering International,* Volume 21, Number 3, 293–309.

Newton, P.K., and K. Aslam (2009). "Monte Carlo Tennis: A Stochastic Markov Chain Model," *Journal of Quantitative Analysis in Sports,* Volume 5, Number 3, Article 7.

New York Times (2009). "For Today's Graduate, Just One Word: Statistics," *New York Times,* August 5.

Nguyen, A., and S.K. Fan (2009). "Ethics and Stopping Rules in a Phase II Clinical Trial," *Chance,* Volume 22, Number 4, 39–44.

Nolan, D., and D.T. Lang (2010). "Computing in the Statistics Curricula," *The American Statistician,* Volume 64, Number 2, 97–107.

Norwood, J.L. (1995). *Organizing to Count: Change in the Federal Statistical System,* Washington, DC: Urban Institute Press.

Norwood, J.L. (2006). *Wise Elder Lecture,* Washington, DC: Bureau of the Census.

Nussbaum, B. (2008). "Statistics Counts at the Environmental Protection Agency," *Amstat News,* November Issue (#377), 23–24.

Nyberg, J. (2010). "Find Your Fit," *Amstat News,* June Issue (#396), 35–36.

O'Brien, R.G. (2000). "Applying for a Job: Your Curriculum Vitae and Cover Letter," *Amstat News,* September Issue (#279), 15–20. Retrieved from http://www.bio.ri.ccf.org/ASA_TSHS/mockCV/mock.html.

O'Brien, R.G. (2002). "The Completely Sufficient Statistician," Keynote Address, *14th Annual Kansas State University Conference on Applied Statistics in Agriculture,* Manhattan, KS.

Ochsenfeld, C.A., and G.R. Olbricht (2009). "Statistical Community Service: What Role Can Students Play?" *Amstat News,* December Issue (#402), 11–12.

O'Neill, R., G. Campbell, and H. Hsu (2008). "Statisticians at the Food and Drug Administration," *Amstat News,* October Issue (#376), 2–3.

Pantula, S. (2010a). "Celebrating Anniversaries and Achievements," *Amstat News,* July Issue (#397), 3.

Pantula, S. (2010b). "Soft Skills Just as Important as Core, Computational Skills When Looking for a Job," *Amstat News,* September Issue (#399), 3–4.

Patterson, B.F. (2009). "AP Statistics: Students Choices After High School," *Amstat News,* May Issue (#383), 6–12.

Peck, R., G. Casella, G. Cobb, R. Hoerl, D, Nolan, R. Starbuck, and H. Stern (Editors) (2006). *Statistics: A Guide to the Unknown,* Fourth Edition, Belmont, CA: Duxbury Press.

Peterson, J.J., R.D. Snee, P.R. McAllister, T.L. Schofield, and A.J. Carella (2009). "Statistics in Pharmaceutical Development and Manufacturing," *Journal of Quality Technology,* Volume 41, Number 2, 111–147.

Petska, T., J. Hobbs, and F. Scheuren, F. (2001). *Statistical Operations and Studies in the SOI Program of the IRS,* Statistics of Income Division, Internal Revenue Service. Retrieved from http://www.oecd.org/dataoecd/57/62/36237450.pdf.

Pharmaceutical Research and Manufacturers of America (PhRMA) (2009). *PhRMA Principles on Conduct of Clinical Trials and Communication of Clinical Trial Results,* accessible via http://www.phrma.org/sites/default/files/105/042009_clinical_trial_principles_final.pdf.

Piccolo, R.S. (2010). "Biostatisticians: Do You Know What They Do?" *Amstat News,* August Issue (#398), 26.

Pierson, S. (2010). "Members Affect ASA Science Policy," *Amstat News,* June Issue (#396), 33–34.

Pregibon, D. (2009). "Statistics @ Google," *Amstat News,* May Issue (#383), 3–5.

Raftery, A.E., M.A. Tanner, and M.T. Wells (Editors) (2001). *Statistics in the 21st Century,* Boca Raton, FL: Chapman & Hall/CRC.

Ratledge, E.C. (2006). "The Anatomy of a Pre-election Poll," Chapter 2 in *Statistics: A Guide to the Unknown,* Fourth Edition, edited by R. Peck, G. Casella, G.W. Cobb, R. Hoerl, and D. Nolan, Belmont, CA: Duxbury Press.

Reamer, A. (2009). "In Dire Straits: The Urgent Need to Improve Economic Statistics," *Amstat News,* March Issue (#381), 33–35.

Rhew, R.D., and P.A. Parker (2007). "A Parametric Geometry Computational Fluid Dynamics (CFD) Study Utilizing Design of Experiments (DOE)," *AIAA 2007-1615, U.S. Air Force Test and Evaluation Days,* Destin, FL.

Ritter, M.A., R.R. Starbuck, and R.V. Hogg (2001). "Advice from Prospective Employers on Training BS Statisticians," *The American Statistician*, Volume 55, Number 1, 14–18.

Rosenbaum, P.R. (2002). *Observational Studies*, Second Edition, New York, NY: Springer.

Rossetti, G. (2010). "Nuclear Engineering to Ultimate Frisbee: A Conversation with an Intern at the Energy Information Administration," *Amstat News*, September Issue (#399), 21–22.

Salsburg, D. (2001). *The Lady Tasting Tea: How Statistics Revolutionized Science in the Twentieth Century*, New York, NY: Holt.

Scheaffer, R.L., and C. Lee (2000). "The Case for Undergraduate Statistics," Presented at the *Joint Statistical Meetings,* Indianapolis, IN. Retrieved from http://www.amstat.org/meetings/jsm/2000/usei/case.html.

Scheaffer, R.L., W. Mendenhall, and R.L. Ott (2005). *Elementary Survey Sampling*, Sixth Edition, Belmont, CA: Duxbury Press.

Schell, M. (2005). *Baseball's All-Time Best Sluggers*, Princeton, NJ: Princeton University Press.

Schenker, N. (2008). "Statistics in Monitoring the Nation's Health," *Amstat News*, September Issue (#375), 2–3.

Scheuren, F. (2007). "The Pro Bono Statistician," *Journal of the American Statistical Association*, Volume 102, Number 477, 1–6.

Schield, M. (2006). "Beware the Lurking Variable," *Stats*, Issue 46 (Fall), 14–18.

Schwartz, S., and S. Shipman (2009). *Using Program Evaluation and Statistical Analysis to Inform Policy Decisions.* (Handout to talk given by N. Kingsbury at 2009 Joint Statistical Meetings.).

Shipp, S., and S. Cohen (2009). "COPAFS Focuses on Statistical Activities," *Amstat News*, December Issue (#390), 17–20.

Shipp, S., and S. Cohen (2010). COPAFS Focuses on Statistical Activities, *Amstat News*, March Issue (#393), 25–29.

Shutes, K. (2009). "A Note on Using Individualised Data Sets for Statistics Coursework," *Technology Innovations in Statistics Education*, Volume 3, Number 2. Retrieved from http://escholarship.org/uc/item/7d02d6hd.

Siegel, J.S (2005). "Pursuing Non-Census Stuff in the Census," *Newsletter of the Sections on Government Statistics and Social Statistics,* June Issue, American Statistical Association. Retrieved from http://www.amstat.org/sections/sgovt/news0605.pdf.

Siegmund, D., and B. Yakir (2007). *The Statistics of Gene Mapping*, New York, NY: Springer.

Sinclair, N. (2009). "A Practitioner's Reflections to Motivate a Career in Statistics," *Amstat News*, July Issue (#385), 56.

Slutsky, J.R., and C.M. Clancy (2009). "AHRQ's Effective Health Care Program: Why Comparative Effectiveness Matters," *American Journal of Medical Quality*, Volume 24, Number 1, 67–70.

Smith, R.L., L.M. Berliner, and P. Guttorp (2010). "Statisticians Comment on Status of Climate Change Science," *Amstat News*, March Issue (#393), 13–17.

Snee, R. (1998). "Non-Statistical Skills that Can Help Statisticians Become More Effective," *Total Quality Management Journal*, Volume 9, Number 8, 711–722.

Snee, R. (1999). "Meeting the Challenge of Improving Business Performance," Presented at the *Joint Statistical Meetings,* Baltimore, MD.

Snee, R. (2008). "W. Edwards Deming's Making Another World: A Holistic Approach to Performance Improvement and the Role of Statistics," *The American Statistician*, Volume 62, Number 3, 251–255.

Snee, R.D., and R.W. Hoerl (2007). "Integrating Lean and Six Sigma—A Holistic Approach," *Six Sigma Forum Magazine*, Volume 6, Number 3, 15–21.

Snider, V. (2009). "Nancy M. Gordon: Informing Decisions," *Amstat News*, September Issue (#387), 11–13.

Spiegelhalter, D., and A. Barnett (2009). "London Murders: A Predictable Pattern," *Significance*, Volume 6, Number 1, 5–8.

Splaver, S. *(Editor)* (1973). *Nontraditional Careers for Women*, New York, NY: Julian Messner.

Spruill, N., and A. Wilson (2009). "Statistics in Defense and National Security: Lesson in Outreach to Policymakers," *Amstat News*, December Issue (#390), 31–33.

Spurrier, J. (2000). *Practice of Statistics: Putting the Pieces Together*, Pacific Grove, CA: Brooks/Cole.

Squire, P. (1988). "Why the 1936 Literary Digest Poll Failed," *Public Opinion Quarterly* Volume 52, Number 1, 125–133.

Starbuck, R. (2009). "Journey to a Career in Industry: Robert Starbuck Tells His Story," *Amstat News*, September Issue (#387), 51–53.

Starbuck, R. (2010). "Fellow Award: Revisited (Again)," *Amstat News*, August Issue (#398), 6–9.

Statistical Journal of the International Association of Official Statistics (2011). Volume 27, Number 1–2.

Steiner, S.H., and R.J. MacKay (2005). *Statistical Engineering: An Algorithm for Reducing Variation in Manufacturing Processes*, Milwaukee, WI: ASQ Quality Press.

Straf, M.L. (2003). "Statistics: The Next Generation," *Journal of the American Statistical Association*, Volume 98, Number 461, 1–6.

Sung, Y. (2008). "It's Easy to Produce Chartjunk Using Microsoft Excel 2007 but Hard to Make Good Graphs," *Computational Statistics and Data Analysis*, Volume 52, Number 10, 4594–4601.

Szabo, F.E. (2004). *Actuaries' Survival Guide: How to Succeed in One of the Most Desirable Professions*, Boston, MA: Academic Press.

Tanenbaum, E. (2010a). "In the Hot Seat: Two Experienced Consultants Discuss Hiring Statisticians," *Amstat News*, July Issue (#397), 27–31.

Tanenbaum, E. (2010b). "Statistical Communications Anonymous," *Amstat News*, April Issue (#394), 37–38.

Tanur, J.M., F. Mosteller, W.H. Kruskal, R.F. Link, R.S. Pieters, and G.R. Rising (Editors) (1972). *Statistics: A Guide to the Unknown*, San Francisco, CA: Holden-Day.

Tarpey, T., C. Acuna, G. Cobb, and R. DeVeaux (2002). "Curriculum Guidelines for Bachelor of Arts Degrees in Statistical Science," *Journal of Statistical Education*, Volume 10, Number 2. Retrieved from http://www.amstat.org/publications/jse/v10n2/tarpey.html.

The College Board (2010). *Statistics Course Description*. Retrieved from http://apcentral.collegeboard.com/apc/public/repository/ap-statistics-course-description.pdf.

Thorn, J., and P. Palmer (1985). *The Hidden Game of Baseball: A Revolutionary Approach to Baseball and its Statistics*, Revised and Updated Edition, Garden City, NY: Dolphin Book.

Tufte, E.R. (1997). *Visual Explanations: Images and Quantities, Evidence and Narrative*, Cheshire, CT: Graphics Press.

Turner, K.R. (2007). *New Drug Development: Design, Methodology and Analysis*, Hoboken, NJ: Wiley.

U.S. Bureau of Labor Statistics (2007). "The Consumer Price Index," Chapter 17 in *BLS Handbook*. Retrieved from http://www.bls.gov/opub/hom/pdf/homch17.pdf.

Utts, J.M. (2005). *Seeing Through Statistics*, Third Edition, Belmont, CA: Thomson Brooks/Cole.

Valentine, K., R. James, and C. Martial (2010). "The Statistician's Role in an Integrated Health Care System," *Amstat News*, November Issue (#401), 17–19.

Valian, V. (1999). *Why So Slow? The Advancement of Women*, Boston, MA: The MIT Press.

Van Belle, G. (2002). *Statistical Rules of Thumb*, New York, NY: Wiley.

Van Den Heuvel, J., R.J.M.M. Does, and S. Bisgaard (2005). "Dutch Hospital Implements Six Sigma," *Six Sigma Forum Magazine*, Volume 4, Number 2, 11–14.

Vardeman, S., and M. Morris (2004). "Statistics and Ethics: Some Advice for Young Statisticians," *The American Statistician*, Volume 57, Number 1, 21–26.

Varian, H. (2010). Quoted in "Data, Data Everywhere," *The Economist*, February 25.

Wallman, K.K. (2008). "Government Statistics: Bold Ventures, Interesting Challenges," *Amstat News*, June Issue (#372), 26–29.

Washington Post (2009). "In D.C., Statisticians Flex Their Strength in Numbers," *Washington Post*, August 5.

Weisberg, H.I. (2010). *Bias and Causation: Models and Judgment for Valid Comparisons*, Hoboken, NJ: Wiley.

Yalta, A.T. (2008). "The Accuracy of Statistical Distributions in Microsoft Excel 2007," *Computational Statistics and Data Analysis*, Volume 52, Number 10, 4579–4586.

Younger, N. (2009). "Is That Clear: Communicating Effectively," *Amstat News*, November Issue (#389), 11–12.

INDEX[1]

Academia
  administrative service, 294–295
  careers in, 269–308
  comparison with business, industry and
     government, 271–272, 305
  department chair, 301
  location within, 277
  teaching challenges, 280–285
  tenure. *see* Tenure
  training for, 298–300
  types of positions, 273–276
  what statisticians do, 270–273
Academic appointment
  adjunct professor, 274–276
  assistant professor, 274
  associate professor, 274
  clinical professor, 274–275
  distinguished chair, 300
  endowed chair, 300
  full professor, 274
  instructor, 274–275
  lecturer, 274
  letter, 276
  2-month, 276
  9-month, 276
  postdoc, 274–275
  temporary, 274
  visiting professor, 275–276
Academic career
  bright sides, 304
  downsides, 303
Academic freedom, 277, 304
Academic leader, 300
Academic year, 273, 276
Accelerated life testing, 25, 202, (footnote)
     221, 226
Acceptance sampling, 23, 76, 127

Accreditation, 2, 15–16
Actuarial science, 2, 15
Adjunct professor, 150, 254
Advanced Placement course in statistics, 10,
     114
Advertising claim, 13, 26, 28, 30, 67, 207,
     211, 217, 249
Agricultural research, 62–63
Agriculture Department, Economic Research
     Service, 46
AIDS epidemic, 261
Allen, R., (footnote) 35
American Association for Public Opinion
     Research, 16
American Chemical Society, 16
American Community Survey (ACS), 43
American Economics Association, 16
American Educational Research
     Association, 16
American Society for Quality (ASQ), 16,
     (footnote) 206, 263, 290, 314
  Statistics Division, 316
American Statistical Association (ASA), 2,
     16, 27, 90, 115, 152, 209, 282, 302,
     310, 316–317
  Biopharmaceutical Section, 27
  Committee on Professional Ethics, 215
  Ethical Guidelines for Statistical
     Practice, 207, 209
  Fellow, 297, 300
  Job Placement Service, 153
  Section on Statistics and the
     Environment, 65
  Section on Statistics in Sports, 69
  Section on Teaching Statistics in Health
     Sciences, 282
  Section on Statistical Education, 282, 298

[1] Also see References for authors of referenced works.

The American Statistician, 312
Amstat News, 14, 17, 53, 64, 108, 138, 142, 153, 158, 166, 205, 312, 317
Analysis of variance, 127–129, (footnote) 220
Analytic study, (footnote) 221
Anderson Cook, C., 92–93
Annals of Applied Statistics, 313
Annals of Statistics, 286, 290, 313
Artificial intelligence, 8
Asher, J., 92
Association of Public Data Users, 50
    Annual Conference, 317
Attribute measurement, 235
Automated tracking, 318

Bacon, F., 4
Banks, D., 100
Baseball, 68
Battelle Memorial Institute, 72
Baudrillard, J., 310
Bayesian methods, 127
Benefits, quantification, 188–190
Berk, S., 304
Bioinformatics, 72, 318
Biometrics, 287
Biometrika, 290
Biostatistician, 59
Biostatistics, 2, 11, 48–49, 57, 59, 90, (footnote) 115, 116, 120, 123, 128–129, 270, 317
Blumberg, C.J., xv, (footnote) 225
Boardman, T.J., 109, 143
Bowker, A.H., 255, 264, 302
Box, G.E.P., 80, 101, 141, 175
Brown, E., 77
Buderer, N., 129
Bureau of Economic Analysis, 37–38, 46, 88
Bureau of Justice Statistics (BJS), 39, 40, 41, 46
Bureau of Labor Statistics (BLS), 37, 38, 39, (footnote) 42, 46, 50, 75, 213
Bureau of the Census, 37, (footnote) 40, 46, 51, 75, 87, 128–130, 213
    Local Employment Dynamics program, 50–51
    State Data Center program, 50
Bureau of Transportation Statistics, 46
Burn-in. *see* Product, burn-in

Canada
    statistical system, 51–52
    Statistics Act, 51–52
Capture-recapture methodology, 107
Career goals, defining, 149–151
Career paths
    academic statistician, 300–302
    statistician 251–264
CareerCast.com, 11
Case-control study, 232
Categorical data, 28, 71, 127, 129, 133
Caucus for Women in Statistics, 93
Causation, 9
Cause and effect relationship, 225–228, 230
Censored data analysis, 130–131
Census, 221–224
Census Bureau. *see* Bureau of the Census
Centers for Disease Control and Protection (CDC), 59
Challenger space shuttle, 64
Chance, 282, 313
Chance News, 282, 310
Cheetham, B., 201
Chemtech, 206
Chernick, M. R., 143, 191, 311
Chief statistician, U.S., 39, 87, 91
Chuang-Stein, C., 92–93
Citation index, 291
Classified information, 212
Cleveland Clinic Foundation, 60
Climate change. *see* Climate science
Climate science, 45, 64–65, 312, 318
Clinical trials, 26–27, 61, 118, 127, 207, 210, 214, (footnote) 220, 243, 318
    ethical considerations, 214, 242
Cluster sampling, 223
Clustering analysis, 130
Cohort study, 232
Coleman, D., 184
Colleges
    admission, 230–231
    comparison of, 121–125
    ranking, 121–122
    selection of, 120–125
    size of statistics program, 123
    2-year, 270, 273, 277
    4-year, 270–273, 277
Communications, 197–205
    ability, 8, 126, 197
    electronic, 200–201, 206

in a global environment, 198
industry, 29, 76
informal, 201
need for, 190
opportunities for, 199–201
Community college, 273–275, 278, 286
Comparative effectiveness research, 61
Comparative investigation, 221
Computer program. *see* Software
Computer scientist, 78
Computer skills, 8, 45, 100, 117, 126
Computing. *see* Statistical computing
Conferences, professional. *see* Conferences, technical
Conferences, technical, 315–318
Confidence interval, 131, 223, 265
Confidential information, 43, (footnote) 163, 211–213, 239, 293
Confidential Information Protection and Statistical Efficiency Act, 213
Confidentiality. *see* Confidential information
Confounding, 210, 227
Consultant. *see* Statistical consultant
Consulting. *see* Statistical consulting
Consumer Price Index (CPI), 37, 38, 39, 42
Consumer Product Safety Commission, 41, 57
Continuing education, 311
Continuous measurement, 235
Control chart, 23, 69, 76, 155, 156, 224
Controlled study, (footnote) 221, 225–226
Cook, L., (footnote) 35, 38, (footnote) 212
Correlation, 226, 228
Cost of Living Adjustments (COLA), 39–41
Council of Professional Associations on Federal Statistics, 49, 317
Cox, G., 90
Credit approval. *see* Loan approval
Credit score, 28, 225
Crosby, P., 24, (footnote) 24
Current Index to Statistics (CIS), 311
Current Population Survey (CPS), 37, 187

Darwin, C., 106
Data detective, 233
Data gathering, 219–247
Data mining, 8, 78, 127, 225, 277
Data quality, 223, 230, 233, 240
De Berk, L., 69

Deadlines, meeting of, 88–89
Decision Sciences Institute, 16
Decision theory, 8, 127
Default risk modeling, 29, 178, 181–182, 188, 194, 225
Degrees in statistics. *see* Statistics, degrees in
Deming Conference on Applied Statistics, 317
Deming, W. E., 24, 76, 93, 113, 141, 178, 258–259, 264, 265, 283, 284
Democratization of statistics, 29, 78–79, 83, 94, 95, 102, 113, 145, 287
Design of experiments, 12, 24, 25, 30, 220–221
Deterministic thinking, 7, 103
Discrete parts assembly, 26
Discrimination claim, assessment, 208
Disease prediction, control and prevention, 59–60
Disputes, assessment of, 208
Disraeli, B., 9
Dissertation, 124, 134–135
advisor, selection of, 134–135
committee, selection of, 136–137
topic, selection of, 136
Dodge, H.F., 76
Doganaksoy, N., 13–14
Domesday Book, 35
Drug approval, 26
Drug Information Association Annual Meeting, 318
Dynamics of work environment, 87, 185

Econometric forecasting, 71
Education, 67
Educational testing, 67
Educational Testing Service (ETS), 67
Electric Power Research Institute, 72
Electronic classroom, 282, 284
Electronic communication. *see* Communications, electronic
Elevator speech, 201–203, 207
Energy Information Administration (EIA), 41, 46, 128, 129, 138, 155
Engineering Statistics Handbook, 310
Enthusiasm, need for, 105
Enumeration, complete, 222
Enumerative study, (footnote) 221
Environment, 64–65

Environmental Protection Agency (EPA), 46, 57, 58–59, 233
  Department of Environmental Information, 46
Epidemiology, 59
Ethical conflicts, up-front avoidance, 209–210
Ethical considerations, 207–216
European Network of Business and Industrial Statisticians (ENBIS), 16, 316
Eurostat, 52
Excel, 133. *see* Microsoft Excel
Exit poll, 66
Experimental design. *see* Design of experiments
Expert witness, 69, 100, 214–215, 217, 253, 293
Explanatory variable, 220, 225, 227–230, 232
Extrapolation, 131

Faculty committee, 294–295
Faculty manual, 274, 276, 279
Failure mode, 226, 242–246
Failure modes and effects analysis (FMEA), 244
Fall Technical Conference (FTC), 317
Fecso, R., (footnote) 35, 41
Federal Bureau of Investigation, 40
Federal Committee on Statistical Methodology, 49, 317
Federal Communications Commission, 57
Federal Highway Administration, 46
Federal Reserve System, 46
Feigenbaum, A., (footnote) 24
Field support, 25–26
Financial services applications, 28–29
Financial Times Top Academic Journals, 291
Firefighting, 24
Fisher, R.A., 62, 141, 184, 220, 231
Flammability test, 235
Flexibility, need for, 102
Food and beverage industry, 26, 28
Food and Drug Administration (FDA), 26, 57, 58, 59, 75, 118, 317
Food and Drug Administration Industry Statistics Workshop, 317
Food Drug and Cosmetics Act, 26
Fort Wayne, Indiana, use of Six Sigma, 77

Fowler, L., 92, 264
Fractional factorial experiment, 162
Fraud detection, 29–31
Ftp site, xiv, 50, 131, 211, 213, 279, 285
Fuzzy logic, 78

Gaines, L., xv, 1, (footnote) 35, 206
Gardner, M., 92, 107, 186–187, 190–191
Garfinkel, L., 255, 256, 264
GE, 77
GE Global Research Center, 79
Geller, N., 188
Genetic predisposition, 231
Genetic research, 60
Gentleman, R., 134
Geographical considerations, in job assessment, 168
Global environment, communicating in. *see* Communications, in a global environment
Globalization, 83–84, 198
  consequences and impact, 83–84
  evolution, 83
Godin, S., 203
Golomski, W., (footnote) 24, 260
Go/no-go measurement, 235
Goodman, A., 23
Goodnight, J., 15, 255, 264
Google, 30, 80
Gossett, W.S., (footnote) 12
Government Accountability Office (GAO), 40–41
Government, job grading. *see* Job grading systems
Graphical methods, 29
Gross Domestic Product (GDP), 37, 38, 52

Hahn, G.J., 12–13
Hahn, J., (footnote) 225
Hand, D.J., 1
Happiness index, 2, 41
Hare, L., xv, (footnote) 4, 10, 109, 191, 259
Health, 59–61, 77
  care, 60–61
  research, 60
High school programs, 114
Hill, W., 109, 191
Hiring process
  New York State, 153
  U. S. government, 153

Hoadley, B., 191
Hoerl. R., 104, 106, 109, 191, 261
Hogg, R., 191
Hollerith, H., (footnote) 43
Honeywell, 77
Hormone replacement therapy, 226–227
Hunter, W., 1

Ihaka, R., 134
Iman, R., 109, 143, 192
In vitro studies, 27
In vivo studies, 27
Incomplete block design, 62
Industrial and Engineering Chemistry, 206
Informatics, 277
Institute for Operations Research and the
    Management Sciences
    (INFORMS), 16, 314
Institute of Electrical and Electronics
    Engineers (IEEE), 16, 314
Institute of Mathematical Statistics
    (IMS), 16, 302, 311, 313, 315
Integrated Postsecondary Education Data
    System, 90
Interagency Council on Statistical Policy, 46
Interdisciplinary program, 277
Internal Revenue Service (IRS), 40
International Association for Official
    Statistics (IAOS), 314
International Association for Statistics
    Education, 282
International Biometric Society, 16
    Eastern North American Region Spring
        Meeting, 317
International Conference on Statistics in
    Business and Industry, 317
International Monetary Fund, 52
International Statistical Institute (ISI), 16,
    51, 209, 216, 262, 263, 313, 316
    Declaration of Professional Ethics, 209
    Sports Statistics Committee, 69
International Statistical Review (ISR), 313
Internet operations, 30
Internet resources, 310–311
Internship, 11, 14, 118, 123–124, 135–140,
    299
Interpersonal skills, 8, 100
Introductory statistics course, 280–283
    applications-area oriented, 281
    liberal arts, 280–281

Introductory statistics course, omnibus, 281
Isolated statistician, 297–298, 309
Iterative learning process, 220

James, W., 68
JMP, 133
Job grading systems, 82–83
Job growth opportunities, assessment of, 166
Job interview, 157–164
    follow up, 164
Job mobility, 257
Job offers, assessment, 166–170
Job search
    approaches, 152–153
    identifying opportunities, 151–154
Joint Statistical Meetings, 64, 69, 91, 142,
    153, 262, 298, 311, 315–316
*Journal of Business and Economic
    Statistics*, 287
*Journal of Computational and Graphical
    Statistics*, 290
*Journal of the American Medical
    Association*, 290
*Journal of the American Statistical
    Association*, 287, 290
Journal of the Royal Statistical Society, 290
The Journal of Statistics Education, 282
Journals
    general interest, 312–313
    professional and technical, 312–314
    specialized, 313–314
    tiering, 290–291
Juran, J., (footnote) 24

Kalman filtering, 78
Kass, R., 77
Kefauver-Harris Amendments, 26, 75
Kettenring, J., 1
Keyes, T., 192
Knowledge
    adaption, skill for, 107
    discovery in databases, 8

Lambert, D., 192
Latin Square design, 62
Leadership skills. *see* Statistical leadership
Lean manufacturing, 77
Legal applications, 69–70
Leon, R., 144
Life data analysis, 127

Lifelong learning, 108, 309–319
Listening, art of, 199
Literary Digest Magazine, 66
Loan approval, 178, 184, 188
Loan default prediction, 181–182, 188
London cholera epidemic, 231
Look, A., (footnote) 8
Los Alamos National Laboratory, 72
Lott, J., (footnote) 35
Lurking variable, 227–228, 232

Makuch, W., 109
Management science, 8
Managers, impact of, 84–85
Manufactured product applications, 23–28
Margin of error, 66, 223
Market research, 25, 66
Markov chain models, 69
Martz, H., 110, 144, 192
Mathematica Policy Research, 47
Mathematical statistician, 47, 48
Mathematics
  skill, 7, 12, 48, 54, 100, 125, 126
  training, 114
Mayo Clinic, 60, 117, 157
McDonald, G., 159
McNulty, S., 1
Measurement
  error, 25
  precision, 236
Meeker, W. Q., xv, (footnote) 206, (footnote)
    269
Meetings
  national societies, 315
  tips for running, 199–200
Michelson, D.K., 91
Michigan, Department of Agriculture, 50
Microsoft Excel, 132–133
Minitab, 133
Missing data, 130, 236
Mixture experiment, 130, 205
Money considerations, in job
    assessment, 168
Monitoring process or product, 27, 28, 41
Moore, D.S., 144, 192, 219
Morgan, C., 93, 110, 144
Morton, S., 15, 39, 64
Motorola, 77
Multiple projects, 89
Multivariate methods, 127, 299

Nanotechnology, 63, 76, 167, 318
National Academy of Sciences, Committee
    on National Statistics, 317
National Aeronautics and Space Agency
    (NASA), 46, 63, 64
National Agricultural Statistics Service
    (NASS), 46, 50, 213
National Center for Education Statistics, 37,
    90, 115
National Center for Health Statistics, 37, 46
National Crime Victimization Survey, 39
National Institute of Standards and
    Technology (NIST), 72–73, 310
National Institute of Statistical Sciences
    (NISS), 72
National Institutes of Health, 59, 82, 233, 288
National laboratories, 71–72
National Opinion Research Center, 47
National Research Council (NRC), 120–121
National Science Foundation, 115, 288,
    (footnote) 290
  Science Research Statistics Division, 46
Needleman, S., 11
Neidermeyer, P., 261
Nelson, L., (footnote) 8, 23, 192
Nelson, W., (footnote) 240
Nemeth, M.A., 110, 144
Neural networks, 8, 78
New England Journal of Medicine, 290
New Testament, 35
New York State
  hiring process (*see* Hiring process, New
    York State)
  job grading (*see* Job grading systems)
  statistician requirements, 48–49
New Zealand Statistical Association, 16
Newsletters, 206
Nightingale, F., 90
Nonlinear estimation, 127
Nonparametric methods, 127, 299
Non-response bias, 223
Nontenure-track position, 274–276
Normal distribution, 131
North American Industry Classification
    System, 44, 130
North Carolina State University, 15
Nortwood, J., 38, 90, 212
Numerical analysis, 299
Nurses Health Study, 226
Nussbaum, B.D., 110, 193

Oakland Athletics, 68
Observational study, 5, 30, 225–233
Office of Management and Budget
    (OMB), 44, 46, 49, 88, 233
Office of Personnel Management, 47, 82
Office of Research, Evaluation and
    Statistics, 46
Official statistics, 35–56, 79, 186
    challenges, 41–45
    compensation, 52–53
    examples, 39–41
    integration of activities, 49–50
    local government activities, 48–51
    major employers, 46–47
    outside the U.S., 51–52
    protecting confidentiality of, 212–213
    required credentials, 47–49
    research opportunities, 45
    scope, 36–38
    some technical challenges, 130
    training for, 130
Old Testament, 35
On-site involvement, 186
Operations research, 8, 28, 48, 61,
    127–128
Ordinal data, 28
Oregon, statistician requirements, 49
OR/MS Today, 314
Orszag, P., 38
Outlier, 131
Outsourcing. *see* Globalization

Pantula, S., 99
Parastatistician, 78, 80
Pardo, S., 197
Pareto Rule, 187
Parr, W.C., 110, 144
Patents, 291
Peer-review, 291, 303
Percentile, 131
Performance
    evaluation, 85
    need for, 187–188
Persistence, need for, 103
Peterson, J.J., 103
Pfeifer, C.G., 111, 193
Pharmaceutical industry, 26–27, 58, 75, 76,
    78, 80, 81
Population, 221–223, 230–231, 243–244
Post-stratification, 70

Practical statistical efficiency, 180
Practitioner, definition, 3
Pre-clinical trial, 26
Prediction, 230
    interval, 131
Presentation
    charts, preparation, 203–204
    effective, 202–205
    hints for giving, 204–205
Principal Federal Economic Indicators, 88
Privacy protection, 213
Proactive role, 6, 10, 24, 106, 252
Probability, 127
Problem assessment skills, 101–102
Problem owners, definition, 5
Process
    control, 63
    improvement, 24, 27, 30–31
    industry, 26
    knowledge, 24
    and product understanding, 184
Product
    burn-in, 28
    design, 24–25
    scale-up, 24–25
    servicing, 25–26
Professional participation. *see* Statistician,
    professional participation
Professional societies, 16
Professor, selection of, 134
Project attractiveness index, 180
Project benefits estimation, 182–183
Project cost estimation, 183–184
Project execution, 184–190
Project implementation, 190
Project initiation, 175–184
    by statistician, 178–179
    statistician recruited during course of
        project, 176–177
    statistician recruited up-front, 177–178
Project report, 201
Project scoping study, 183
Project selection, 179–180
Proposal
    development, 288–289
    process, 180–184
    request for, 180
    typical requirements, 180
    within an organization, 180–184
Public opinion poll, 65–67, 223, 232

Publicizing statistics and statisticians, 205–207

Quality and Productivity Research Conference, 317
Quality assurance, 127–128, 190, 255, 263
Quality control. *see* Quality assurance
Quality improvement, (footnote) 24, 61, 76–77
Quality Progress, (footnote) 206, 314

R programming environment, 133–134
RAND Corporation, 72
Random sampling, 65, 221–226, 241
Randomization, 220, 225
Randomized trial, 225, 233
Ranney, G., 65, 193
Rao, C.R., 1
Realistic attitude, need for, 104
Recruiting process, insights into, 158
References, for job applications, 165–166
Regression analysis, 127–129, 133, 281
Regulatory activities, 57–59
Reliability, 24–25, 26, 29, (footnote) 221
Reliability, Availability and Maintainability Symposium, 318
Research. *see* Statistical research
grant, 288–289
institutes, 72
university, 270, 275
Research Triangle Institute (RTI), 47, 72
Response variable, 220–221, 227–228, 230
Résumé writing, 154–156
Reverse interview, 169
Risk assessment, 29
Robust products, 25
Role. *see* Statistician, roles
Romig, H., 76
Rothamsted Agriculture Experimental Station, 62
Royal Statistical Society, 16, 90, 290, (footnote) 297, 312
RSPLIDA, 134

S programming environment, 134
Sabbatical, 275, 305
Sabermetrics, 68
Sall, J., 15
Sample size, 144
Sampling frame, 65, 223

Sampling studies, 5, 37, 41, 43, 65–67, 89, 127, 137, 185, 195, 210, 221–224, 241, 293, 299
response rate, 66
Sampson, C.B., 111, 193
SAS, 15, 133
Schmee, J., xv, (footnote) 269
Schwab, C., 197
Scientific method, 4–7, 11
Self-confidence, need for, 105
Sematech, 310
Semiconductor industry, 26, 27, 63, 76, 310
Senior academician, 300
Service business, 28–30
Shainin, D., (footnote) 24
Shewhart, W., 23, 76, 141
Short course, 205–206, 311
Significance, 282, 312
Simple random sampling, 223
Simpson's Paradox, 68
Simulation, 127, 245, 299
Singapore, Statistics Act, 52
Singpurwalla, N.D., 111, 161
Six Sigma, 12, 24, 61, 76–77, 80, 93, 94, 141, 184, 206, 263, 264
Six Sigma Forum Magazine, 77
Small Area Health Insurance Estimates, 37
Smoking and lung cancer, 231
Snee, R.D., 99
Snow, J., 231
Social and behavioral sciences, 65–71
Social Security Administration, 128
Society for Clinical Trials, 16
Society for Clinical Trials Annual Meeting, 318
Society for Industrial and Applied Mathematics (SIAM), 314
Software, 29–30, 131–135
development and support of, 29–30
general purpose statistical, 133
general purpose with statistical features, 132–133
specialized statistical, 133
statistical, 131–135
programming environment, 133–134
Southwest Research Institute, 72
Space exploration, 63–64
Spar, E.J., 151
Spatial statistics, 63, 127
SPLIDA, 134

Split plot design, 62
S-Plus programming environment, 134
Sports strategy and assessment, 67–69
SPSS, 133
Standard Industrial Classification (SIC)
    System, 44, 130
Stata, 133
STATCOM, 140
Statgraphics, 133
Statistical advocate, 6
Statistical analyst, 251, 254, 256–257
Statistical computing, 127, 299
Statistical consultant, 251–254, 257
Statistical consulting, 126–129
    center, 140
    external, 139, 273, 292–293
    internal, 292
Statistical engineering, 77
Statistical genetics, 318
Statistical group, 23
Statistical Journal of the IAOS, 314
Statistical journals, tiering. *see* Journals,
    tiering
Statistical leadership, 24
Statistical literacy, 9, 280
Statistical manager, 252, 254
Statistical precision, 240
Statistical quality control. *see* Quality
    assurance
Statistical research, 286–292
    assessment of quality, 289
    grants for, 288–289
    identifying projects, 287–288
    publishing, 289
    types of, 286–287
Statistical science, 77
Statistical significance, 131, 211
Statistical Society of Canada, 16
    code of ethical statistical practice, 209
Statistical software. *see* Software
Statistical theory, 5
Statistical thinking, 7, 255
Statistical uncertainty, 223–224, 235
Statistician
    in academia, 269–305
    applied, moving into application
        area, 255, 256–257
    applied, moving into management, 255,
        257
    in business and industry, 23–33

career paths (*see* Career paths, statistician)
contribution beyond the
    workplace, 259–261
contributions to community and
    society, 260–261
definition, 2–3
in government, 35–56
place in an organization, 79–80
preparing for a successful career
    as, 110–185
professional participation, 260
receptiveness to, 75–76
roles, 251–258
    aspirations, 150
    as technical arbitrator, 89–90
senior, 251–252, 254, 257, 262
technical contributions, 259–260
traits for success, 99–111
work environment, 75–93
work of, 37–109
Statistics
    bachelor's degree in, 114–118, 120, 122,
        125
    beyond the workplace, 8–9
    Canada, 51–52
    career downsides, 9–10
    career, excitement of, 10–11
    definition, 1–2
    degrees in, 90–91, 114–116, 256, 298
    department offering degree in, 122–123
    education, 113–145
        limits of, 142–143
    entering from other fields, 141
    groups, 79–80
    introductory course, 10
    job rating, 11
    master degree in, 114–120, 122, 124, 125,
        133
    Ph.D. degree in, 114–120, 122, 124,
        134–136, 138, 142
    receptiveness to, 10, 75–79
    recognition (*see* Statistics, receptiveness to)
    Roundtable, (footnote) 206
    selection of courses, 126–131
    without Borders, 261
    women in (*see* Women in statistics)
Statistics Education Research Journal, 282
STATS, (footnote) 312
Stochastic processes, 127
Stratified sampling, 223

Student. *see* Gossett, W.S.
Student questionnaire, 279, 284–285
Study planning, 219
Subject selection, 221
Survey sampling. *see* Sampling studies
Survival data analysis, 127
Swedish Statistical Association, 16
Synthetic data, 51, 131, 163
SYSTAT, 133
Systems development study, 224

Taguchi methods, 78
Target population, 223
Teaching
  evaluation of, 285–286
  experience, 299
  in non-academic settings, 71
  load, 273, 276
Teaching Statistics, 282
Team
  member, 6, 81
  skills, 106
Technology Innovations in Statistics
    Education, 282
Technometrics, 287, 290
Tenure, 273–276, 277–279
  criteria for, 278–279
  granting of, 274
  life beyond, 279–280
  track position, 273–275, 278–279
Testing
  component, 243–244
  in-house, 243–244
  subassembly, 243–244
Theoretical statistics. *see* Statistical theory
Thesis. *see* Dissertation
Time accounting, 86
Time management ability, 105–106
Time series, 127, 130, 281
  data, seasonal adjustments, 45
Toastmasters, 202
Tolerance interval, 131
Traceability, 239
Travel safety, 8
Treasury Department, Statistics of Income
    Division, 46

Treatment, 220, 225
Truman, H. S., 66
Tukey, J., 10, 225
Twain, M., 9

Unbiased estimate, 223, 228
Unfavorable results, conveying of, 211
United Kingdom
  Office for National Statistics, 52
  Statistics Board, 52
United Nations, 35, 52, 64
U.S. citizenship requirement, 48
U.S. Geological Survey, 46
U.S. government agencies. *see* Individual
    agency name
U.S. News & World Report, college
    rankings, 120–121, 123
U.S. Supreme Court, 70
USAJobs, 153, (footnote) 156
Use-rate acceleration, 244

Validation, 24, 25
Value-added modeling, 74
Variability, 7, 18, 24, 25, 28, 31, 137, 177,
    224, 229, 255
  limited, 229
Varian. H., 318
Volunteer activities. *see* Statistician,
    contributions to community and
    society

Wallman, K., (footnote) 35, 91
Walsh, K. C., 87, 222
Washing machine reliability, 240
Washington Statistical Society, (footnote) 47
Wendelberger, J., 92, 105
Westat, 47
Wludyka, P., 184
Women in statistics, 90–93
World Bank, 52
World Statistics Congress, 16, 316–317
World Trade Organization, Harmonized
    System, 52

Youden, J., 206
Young Investigator Programs, 288